DISCARDED,

NASA SP-4012

NASA HISTORICAL DATA BOOK
Volume V

NASA Launch Systems, Space Transportation,
Human Spaceflight, and Space Science
1979–1988

Judy A. Rumerman

The NASA History Series

National Aeronautics and Space Administration
NASA History Office
Office of Policy and Plans
Washington, D.C.   1999

Library of Congress Cataloguing-in-Publication Data
(Revised for vol. 5)

NASA historical data book.

  (The NASA historical series) (NASA SP ; 4012)
  Vol. 1 is a republication of: NASA historical data book, 1958–1968./ Jane Van Nimmen and Leonard C. Bruno.
  Vol. 5 in series: The NASA history series.
  Includes bibliographical references and indexes.
  Contents: v. 1 NASA resources, 1958–1968 / Jane Van Nimmen and Leonard C. Bruno — v. 2. Programs and projects, 1958–1968 / Linda Neuman Ezell — v. 3. Programs and projects, 1969—1978 / Linda Neuman Ezell — v. 4. NASA resources, 1969–1978 / Ihor Gawdiak with Helen Fedor — v. 5. NASA launch systems, space transportation, human spaceflight, and space science, 1979–1988 / Judy A. Rumerman.
  1. United States. National Aeronautics and Space Administration—History. I. Van Nimmen, Jane. II. Bruno, Leonard C. III. Ezell, Linda Neuman. IV. Gawdiak, Ihor. V. Rumerman, Judy A.
  VI. Series. VII. Series. VIII. Series: NASA SP ; 4012.

For sale by the U.S. Government Printing Office
Superintendent of Documents, Mail Stop: SSOP, Washington, DC 20402-9328
ISBN 0-16-050030-3

CONTENTS

List of Figures and Tables v

Preface and Acknowledgments xi

Chapter One: Introduction 1

Chapter Two: Launch Systems 11

Chapter Three: Space Transportation/Human Spaceflight 105

Chapter Four: Space Science 361

Index 527

About the Compiler 535

The NASA History Series 537

LIST OF FIGURES AND TABLES

Chapter One: Introduction

Figure 1–1   Program Office Functional Areas   7

Chapter Two: Launch Systems

Figure 2–1	NASA Space Transportation System (1988)	14
Figure 2–2	Top-Level Launch Vehicle Organizational Structure	16
Figure 2–3	Office of Space Transportation (as of October 1979)	16
Figure 2–4	Code M/Code O Split (as of February 1980)	17
Figure 2–5	Code M Merger (as of October 1982)	18
Figure 2–6	Office of Space Flight 1986 Reorganization	20
Figure 2–7	Expendable Launch Vehicle Success Rate	24
Figure 2–8	Atlas-Centaur Launch Vehicle	30
Figure 2–9	Delta 3914	31
Figure 2–10	Delta 3920/PAM-D	31
Figure 2–11	Scout-D Launch Vehicle (Used in 1979)	32
Figure 2–12	External Tank	38
Figure 2–13	Solid Rocket Booster	41
Figure 2–14	Solid Rocket Motor Redesign Schedule	44
Figure 2–15	Inertial Upper Stage	48
Figure 2–16	Transfer Orbit Stage	50
Figure 2–17	Orbital Maneuvering Vehicle	55
Table 2–1	Appropriated Budget by Launch Vehicle and Launch-Related Component	59
Table 2–2	Atlas E/F Funding History	63
Table 2–3	Atlas-Centaur Funding History	64
Table 2–4	Delta Funding History	65
Table 2–5	Scout Funding History	66
Table 2–6	Space Shuttle Main Engine Funding History	67
Table 2–7	Solid Rocket Boosters Funding History	69
Table 2–8	External Tank Funding History	71
Table 2–9	Upper Stages Funding History	73
Table 2–10	Orbital Maneuvering Vehicle Funding History	75
Table 2–11	Tethered Satellite System Funding History	76
Table 2–12	Advanced Programs/Planning Funding History	77
Table 2–13	ELV Success Rate by Year and Launch Vehicle for NASA Launches	78
Table 2–14	NASA Atlas E/F Vehicle Launches	79
Table 2–15	Atlas E/F Characteristics	80
Table 2–16	NASA Atlas-Centaur Vehicle Launches	82
Table 2–17	Atlas-Centaur Characteristics	83

Table 2–18	Chronology of Delta Vehicle Launches	84
Table 2–19	Delta 2914 Characteristics	86
Table 2–20	Delta 3910/3914 Characteristics	87
Table 2–21	Delta 3920/3924 Characteristics	88
Table 2–22	NASA Scout Launches	89
Table 2–23	Scout Characteristics (G-1)	90
Table 2–24	STS-Launched Missions	91
Table 2–25	Space Shuttle Main Engine Characteristics	93
Table 2–26	Main Engine Development and Selected Events	94
Table 2–27	Space Shuttle External Tank Characteristics	95
Table 2–28	External Tank Development and Selected Events	96
Table 2–29	Space Shuttle Solid Rocket Booster Characteristics	97
Table 2–30	Chronology of Selected Solid Rocket Booster Development Events	98
Table 2–31	Upper Stage Development	101
Table 2–32	Transfer Orbit Stage Characteristics	103

Chapter Three: Space Transportation/Human Spaceflight

Figure 3–1	NSTS Organization	110
Figure 3–2	Safety, Reliability, and Quality Assurance Office Organization	113
Figure 3–3	Space Station Program Management Approach	116
Figure 3–4	Office of Space Station Organization (December 1986)	117
Figure 3–5	Space Shuttle Orbiter	124
Figure 3–6	Typical STS Flight Profile	132
Figure 3–7	Types of Intact Aborts	138
Figure 3–8	Pallet Structure and Panels	149
Figure 3–9	Spacelab Igloo Structure	149
Figure 3–10	Insulating Materials	159
Figure 3–11	STS-1 Entry Flight Profile	163
Figure 3–12	Continuous Flow Electrophoresis System Mid-deck Gallery Location	170
Figure 3–13	STS-5 Payload Configuration	171
Figure 3–14	Payload Flight Test Article	173
Figure 3–15	Manned Maneuvering Unit	175
Figure 3–16	Solar Max On-Orbit Berthed Configuration	176
Figure 3–17	Long Duration Exposure Facility Configuration	177
Figure 3–18	STS 51-A Cargo Configuration	179
Figure 3–19	STS 61-A Cargo Configuration	183
Figure 3–20	EASE/ACCESS Configuration	184
Figure 3–21	Integrated MSL-2 Payload	185
Figure 3–22	Tracking and Data Relay Satellite On-Orbit Configuration	186
Figure 3–23	STS 51-L Data and Design Analysis Task Force	211
Figure 3–24	Space Shuttle Return to Flight	213
Figure 3–25	Space Shuttle Return to Flight Milestones	220
Figure 3–26	Field Joint Redesign	221

Figure 3–27	Extendible Rod Escape System	226
Figure 3–28	Availability of Fourth Orbiter	228
Figure 3–29	System Integrity Assurance Program	229
Figure 3–30	Major Orbiter Modifications	230
Figure 3–31	Dual Keel Final Assembly Configuration	244
Figure 3–32	Revised Baseline Configuration (1987), Block I	245
Figure 3–33	Enhanced Configuration, Block II	245
Figure 3–34	Habitation Module	247
Figure 3–35	Flight Telerobotic Servicer	249
Figure 3–36	Photovoltaic Module	249
Figure 3–37	Mobile Servicing System and Special Purpose Dexterous Manipulator	250
Figure 3–38	Columbus Attached Laboratory	251
Figure 3–39	Columbus Free-Flying Laboratory	252
Figure 3–40	Columbus Polar Platform	253
Figure 3–41	Japanese Experiment Module	253
Table 3–1	Total Human Spaceflight Funding History	256
Table 3–2	Programmed Budget by Budget Category	259
Table 3–3	Orbiter Funding History	260
Table 3–4	Orbiter Replacement Funding History	261
Table 3–5	Launch and Mission Support Funding History	262
Table 3–6	Launch and Landing Operations Funding History	264
Table 3–7	Spaceflight Operations Program Funding History	265
Table 3–8	Flight Operations Funding History	266
Table 3–9	Spacelab Funding History	267
Table 3–10	Space Station Funding History	268
Table 3–11	Orbiter Characteristics	269
Table 3–12	Typical Launch Processing/Terminal Count Sequence	271
Table 3–13	Space Shuttle Launch Elements	272
Table 3–14	Mission Command and Control Positions and Responsibilities	273
Table 3–15	Shuttle Extravehicular Activity	274
Table 3–16	STS-1–STS-4 Mission Summary	275
Table 3–17	STS-1 Mission Characteristics	277
Table 3–18	STS-2 Mission Characteristics	279
Table 3–19	STS-3 Mission Characteristics	281
Table 3–20	STS-4 Mission Characteristics	283
Table 3–21	STS-5–STS-27 Mission Summary	285
Table 3–22	STS-5 Mission Characteristics	294
Table 3–23	STS-6 Mission Characteristics	296
Table 3–24	STS-7 Mission Characteristics	298
Table 3–25	STS-8 Mission Characteristics	300
Table 3–26	STS-9 Mission Characteristics	302
Table 3–27	STS 41-B Mission Characteristics	303
Table 3–28	STS 41-C Mission Characteristics	306
Table 3–29	STS 41-D Mission Characteristics	307
Table 3–30	STS 41-G Mission Characteristics	309

Table 3–31	STS 51-A Mission Characteristics	312
Table 3–32	STS 51-C Mission Characteristics	313
Table 3–33	STS 51-D Mission Characteristics	314
Table 3–34	STS 51-B Mission Characteristics	317
Table 3–35	STS 51-G Mission Characteristics	318
Table 3–36	STS 51-F Mission Characteristics	321
Table 3–37	STS 51-I Mission Characteristics	323
Table 3–38	STS 51-J Mission Characteristics	324
Table 3–39	STS 61-A Mission Characteristics	325
Table 3–40	STS 61-B Mission Characteristics	326
Table 3–41	STS 61-C Mission Characteristics	328
Table 3–42	STS 51-L Mission Characteristics	333
Table 3–43	STS-26 Mission Characteristics	334
Table 3–44	STS-27 Mission Characteristics	337
Table 3–45	Return to Flight Chronology	338
Table 3–46	Sequence of Major Events of the *Challenger* Accident	342
Table 3–47	Chronology of Events Prior to Launch of *Challenger* (STS 51-L) Related to Temperature Concerns	345
Table 3–48	Schedule for Implementation of Recommendations (as of July 14, 1986)	354
Table 3–49	Revised Shuttle Manifest (as of October 3, 1986)	356
Table 3–50	Space Station Work Packages	359
Table 3–51	Japanese Space Station Components	360

Chapter Four: Space Science

Figure 4–1	Office of Space Science (Through November 1981)	369
Figure 4–2	Office of Space Science and Applications (Established November 1981)	370
Figure 4–3	HEAO High-Spectral Resolution Gamma Ray Spectrometer	376
Figure 4–4	HEAO Isotopic Composition of Primary Cosmic Rays	376
Figure 4–5	HEAO Heavy Nuclei Experiment	377
Figure 4–6	Solar Maximum Instruments	378
Figure 4–7	Solar Mesospheric Explorer Satellite Configuration	380
Figure 4–8	Altitude Regions to Be Measured by Solar Mesospheric Explorer Instruments	381
Figure 4–9	Infrared Astronomy Satellite Configuration	382
Figure 4–10	Exploded View of the European X-Ray Observatory Satellite	385
Figure 4–11	Distortion of Earth's Magnetic Field	387
Figure 4–12	Spartan 1	389
Figure 4–13	Plasma Diagnostics Package Experiment Hardware	390
Figure 4–14	Spartan Halley Configuration	391
Figure 4–15	San Marco D/L Spacecraft	393
Figure 4–16	Spacelab 1 Module Experiment Locations (Port Side)	395

Figure 4–17	Spacelab 1 Module Experiment Locations (Starboard Side)	396
Figure 4–18	Spacelab 1 Pallet Experiment Locations	397
Figure 4–19	Spacelab 3 Experiment Module Layout (Looking Down From the Top)	398
Figure 4–20	Spacelab 2 Configuration	398
Figure 4–21	OSS-1 Payload Configuration	400
Figure 4–22	Hubble Space Telescope	404
Figure 4–23	Compton Gamma Ray Observatory Configuration	405
Figure 4–24	Extreme Ultraviolet Explorer Observatory	407
Figure 4–25	Two Phases of the Extreme Ultraviolet Explorer Mission	408
Figure 4–26	ROSAT Flight Configuration	409
Figure 4–27	Cosmic Background Explorer Observatory (Exploded View)	411
Figure 4–28	Cosmic Background Explorer Orbital Alignments	412
Figure 4–29	Magellan Spacecraft Configuration	417
Figure 4–30	Magellan Orbit	418
Figure 4–31	Galileo Mission	419
Figure 4–32	Galileo Spacecraft	420
Figure 4–33	Ulysses Spacecraft Configuration	421
Table 4–1	Total Space Science Funding History	422
Table 4–2	Programmed Budget by Budget Category	425
Table 4–3	High Energy Astronomy Observatories Development Funding History	426
Table 4–4	Solar Maximum Mission Development Funding History	426
Table 4–5	Space Telescope Development Funding History	426
Table 4–6	Solar Polar Mission Development Funding History	427
Table 4–7	Gamma Ray Observatory Development Funding History	427
Table 4–8	Shuttle/Spacelab Payload Development Funding History	428
Table 4–9	Explorer Development Funding History	429
Table 4–10	Physics and Astronomy Mission Operations and Data Analysis Funding History	429
Table 4–11	Physics and Astronomy Research and Analysis Funding History	430
Table 4–12	Physics and Astronomy Suborbital Programs Funding History	430
Table 4–13	Space Station Planning Funding History	431
Table 4–14	Jupiter Orbiter/Probe and Galileo Programs Funding History	431
Table 4–15	Venus Radar Mapper/Magellan Funding History	431
Table 4–16	Global Geospace Science Funding History	432
Table 4–17	International Solar Polar Mission/Ulysses Development Funding History	432

Table 4–18	Mars Geoscience/Climatology Orbiter Program Funding History	432
Table 4–19	Lunar and Planetary Mission Operations and Data Analysis Funding History	433
Table 4–20	Lunar and Planetary Research and Analysis Funding History	433
Table 4–21	Life Sciences Flight Experiments Program Funding History	434
Table 4–22	Life Sciences/Vestibular Function Research Funding History	434
Table 4–23	Life Sciences Research and Analysis Funding History	435
Table 4–24	Science Missions (1979–1988)	436
Table 4–25	Spacecraft Charging at High Altitudes Characteristics	437
Table 4–26	UK-6 (Ariel) Characteristics	439
Table 4–27	HEAO-3 Characteristics	441
Table 4–28	Solar Maximum Mission	442
Table 4-29	Dynamics Explorer 1 and 2 Characteristics	444
Table 4–30	Solar Mesospheric Explorer Instrument Characteristics	446
Table 4–31	Solar Mesospheric Explorer Characteristics	447
Table 4–32	Infrared Astronomy Satellite Characteristics	449
Table 4–33	European X-Ray Observatory Satellite Characteristics	451
Table 4–34	Shuttle Pallet Satellite-01 Characteristics	452
Table 4–35	Hilat Characteristics	453
Table 4–36	Charge Composition Explorer Characteristics	454
Table 4–37	Ion Release Module Characteristics	455
Table 4–38	United Kingdom Subsatellite Characteristics	456
Table 4–39	Spartan 1 Characteristics	457
Table 4–40	Plasma Diagnostics Package Characteristics	458
Table 4–41	Spartan 203 Characteristics	459
Table 4–42	Polar BEAR Characteristics	460
Table 4–43	San Marco D/L Characteristics	461
Table 4–44	Chronology of Spacelab Development	462
Table 4–45	Spacelab 1 Experiments	480
Table 4–46	Spacelab 3 Experiments	499
Table 4–47	Spacelab 2 Experiments	505
Table 4–48	Spacelab D-1 Experiments	512
Table 4–49	OSS-1 Investigations	516
Table 4–50	Hubble Space Telescope Development	518
Table 4–51	Ulysses Historical Summary	525

PREFACE AND ACKNOWLEDGMENTS

In 1973, NASA published the first volume of the *NASA Historical Data Book*, a hefty tome containing mostly tabular data on the resources of the space agency between 1958 and 1968. There, broken into detailed tables, were the facts and figures associated with the budget, facilities, procurement, installations, and personnel of NASA during that formative decade. In 1988, NASA reissued that first volume of the data book and added two additional volumes on the agency's programs and projects, one each for 1958–1968 and 1969–1978. NASA published a fourth volume in 1994 that addressed NASA resources for the period between 1969 and 1978.

This fifth volume of the *NASA Historical Data Book* is a continuation of those earlier efforts. This fundamental reference tool presents information, much of it statistical, documenting the development of four critical areas of NASA responsibility for the period between 1979 and 1988. This volume includes detailed information on the development and operation of launch systems, space transportation, human spaceflight, and space science during this era. As such, it contains in-depth statistical information about the early Space Shuttle program through the return to flight in 1988, the early efforts to build a space station, the development of new launch systems, and the launching of seventeen space science missions.

A companion volume will appear late in 1999, documenting the space applications, support operations, aeronautics, and resources aspects of NASA during the period between 1979 and 1988.

There are numerous people at NASA associated with historical study, technical information, and the mechanics of publishing who helped in myriad ways in the preparation of this historical data book. Stephen J. Garber helped in the management of the project and handled final proofing and publication. M. Louise Alstork edited and prepared the index of the work. Nadine J. Andreassen of the NASA History Office performed editorial and proofreading work on the project; and the staffs of the NASA Headquarters Library, the Scientific and Technical Information Program, and the NASA Document Services Center provided assistance in locating and preparing for publication the documentary materials in this work. The NASA Headquarters Printing and Design Office developed the layout and handled printing. Specifically, we wish to acknowledge the work of Jane E. Penn, Jonathan L. Friedman, Joel Vendette, Patricia M. Talbert, and Kelly L. Rindfusz for their editorial and design work. In addition, Michael Crnkovic, Stanley Artis, and Jeffrey Thompson saw the book through the publication process. Thanks are due them all.

CHAPTER ONE
INTRODUCTION

CHAPTER ONE
INTRODUCTION

NASA began its operations as the nation's civilian space agency in 1958 following the passage of the National Aeronautics and Space Act. It succeeded the National Advisory Committee for Aeronautics (NACA). The new organization was charged with preserving the role of the United States "as a leader in aeronautical and space science and technology" and in its application, with expanding our knowledge of the Earth's atmosphere and space, and with exploring flight both within and outside the atmosphere.

By the 1980s, NASA had established itself as an agency with considerable achievements on record. The decade was marked by the inauguration of the Space Shuttle flights and haunted by the 1986 *Challenger* accident that temporarily halted the program. The agency also enjoyed the strong support of President Ronald Reagan, who enthusiastically announced the start of both the Space Station program and the National Aerospace Plane program.

Overview of the Agency

NASA is an independent federal government agency that, during the 1980s, consisted of 10 field installations located around the United States, the Jet Propulsion Laboratory (a government-owned facility staffed by the California Institute of Technology), and a Headquarters located in Washington, D.C. Headquarters was divided into a number of program and staff offices that provided overall program management and handled administrative functions for the agency. Each program office had responsibility for particular program areas (see Figure 1–1). Headquarters also interacted with Congress and the Executive Branch.

NASA's structure was quite decentralized. Although Headquarters had overall program responsibility, each installation was responsible for the day-to-day execution and operations of its projects, managed its own facility, hired its own personnel, and awarded its own procurements. Each installation also focused on particular types of projects and discipline areas.

Program and Project Development

NASA called most of its activities programs or projects. The agency defined a *program* as "a related series of undertakings which are funded for the most part from NASA's R&D appropriation, which continue over a period of time (normally years), and which are designed to pursue a broad scientific or technical goal." A *project* is "a defined, time-limited activity with clearly established objectives and boundary conditions executed to gain knowledge, create a capability, or provide a service.... A project is normally an element of a program."[1]

NASA's flight programs and projects followed prescribed phases (with associated letter designators) in their development and execution. This sequence of activities consisted of concept development (Pre-Phase A), mission analysis (Phase A), definition or system design (Phase B), execution (design, development, test, and evaluation) (Phase C/D), launch and deployment operations (Phase E), and mission operations, maintenance, and disposal (Phase F). Although most concepts for missions originated within a field installation, Headquarters retained project responsibility through Phase B. Once a program or project was approved and funded by Congress, the principal responsibility for program or project implementation shifted to the field installation. Internal agency reviews were held during and between each phase of a project. Before moving to Phase C/D, NASA held a major agency review, and approval and funding by Congress were required. Particular activities never moved beyond Phase B, nor were they meant to. For instance, many aeronautics activities were designed as research efforts and were intended to be turned over to the private sector or to other government agencies once Phase B concluded.

NASA's Budget Process

NASA's activities relied on getting a reasonable level of funding from Congress. The federal budget process was quite complex, and a brief description as it relates to NASA is presented here. Additional information can be found in Chapter 8, "Finances and Procurement," in Volume VI of the *NASA Historical Data Book*.

NASA operated on a fiscal year (FY) that ran from October 1 through September 30 of the following year. Through FY 1983, the agency budget was broken into three accounts or appropriation categories: Research and Development (R&D), Research and Program Management (R&PM), and Construction of Facilities (C of F). An additional appropriation, Space Flight, Control, and Data Communications (SFC&DC) was added in FY 1984 for ongoing Shuttle-related and tracking and data acquisition activities. Although a program office could administer activities from

---

[1] NASA Management Instruction 7120.3, "Space Flight Program and Project Management," February 6, 1985.

more than one appropriation category, such as the Office of Space Flight, which managed both R&D and SFC&DC activities, all funds were designated for particular appropriation categories and could not be transferred between accounts without congressional approval.

Congress appropriated operating funds each year. These appropriations were the culmination of a series of activities that required at least two years of effort by the installations and Headquarters.

Two years before a budget year began, Headquarters sent guidelines to each installation that contained programmatic and budget information based on its long-range plans and the budget forecasts from the Office of Management and Budget (OMB). Each installation then prepared a detailed budget, or Program Operating Plan (POP), for the fiscal year that would begin two years in the future. The installation also refined the budget for the remainder of the current fiscal year and the next fiscal year that it had already submitted and had approved, and it provided less detailed budget figures for later years. Upon approval from each installation's comptroller and director, this budget was forwarded to the appropriate Headquarters-level program office, to the NASA comptroller's office, and the NASA administrator.

Headquarters reviewed the budget requests from each installation, held discussions with the installations, and negotiated with OMB to arrive at a budget that looked realistic and had a fair chance of passage by Congress. Following these negotiations, NASA formally submitted its budget requests to OMB. This became part of the administration's budget that went to Congress in January of each year.

When Congress received the budget, NASA's proposed budget first went to the House and Senate science committees that were charged with authorizing the agency's budget. Each committee held hearings, usually with NASA administrators; reviewed the submission in great detail; debated, revised, and approved the submitted budget; and sent it to the full House or Senate for approval. The authorization committees could limit how much could be appropriated and often set extensive conditions on how the funds were to be spent. Each house approved its own authorization bill, which was then submitted to a House-Senate conference committee to resolve any differences. After this took place, the compromise bill was passed by the full House and Senate and submitted to the President for his signature.

The process to appropriate funds was similar, with the bills going to the proper appropriations committees for discussion, revision, and approval. However, in practice, the appropriations committees usually did not review the proposed budget in as great detail as the authorization committees. Upon committee approval, the appropriations bills went to the full House and Senate, back to a conference committee if necessary, and finally to the President. After approval by the President, OMB established controls on the release of appropriated funds to the various agencies, including NASA.

Once NASA received control over its appropriated funds, it earmarked the funds for various programs, projects, and facilities, each of which had an "account" with the agency established for it. Funds were then committed, obligated, costed, and finally disbursed according to the progression of activities, which hopefully coincided with the timing of events spelled out in the budget. NASA monitored all of its financial activities scrupulously, first at the project and installation level and then at the Headquarters level. Its financial transactions were eventually reviewed by the congressional General Accounting Office to ensure that they were legal and followed prescribed procedures.

In the budget tables that follow in each chapter, the "request" or "submission" column contains the amount that OMB submitted to Congress. It may not be the initial request that NASA submitted to OMB. The "authorization" is the ceiling set by the authorization committees in their bill. The "appropriation" is the amount provided to the agency. The "programmed" column shows the amount the agency actually spent during the fiscal year for a particular program.

INTRODUCTION

	R&D Programs				
1979	1980	1981	1982	1983	
Space Transportation Systems • Space Shuttle • Space Flight Operations • Expendable Launch Vehicles					
Space Science • Physics and Astronomy • Lunar and Planetary • Life Sciences	Space Science • Physics and Astronomy • Planetary Exploration • Life Sciences			Space Science and Applications • Physics and Astronomy • Planetary Exploration • Life Sciences • Space Applications • Technology Utilization	
Space and Terrestrial Applications • Space Applications • Technology Utilization					
Aeronautics and Space Technology • Aeronautical Research and Technology • Space Research and Technology • Energy Technology				Aeronautics and Space Technology • Aeronautical Research and Technology • Space Research and Technology	
Space Tracking and Data Systems • Tracking and Data Acquisition					

Figure 1–1. Program Office Functional Areas

R&D Programs

1984	1985	1986	1987	1988
Space Transportation Systems • Space Transportation and Capability • Space Transportation Operations	Space Flight • Space Transportation Capability Development			
	Space Station Task Force • Space Station	Space Station • Space Station		Space Station • Space Station • Industrial Space Facility
Space Science and Applications • Physics and Astronomy • Planetary Exploration • Life Sciences • Space Applications				
External Relations • Technology Utilization		Commercial Programs • Technology Utilization • Commercial Use of Space		
Aeronautics and Space Technology • Aeronautical Research and Technology • Space Research and Technology			Aeronautics and Space Technology • Aeronautical Research and Technology • Transatmospheric Research and Technology • Space Research and Technology	
			Safety, Reliability, Maintainability, and Quality Assurance • Safety, Reliability, and Quality Assurance	
Space Tracking and Data Systems • Tracking and Data Acquisition	Space Tracking and Data Systems • Tracking and Data Advanced Systems			

Figure 1–1 continued

INTRODUCTION

SFC&DC Programs

1984	1985	1986	1987	1988
Space Transportation Systems • Space Transportation and Capability Development • Space Transportation Operations	Space Flight • Space Production and Operational Capability • Space Transportation Operations		Space Flight • Shuttle Production and Operational Capability • Space Transportation Operations	Space Flight • Shuttle Production and Operational Capability • Space Transportation Operations • Expendable Launch Vehicles
Space Tracking and Data Systems • Tracking and Data Acquisition		Space Tracking and Data Systems • Space and Ground Network Communications and Data Systems		

Source: *NASA Chronological Histories, Budget Submissions, 1979–1988*

Figure 1–1 continued

CHAPTER TWO
LAUNCH SYSTEMS

CHAPTER TWO
LAUNCH SYSTEMS

Introduction

Launch systems provide access to space, obviously a necessary component of all spaceflights. The elements of launch systems include the various vehicles, engines, boosters, and other propulsive and launch devices that help propel a spacecraft into space and position it properly. From 1979 through 1988, NASA used both expendable launch vehicles (ELVs)—those that can be used only once—and reusable launch vehicles. This chapter addresses both types of vehicles, as well as other launch system-related elements.

NASA used three families of ELVs (Scout, Delta, and Atlas) and one reusable launch vehicle (Space Shuttle) from 1979 through 1988 (Figure 2–1). Each family of ELVs had several models, which are described in this chapter. For the Space Shuttle, or Space Transportation System (STS), the solid rocket booster, external tank, and main engine elements comprised the launch-related elements and are addressed. The orbital maneuvering vehicle and the various types of upper stages that boosted satellites into their desired orbit are also described.

This chapter includes an overview of the management of NASA's launch vehicle program and summarizes the agency's launch vehicle budget. In addition, this chapter addresses other launch vehicle development, such as certain elements of advanced programs.

Several trends that began earlier in NASA's history continued in this decade (1979–1988). The trend toward acquiring launch vehicles and services from the commercial sector continued, as did the use of NASA-launched vehicles for commercial payloads. President Reagan's policy directive of May 1983 reiterated U.S. government support for commercial ELV activities and the resulting shift toward commercialization of ELV activities. His directive stated that the "U.S. government fully endorses and will facilitate commercialization of U.S. Expendable Launch Vehicles." His directive said that the United States would encourage use of its national ranges for commercial ELV operations and would "make available, on a reimbursable basis, facilities, equipment, tooling, and services that are required to support the production and operation of

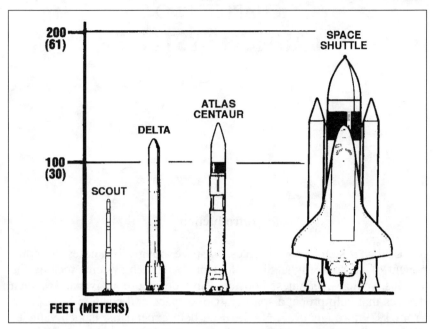

Figure 2–1. NASA Space Transportation System (1988)

U.S. commercial ELVs." Use of these facilities would be priced to encourage "viable commercial ELV launch activities."[1]

The policy also stated the government's intention of replacing ELVs with the STS as the primary launch system for most spaceflights. (Original plans called for a rate flight of up to fifty Space Shuttle flights per year.) However, as early as FY 1984, Congress recognized that relying exclusively on the Shuttle for all types of launches might not be the best policy. Congress stated in the 1984 appropriations bill that "the Space Shuttle system should be used primarily as a launch vehicle for government defense and civil payloads only" and "commercial customers for communications satellites and other purposes should begin to look to the commercialization of existing expendable launch vehicles."[2] The *Challenger* accident, which delayed the Space Shuttle program, also con-

---

[1]*Announcement of U.S. Government Support for Commercial Operations by the Private Sector, May 16, 1983,* from National Archives and Records Service's *Weekly Compilation of Presidential Documents* for May 16, 1983, pp. 721–23.

[2]House Committee on Appropriations, *Department of Housing and Urban Development-Independent Agencies Appropriation Bill, 1984, Report to Accompany H.R. 3133,* 98th Cong., 1st sess., 1983, H. Rept. 98— (unnumbered).

tributed to the development of a "mixed fleet strategy," which recommended using both ELVs and the Shuttle.[3]

Management of the Launch Vehicle Program

Two NASA program offices shared management responsibility for the launch vehicle program: Code M (at different times called the Office of Space Transportation, the Office of Space Transportation Acquisition, and the Office of Space Flight) and Code O (the Office of Space Transportation Operations). Launch system management generally resided in two or more divisions within these offices, depending on what launch system elements were involved.

The organizational charts that follow illustrate the top-level structure of Codes M and O during the period 1979–1988. As in other parts of this chapter, there is some overlap between the management-related material presented in this chapter and the material in Chapter 3, "Space Transportation and Human Spaceflight."

Also during the period 1979 through 1988, two major reorganizations in the launch vehicle area occurred (Figure 2–2): the split of the Office of Space Transportation into Codes M and O in 1979 (Phase I) and the merger of the two program offices into Code M in 1982 (Phase II). In addition, the adoption of the mixed fleet strategy following the loss of the Challenger reconfigured a number of divisions (Phase III). These management reorganizations reflected NASA's relative emphasis on the Space Shuttle or on ELVs as NASA's primary launch vehicle, as well as the transition of the Shuttle from developmental to operational status.

Phase I: Split of Code M Into Space Transportation Acquisition (Code M) and Space Transportation Operations (Code O)

John F. Yardley, the original associate administrator for the Office of Space Transportation Systems since its establishment in 1977, continued in that capacity, providing continuous assessment of STS development, acquisition, and operations status. In October 1979, Charles R. Gunn assumed the new position of deputy associate administrator for STS (Operations) within Code M, a position designed to provide transition management in anticipation of the formation of a new program office planned for later that year (Figure 2–3).

---

[3]NASA Office of Space Flight, *Mixed Fleet Study,* January 12, 1987. The NASA Advisory Council had also established a Task Force on Issues of a Mixed Fleet in March 1987 to study the issues associated with the employment of a mixed fleet of launch vehicles and endorsed the Office of Space Flight study results in its *Study of the Issues of a Mixed Fleet.* Further references to a mixed fleet are found in remarks made by NASA Administrator James C. Fletcher on May 15, 1987.

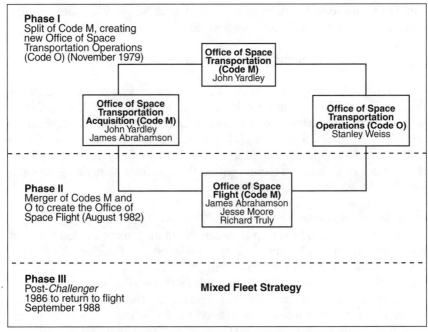

Figure 2–2. Top-Level Launch Vehicle Organizational Structure

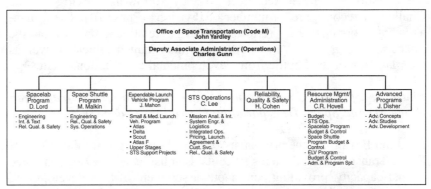

Figure 2–3. Office of Space Transportation (as of October 1979)

The formal establishment of the new Office of Space Operations (Code O) occurred in November 1979, and Dr. Stanley I. Weiss became its first permanent associate administrator in July 1980. Code O was the principal interface with all STS users and assumed responsibilities for Space Shuttle operations and functions, including scheduling, manifesting, pricing, launch service agreements, Spacelab, and ELVs, except for the development of Space Shuttle upper stages. The ELV program—Atlas, Centaur, Delta, Scout, and Atlas F—moved to Code O and was managed by Joseph B. Mahon, who had played a significant role in launch vehicle management during NASA's second decade.

Yardley remained associate administrator for Code M until May 1981, when L. Michael Weeks assumed associate administrator responsibilities.

Two new divisions within Code M were established in May 1981. The Upper Stage Division, with Frank Van Renssalaer as director, assumed responsibility for managing the wide-body Centaur, the Inertial Upper Stage (IUS), the Solid Spinning Upper Stage (SSUS), and the Solar-Electric Propulsion System. The Solid Rocket Booster and External Tank Division, with Jerry Fitts as director, was also created. In November 1981, Major General James A. Abrahamson, on assignment from the Air Force, assumed duties as permanent associate administrator of Code M (Figure 2–4).

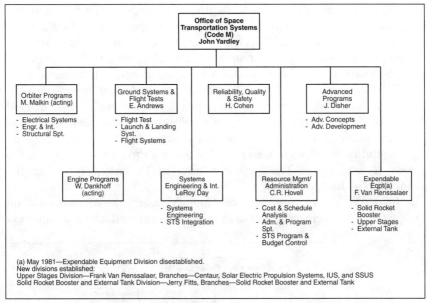

Figure 2–4. Code M/Code O Split (as of February 1980) (1 of 2)

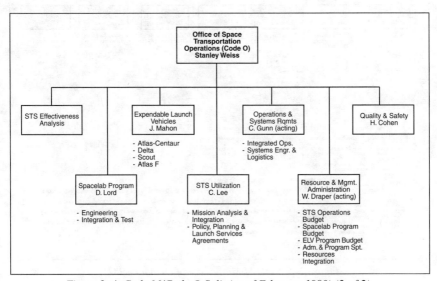

Figure 2–4. Code M/Code O Split (as of February 1980) (2 of 2)

Phase II: Merger of Codes M and O Into the Office of Space Flight

In preparation for Space Shuttle operations, Codes M and O merged in 1982 into the Office of Space Flight, Code M, with Abrahamson serving as associate administrator (Figure 2–5). Weiss became NASA's chief engineer. Code M was responsible for the fourth and final developmental Shuttle flight, the operational flights that would follow, future Shuttle procurements, and ELVs. The new office structure included the Special Programs Division (responsible for managing ELVs and upper stages), with Mahon continuing to lead that division, the Spacelab Division, the Customer Services Division, the Space Shuttle Operations Office, and the Space Station Task Force. This task force, under the direction of John D. Hodge, developed the programmatic aspects of a space station, including mission analysis, requirements definition, and program management. In April 1984, an interim Space Station Program Office superseded the Space Station Task Force and, in August 1984, became the permanent Office of Space Station (Code S), with Philip E. Culbertson serving as associate administrator. In the second quarter of 1983, organizational responsibility for ELVs moved from the Special Programs Division to the newly formed Space Transportation Support Division, still under the leadership of Joseph Mahon.

Jesse W. Moore took over as Code M associate administrator on August 1, 1984, replacing Abrahamson, who accepted a new assignment

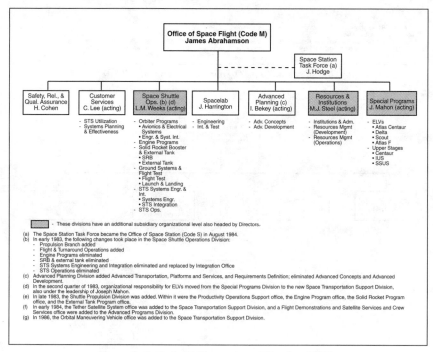

Figure 2–5. Code M Merger (as of October 1982)

in the Department of Defense (DOD). Moore was succeeded by Rear Admiral Richard H. Truly, a former astronaut, on February 20, 1986.

Phase III: Post-Challenger Launch Vehicle Management

From the first Space Shuttle orbital test flight in April 1981 through STS 61-C on January 12, 1986, NASA flew twenty-four successful Shuttle missions, and the agency was well on its way to establishing the Shuttle as its only launch vehicle. The loss of the *Challenger* (STS 51-L) on January 26, 1986, grounded the Shuttle fleet for thirty-two months. When flights resumed with STS-26 in September 1988, NASA planned a more conservative launch rate of twelve launches per year. The reduction of the planned flight rate forced many payloads to procure ELV launch services and forced NASA to plan to limit Shuttle use to payloads that required a crewed presence or the unique capabilities of the Shuttle. It also forced NASA to recognize the inadvisability of relying totally on the Shuttle. The resulting adoption of a "mixed fleet strategy" included increased NASA-DOD collaboration for the acquisition of launch vehicles and the purchase of ELV launch services. This acquisition strategy consisted of competitive procurements of the vehicle, software, and engineering and logistical work, except for an initial transitional period through 1991, when procurements would be noncompetitive if it was shown that it was in the government's best interest to match assured launch vehicle availability with payloads and established mission requirements.

The mixed fleet strategy was aimed at a healthy and affordable launch capability, assured access to space, the utilization of a mixed fleet to support NASA mission requirements, a dual-launch capability for critical payloads, an expanded national launch capability, the protection of the Shuttle fleet, and the fostering of ELV commercialization. This last goal was in accordance with the Reagan administration's policy of encouraging the growth of the fledgling commercial launch business whenever possible. The Office of Commercial Programs (established in 1984) was designated to serve as an advocate to ensure that NASA's internal decision-making process encouraged and facilitated the development of a domestic industrial base to provide access to space.

During this regrouping period, the ELV program continued to be managed at Headquarters within the Office of Space Flight, through the Space Transportation Support Division, with Joseph Mahon serving as division director and Peter Eaton as chief of ELVs, until late 1986. During this period, the Tethered Satellite System and the Orbital Maneuvering Vehicle also became responsibilities of this division. In late 1986, Code M reorganized into the basic configuration that it would keep through 1988 (Figure 2–6). This included a new management and operations structure for the National Space Transportation System (NSTS). Arnold J. Aldrich was named director of the NSTS at NASA Headquarters. A new Flight Systems Division, still under the leadership of Mahon, consisted of divisions for ELVs and upper stages, as well as divisions for advanced programs and

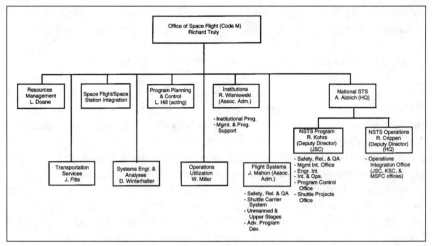

Figure 2–6. Office of Space Flight 1986 Reorganization

Space Shuttle carrier systems. The Propulsion Division was eliminated as part of the NSTS's move to clarify the points of authority and responsibility in the Shuttle program and to establish clear lines of communication in the information transfer and decision-making processes.

Money for NASA's Launch Systems

From 1979 through 1983, all funds for NASA's launch systems came from the Research and Development (R&D) appropriation. Beginning in FY 1984, Congress authorized a new appropriation, Space Flight, Control, and Data Communications (SFC&DC), to segregate funds for ongoing Space Shuttle-related activities. This appropriation was in response to an October 1983 recommendation by the NASA Advisory Council, which stated that the operating budgets, facilities, and personnel required to support an operational Space Shuttle be "fenced" from the rest of NASA's programs. The council maintained that such an action would speed the transition to more efficient operations, help reduce costs, and ease the transfer of STS operations to the private sector or some new government operating agency, should such a transfer be desired.[4] SFC&DC was used for Space Shuttle production and capability development, space transportation operations (including ELVs), and space and ground network communications and data systems activities.

Most data in this section came from two sources. Programmed (actual) figures came from the yearly budget estimates prepared by NASA's Budget Operations Division, Office of the Comptroller. Data on NASA's submissions and congressional action came from the chronological history budget submissions issued for each fiscal year.

---

[4]NASA, *Fiscal Year 1985 Budget Submission, Chronological History,* House Authorization Committee Report, issued April 22, 1986, p. 15.

Table 2–1 shows the total appropriated amounts for launch vehicles and launch-related components. Tables 2–2 through 2–12 show the requested amount that NASA submitted to Congress, the amount authorized for each item or program, the final appropriation, and the programmed (or actual) amounts spent for each item or program. The submission represented the amount agreed to by NASA and OMB, not necessarily the initial request NASA made to the President's budget officer. The authorized amount was the ceiling set by Congress for a particular purpose. The appropriated amount reflected the amount that Congress actually allowed the Treasury to provide for specific purposes.[5]

As is obvious from examining the tables, funds for launch vehicles and other launch-related components were often rolled up into the total R&D or SFC&DC appropriation or other major budget category ("undistributed" funds). This made tracking the funding levels specifically designated for launch systems difficult. However, supporting congressional committee documentation clarified some of Congress's intentions. In the late 1970s and early 1980s, Congress intended that most space launches were to move from ELVs to the Space Shuttle as soon as the Shuttle became operational. This goal was being rethought by 1984, and it was replaced by a mixed fleet strategy after 1986. However, even though the government returned to using ELVs for many missions, it never again took prime responsibility for most launch system costs. From 1985 through 1987, Congress declared that the NASA ELV program would be completely funded on a reimbursable basis. Launch costs would be paid by the customer (for example, commercial entities, other government agencies, or foreign governments). Not until 1988 did Congress provide direct funding for two Delta II launch vehicles that would be used for NASA launches in the early 1990s. Although the federal government funded the Shuttle to a much greater degree, it was also to be used, when possible, for commercial or other government missions in which the customer would pay part of the launch and payload costs.

In some fiscal years, ELVs, upper stages, Shuttle-related launch elements, and advanced programs had their own budget lines in the congressional budget submissions. However, no element always had its own budget line. To follow the changes that took place, readers should consult the notes that follow each table as well as examine the data in each table. Additional data relating to the major Space Shuttle budget categories can be found in the budget tables in Chapter 3.

NASA's budget structure changed from one year to the next depending on the status of various programs and budget priorities. From 1979 through 1983, all launch-related activities fell under the R&D appropriation.

---

[5]The term "appropriation" is used in two ways. It names a major budget category (for instance, R&D or SFC&DC). It is also used to designate an amount that Congress allows an agency to spend (for example, NASA's FY 1986 appropriation was $7,546.7 million).

Launch elements were found in the Space Flight Operations program, the Space Shuttle program, and the ELV program. The Space Flight Operations program included the major categories of space transportation systems operations capability development, space transportation system operations, and advanced programs (among others not relevant here). Upper stages were found in two areas: space transportation systems operations capability development included space transportation system upper stages, and space transportation system operations included upper stage operations.

The Space Shuttle program included design, development, test, and evaluation (DDT&E), which encompassed budget items for the orbiter, main engine, external tank, solid rocket booster (SRB), and launch and landing. The DDT&E category was eliminated after FY 1982. The production category also was incorporated into the Space Shuttle program. Production included budget line items for the orbiter, main engine, and launch and landing.

The ELV program included budget items for the Delta, Scout, Centaur, and Atlas F. (FY 1982 was the last year that the Atlas F appeared in the budget.)

FY 1984 was a transition year. Budget submissions (which were submitted to Congress as early as FY 1982) and authorizations were still part of the R&D appropriation. By the time the congressional appropriations committee acted, the SFC&DC appropriation was in place. Two major categories, Shuttle production and operational capability and space transportation operations, were in SFC&DC. Shuttle production and operational capability contained budget items for the orbiter, launch and mission support, propulsion systems (including the main engine, solid rocket booster, external tank, and systems support), and changes and systems upgrading. Space transportation operations included Shuttle operations and ELVs. Shuttle operations included flight operations, flight hardware (encompassing the orbiter, solid rocket booster, and external tank), and launch and landing. ELVs included the Delta and Scout. (FY 1984 was the last year that there was a separate ELV budget category until the FY 1988 budget.) R&D's Space Transportation Capability Development program retained upper stages, advanced programs, and the Tethered Satellite System.

Beginning in FY 1985, most launch-related activities moved to the SFC&DC appropriation. In 1987, NASA initiated the Expendable Launch Vehicles/Mixed Fleet program to provide launch services for selected NASA payloads not requiring the Space Shuttle's capabilities.

Space Shuttle Funding

Funds for the Space Shuttle Main Engine (SSME) were split into a DDT&E line item and a production line item from 1979 through 1983. Funds for the external tank and SRB were all designated as DDT&E. Beginning with FY 1984, SSME, external tank, and SRB funds were located in the capability development/flight hardware category and in the Propulsion System program. Capability development included continuing capability development tasks for the orbiter, main engine, external tank, and SRB and the development of the filament wound case SRB. Congress defined propulsion systems as systems that provided "for the production of the SSME, the implementation of the capability to support operational requirements, and the anomaly resolution for the SSME, SRB, and external tank."

Some Space Shuttle funds were located in the flight hardware budget category. Flight hardware provided for the procurement of the external tank, the manufacturing and refurbishment of SRB hardware and motors, and space components for the main engine; orbiter spares, including external tank disconnects, sustaining engineering, and logistics support for external tank, SRB, and main engine flight hardware elements; and maintenance and operation of flight crew equipment.

Tables 2–1 through 2–9 provide data for the launch-related elements of the Space Shuttle and other associated items. Budget data for additional Shuttle components and the major Shuttle budget categories are found in the Chapter 3 budget tables.

Characteristics

The following sections describe the launch vehicles and launch-related components used by NASA during the period 1979 through 1988. A chronology of each vehicle's use and its development is also presented, as well as the characteristics of each launch vehicle and launch-related component.

In some cases, finding the "correct" figures for some characteristics was difficult. The specified height, weight, or thrust of a launch vehicle occasionally differed among NASA, contractor, and media sources. Measurements, therefore, are approximate. Height or length was measured in several different ways, and sources varied on where a stage began and ended for measuring purposes. The heights of individual stages were generally without any payload. However, the overall height of the assembled launch vehicle may include the payload. Source material did not always indicate whether the overall length included the payload, and sometimes one mission operations report published two figures for the height of a launch vehicle within the same report.

Thrust was also expressed in more than one way. Source material referred to thrust "in a vacuum," "at sea level," "average," "nominal," and "maximum." Thrust levels vary during a launch and were sometimes presented as a range of values or as a percentage of "rated thrust." Frequently, there was no indication of which definition of thrust was being used.

This chapter uses the following abbreviations for propellants: LH_2 = liquid hydrogen, LOX = liquid oxygen, N_2H_2 = hydrazine, N_2O_4 = nitrogen tetroxide, RJ-1 = liquid hydrocarbon, and RP-1 = kerosene.

Expendable Launch Vehicles

From 1979 through 1988, NASA attempted seventy-four launches with a 94.6-percent success rate using the expendable Atlas E/F, Atlas-Centaur, Delta, or all-solid-fueled Scout vehicle—all vehicles that had been used during NASA's second decade. During this time, the agency continued to built Deltas and maintained its capability to build Scouts and Atlases on demand. It did not emphasize ELV development but rather focused on Space Shuttle development and the start of STS operational status. However, the adoption of the mixed fleet strategy returned some attention to ELV development

The following section summarizes ELV activities during the decade from 1979 through 1988. Figure 2–7 and Table 2–13 present the success rate of each launch vehicle.

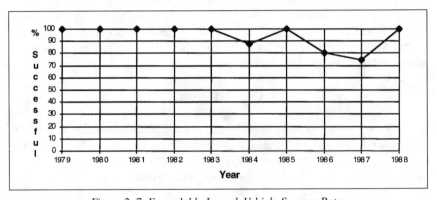

Figure 2–7. Expendable Launch Vehicle Success Rate

1979

NASA conducted nine launches during 1979, all successful. These used the Scout, the Atlas E/F, the Atlas-Centaur, and the Delta. Of the nine launches, three launched NASA scientific and application payloads, and six supported other U.S. government and nongovernment reimbursing customers.[6]

A Scout vehicle launched the NASA Stratospheric Aerosol and Gas Experiment (SAGE), a NASA magnetic satellite (Magsat), and a reimbursable United Kingdom scientific satellite (UK-6/Ariel). An Atlas-Centaur launched a FltSatCom DOD communications satellite and a NASA scientific satellite (HEAO-3). Three launches used the Delta: one domestic communications satellite for Western Union, another for RCA, and an experimental satellite, called SCATHA, for DOD. A weather satellite was launched on an Atlas F by the Air Force for NASA and the National Oceanic and Atmospheric Administration (NOAA).

---

[6]*Aeronautics and Space Report of the President, 1979* (Washington, DC: U.S. Government Printing Office (GPO), 1980), p. 39.

1980

Seven ELV launches took place in 1980: three on Deltas, three on Atlas-Centaurs, and one on an Atlas F. Of the seven, one was for NASA; the other six were reimbursable launches for other U.S. government, international, and domestic commercial customers that paid NASA for the launch and launch support costs.[7]

A Delta launched the Solar Maximum Mission, the single NASA mission, with the goal of observing solar flares and other active Sun phenomena and measuring total radiative output of the Sun over a six-month period. A Delta also launched GOES 4 (Geostationary Operational Environmental Satellite) for NOAA. The third Delta launch, for Satellite Business Systems (SBS), provided integrated, all-digital, interference-free transmission of telephone, computer, electronic mail, and videoconferencing to clients.

An Atlas-Centaur launched FltSatCom 3 and 4 for the Navy and DOD. An Atlas-Centaur also launched Intelsat V F-2. This was the first in a series of nine satellites launched by NASA for Intelsat and was the first three-axis stabilized Intelsat satellite. An Atlas F launched NOAA-B, the third in a series of Sun-synchronous operational environmental monitoring satellites launched by NASA for NOAA. A booster failed to place this satellite in proper orbit, causing mission failure.

1981

During 1981, NASA launched missions on eleven ELVs: one on a Scout, five using Deltas (two with dual payloads), four on Atlas-Centaurs, and one using an Atlas F. All but two were reimbursable launches for other agencies or commercial customers, and all were successful.[8]

A Scout vehicle launched the DOD navigation satellite, NOVA 1. In five launches, the Delta, NASA's most-used launch vehicle, deployed seven satellites. Two of these launches placed NASA's scientific Explorer satellites into orbit: Dynamics Explorer 1 and 2 on one Delta and the Solar Mesosphere Explorer (along with Uosat for the University of Surrey, England) on the other. The other three Delta launches had paying customers, including the GOES 5 weather satellite for NOAA and two communications satellites, one for SBS and one for RCA.

An Atlas-Centaur, which was the largest ELV being used by NASA, launched four missions: Comstar D-4, a domestic communications satellite for Comsat; two Intelsat V communications satellites for Intelsat; and the last in the current series of FltSatCom communications satellites for DOD. An Atlas F launched the NOAA 7 weather satellite for NOAA.

---

[7]*Aeronautics and Space Report of the President, 1980* (Washington, DC: GPO, 1981).

[8]*Aeronautics and Space Report of the President, 1981* (Washington, DC: GPO, 1982).

In addition, ELVs continued to provide backup support to STS customers during the early development and transition phase of the STS system.

1982

NASA launched nine missions on nine ELVs in 1982, using seven Deltas and two Atlas-Centaurs. Of the nine, eight were reimbursable launches for other agencies or commercial customers, and one was a NASA applications mission.[9]

The Delta supported six commercial and international communications missions for which NASA was fully reimbursed: RCA's Satcom 4 and 5, Western Union's Westar 4 and 5, India's Insat 1A, and Canada's Telesat G (Anik D-1). In addition, a Delta launched Landsat 4 for NASA. The Landsat and Telesat launches used improved, more powerful Deltas. An Aerojet engine and a tank with a larger diameter increased the Delta weight-carrying capability into geostationary-transfer orbit by 140 kilograms. An Atlas-Centaur launched two communications satellites for the Intelsat.

1983

During 1983, NASA launched eleven satellites on eleven ELVs, using eight Deltas, one Atlas E, one Atlas-Centaur, and one Scout. A Delta launch vehicle carried the European Space Agency's EXOSAT x-ray observatory to a highly elliptical polar orbit. Other 1983 payloads launched into orbit on NASA ELVs were the NASA-Netherlands Infrared Astronomy Satellite (IRAS), NOAA 8 and GOES 6 for NOAA, Hilat for the Air Force, Intelsat VF-6 for Intelsat, Galaxy 1 and 2 for Hughes Communications, Telstar 3A for AT&T, and Satcom 1R and 2R for RCA; all except IRAS were reimbursable.[10]

The increased commercial use of NASA's launch fleet and launch services conformed to President Reagan's policy statement on May 16, 1983, in which he announced that the U.S. government would facilitate the commercial operation of the ELV program.

1984

During 1984, NASA's ELVs provided launch support to seven satellite missions using four Deltas, one Scout, one Atlas-Centaur, and one Atlas E. During this period, the Delta vehicle completed its forty-third consecutive successful launch with the launching of the NATO-IIID satellite in November 1984. In addition, a Delta successfully launched Landsat 5 for NOAA in March (Landsat program management had trans-

---

[9]*Aeronautics and Space Report of the President, 1982* (Washington, DC: GPO, 1983), p. 19.

[10]*Aeronautics and Space Report of the President, 1983* (Washington, DC: GPO, 1984), p. 17.

ferred to NOAA in 1983); AMPTE, a joint American, British, and German space physics mission involving three satellites, in August; and Galaxy-C in September. Other payloads launched during 1984 by NASA ELVs included a Navy navigation satellite by a Scout, an Intelsat communications satellite by an Atlas-Centaur, and a NOAA weather satellite by an Atlas F vehicle. The launch of the Intelsat satellite experienced an anomaly in the launch vehicle that resulted in mission failure. All missions, except the NASA scientific satellite AMPTE, were reimbursable launches for other U.S. government, international, and domestic commercial missions that paid NASA for launch and launch support.[11]

In accordance with President Reagan's policy directive to encourage commercialization of the launch vehicle program, Delta, Atlas-Centaur, and Scout ELVs were under active consideration during this time by commercial operators for use by private industry. NASA and Transpace Carriers, Inc. (TCI), signed an interim agreement for exclusive rights to market the Delta vehicle, and negotiations took place with General Dynamics on the Atlas-Centaur. A *Commerce Business Daily* announcement, published August 8, 1984, solicited interest for the private use of the Scout launch vehicle. Ten companies expressed interest in assuming a total or partial takeover of this vehicle system.

Also in August 1984, President Reagan approved a National Space Strategy intended to implement the 1983 National Space Policy. This strategy called for the United States to encourage and facilitate commercial ELV operations and minimize government regulation of these operations. It also mandated that the U.S. national security sector pursue an improved assured launch capability to satisfy the need for a launch system that complemented the STS as a hedge against "unforeseen technical and operational problems" and to use in case of crisis situations. To accomplish this, the national security sector should "pursue the use of a limited number of ELVs."[12]

1985

In 1985, NASA's ELVs continued to provide launch support during the transition of payloads to the Space Shuttle. Five launches took place using ELVs. Two of these were DOD satellites launched on Scouts—one from the Western Space and Missile Center and the other from the Wallops Flight Facility. Atlas-Centaurs launched the remaining three missions for Intelsat on a reimbursable basis.[13]

---

[11]*Aeronautics and Space Report of the President, 1984* (Washington, DC: GPO, 1985), p. 23

[12]White House Fact Sheet, "National Space Strategy," August 15, 1984.

[13]*Aeronautics and Space Report of the President, 1985* (Washington, DC: GPO, 1986).

1986

In 1986, NASA's ELVs launched five space application missions for NOAA and DOD. A Scout launched the Polar Beacon Experiments and Auroral Research satellite (Polar Bear) from Vandenberg Air Force Base; an Atlas-Centaur launched a FltSatCom satellite in December; an Atlas E launched a NOAA satellite; and two Delta vehicles were used—one to launch a NOAA GOES satellite and the other to launch a DOD mission. One of the Delta vehicles failed during launch and was destroyed before boosting the GOES satellite into transfer orbit. An investigation concluded that the failure was caused by an electrical short in the vehicle wiring. Wiring modifications were incorporated into all remaining Delta vehicles. In September, the second Delta vehicle successfully launched a DOD mission.[14]

Partly as a result of the *Challenger* accident, NASA initiated studies in 1986 on the need to establish a Mixed Fleet Transportation System, consisting of the Space Shuttle and existing or new ELVs. This policy replaced the earlier stated intention to make the Shuttle NASA's sole launch vehicle.

1987

In 1987, NASA launched four spacecraft missions using ELVs. Three of these missions were successful: a Delta launch of GOES 7 for NOAA into geostationary orbit in February; a Delta launch of Palapa B-2, a communications satellite for the Indonesian government, in March; and a Scout launch of a Navy Transit satellite in September. In March, an Atlas-Centaur launch attempt of FltSatCom 6, a Navy communications satellite, failed when lightning in the vicinity of the vehicle caused the engines to malfunction. The range safety officer destroyed the vehicle approximately fifty-one seconds after launch.[15]

1988

The ELV program had a perfect launch record in 1988 with six successful launches. In February, a Delta ELV lifted a classified DOD payload into orbit. This launch marked the final east coast Delta launch by a NASA launch team. A NASA-Air Force agreement, effective July 1, officially transferred custody of Delta Launch Complex 17 at Cape Canaveral Air Force Station to the Air Force. Over a twenty-eight-year period, NASA had launched 143 Deltas from the two Complex 17 pads. A similar transaction transferred accountability for Atlas/Centaur Launch Complex 36 to the Air Force.[16]

---

[14]*Aeronautics and Space Report of the President, 1986* (Washington, DC: GPO, 1987).

[15]*Aeronautics and Space Report of the President, 1987* (Washington, DC: GPO, 1988).

[16]*Aeronautics and Space Report of the President, 1988* (Washington, DC: GPO, 1989).

Also in 1988, a Scout launched San Marcos DL from the San Marco launch facility in the Indian Ocean, a NASA-Italian scientific mission, during March. Its goal was to explore the relationship between solar activity and meteorological phenomena by studying the dynamic processes that occur in the troposphere, stratosphere, and thermosphere. In April, another Scout deployed the SOOS-3, a Navy navigation satellite. In June, a third Scout carried the NOVA-II, the third in a series of improved Navy Transit navigation satellites, into space. The final Scout launch of the year deployed a fourth SOOS mission in August. In September, an Atlas E launched NOAA H, a National Weather Service meteorological satellite funded by NOAA, into Sun-synchronous orbit. This satellite payload included on-board search-and-rescue instruments.

In addition to arranging for the purchase of launch services from the commercial sector, NASA took steps to divest itself of an adjunct ELV capability and by making NASA-owned ELV property and services available to the private sector. During 1988, NASA finalized a barter agreement with General Dynamics that gave the company ownership of NASA's Atlas-Centaur flight and nonflight assets. In exchange, General Dynamics agreed to provide the agency with two Atlas-Centaur launches at no charge. An agreement was signed for the first launch service—supporting the FltSatCom F-8 Navy mission. NASA and General Dynamics also completed a letter contract for a second launch service to support the NASA-DOD Combined Release and Radiation Effects Satellite (CRRES) mission. In addition, NASA transferred its Delta vehicle program to the U.S. Air Force. Finally, enabling agreements were completed to allow ELV companies to negotiate directly with the appropriate NASA installation. During 1988, NASA Headquarters signed enabling agreements with McDonnell Douglas, Martin Marietta, and LTV Corporation. The Kennedy Space Center and General Dynamics signed a subagreement in March to allow General Dynamics to take over maintenance and operations for Launch Complex 36.

ELV Characteristics

The Atlas Family

The basic Atlas launch vehicle was a one-and-a-half stage stainless steel design built by the Space Systems Division of General Dynamics. It was designed as an intercontinental ballistic missile (ICBM) and was considered an Air Force vehicle. However, the Atlas launch vehicle was also used successfully in civilian space missions dating from NASA's early days. The Atlas launched all three of the unmanned lunar exploration programs (Ranger, Lunar Orbiter, and Surveyor). Atlas vehicles also launched the Mariner probes to Mars, Venus, and Mercury and the Pioneer probes to Jupiter, Saturn, and Venus.

NASA used two families of Atlas vehicles during the 1979–1988 period: the Atlas E/F series and the Atlas-Centaur series. The Atlas E/F launched seven satellites during this time, six of them successful (Table 2–14). The Atlas E/F space booster was a refurbished ICBM. It burned kerosene (RP-1) and liquid oxygen in its three main engines, two Rocketdyne MA-3 booster engines, and one sustainer engine. The Atlas E/F also used two small vernier engines located at the base of the RP-1 tank for added stability during flight (Table 2–15). The Atlas E/F was designed to deliver payloads directly into low-Earth orbit without the use of an upper stage.

The Atlas-Centaur (Figure 2–8) was the nation's first high-energy launch vehicle propelled by liquid hydrogen and liquid oxygen. Developed and launched under the direction of the Lewis Research Center, it became operational in 1966 with the launch of Surveyor 1, the first U.S. spacecraft to soft-land on the Moon's surface. Beginning in 1979, the Centaur stage was used only in combination with the Atlas booster, but it had been successfully used earlier in combination with the Titan III booster to launch payloads into interplanetary trajectories, sending two Helios spacecraft toward the Sun and two Viking spacecraft toward Mars.[17] From 1979 through 1988, the Atlas-Centaur launched 18 satellites with only two failures (Table 2–16).

The Centaur stage for the Atlas booster was upgraded in 1973 and incorporated an integrated electronic system controlled by a digital computer. This flight-proven "astrionics" system checked itself and all other systems prior to and during the launch phase; during flight, it controlled all events after the liftoff. This system was located on the equipment module on the forward end of the Centaur stage. The 16,000-word capacity computer replaced the original 4,800-word capacity computer and enabled it to take over many of the functions previously handled by separate mechanical and electrical systems. The new Centaur system handled navigation, guidance tasks, control pressurization, propellant management, telemetry formats and transmission, and initiation of vehicle events (Table 2–17).

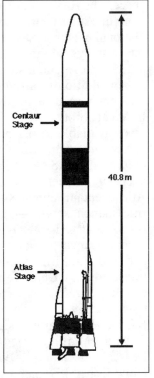

Figure 2–8. Atlas-Centaur Launch Vehicle

---

[17]For details, see Linda Neuman Ezell, *NASA Historical Data Book, Volume III: Programs and Projects, 1969–1978* (Washington, DC: NASA SP-4012, 1988).

LAUNCH SYSTEMS

The Delta Family

NASA has used the Delta launch vehicle since the agency's inception. In 1959, NASA's Goddard Space Flight Center awarded a contract to Douglas Aircraft Company (later McDonnell Douglas) to produce and integrate twelve launch vehicles. The Delta, using components from the Air Force's Thor intermediate range ballistic missile (IRBM) program and the Navy's Vanguard launch program, was available eighteen months later. The Delta has evolved since that time to meet the increasing demands of its payloads and has been the most widely used launch vehicle in the U.S. space program, with thirty-five launches from 1979 through 1988 and thirty-four of them successful (Table 2–18).

The Delta configurations of the late 1970s and early 1980s were designated the 3900 series. Figure 2–9 illustrates the 3914, and Figure 2–10 shows the 3920 with the Payload Assist Module (PAM) upper stage. The 3900 series resembled the earlier 2900 series (Table 2–19), except for the replacement of the Castor II solid strap-on motors with nine larger and more powerful Castor IV solid motors (Tables 2–20 and 2–21).

The RS-27 engine, manufactured by the Rocketdyne Division of Rockwell International, powered the first stage of the Delta. It was a single-start power plant, gimbal-mounted and operated on a combination of liquid oxygen and kerosene (RP-1). The thrust chamber was regeneratively

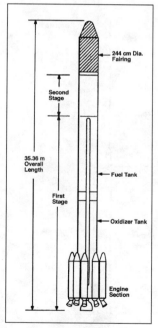

Figure 2–9.
Delta 3914

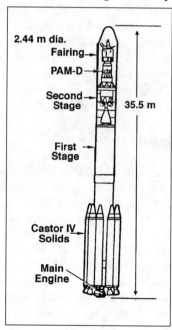

Figure 2–10.
Delta 3920/PAM-D

cooled, with the fuel circulating through 292 tubes that comprised the inner wall of the chamber.

The following four-digit code designated the type of Delta launch vehicle:
- 1st digit designated the type of strap-on engines:
  - 2 = Castor II, extended long tank Thor with RS-27 main engine
  - 3 = Castor IV, extended long tank Thor with RS-27 main engine
- 2nd digit designated the number of strap-on engines
- 3rd digit designated the type of second stage and manufacturer:
  - 1 = ninety-six-inch manufactured by TRW (TR-201)
  - 2 = ninety-six-inch stretched tank manufactured by Aerojet (AJ10-118K)
- 4th digit designated the type of third stage:
  - 0 = no third stage
  - 3 = TE-364-3
  - 4 = TE-364-4

For example, a model designation of 3914 indicated the use of Castor IV strap-on engines, extended long tank with an RS-27 main engine; nine strap-ons; a ninety-six-inch second stage manufactured by TRW; and a TE-364-4 third stage engine. A PAM designation appended to the last digit indicated the use of a McDonnell-Douglas PAM.

Scout Launch Vehicle

The standard Scout launch vehicle (Scout is an acronym for Solid Controlled Orbital Utility Test) was a solid propellant four-stage booster system. It was the world's first all-solid propellant launch vehicle and was one of NASA's most reliable launch vehicles. The Scout was the smallest of the basic launch vehicles used by NASA and was used for orbit, probe, and reentry Earth missions (Figure 2–11).

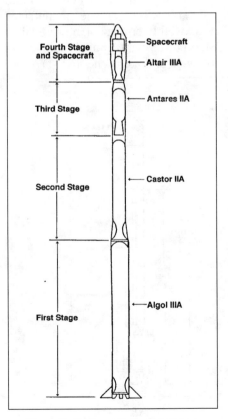

Figure 2–11. Scout-D Launch Vehicle (Used in 1979)

The first Scout launch took place in 1960. Since that time, forty-six NASA Scout launches have taken place, including fourteen between 1979 and 1988, when every launch was successful (Table 2–22). In addition to NASA payloads, Scout clients included DOD, the European Space Research Organization, and several European governments. The Scout was used for both orbital and suborbital missions and has participated in research in navigation, astronomy, communications, meteorology, geodesy, meteoroids, reentry materials, biology, and Earth and atmospheric sensing. It was the only U.S. ELV launched from three launch sites: Wallops on the Atlantic Ocean, Vandenberg on the Pacific Ocean, and the San Marco platform in the Indian Ocean. It could also inject satellites into a wider range of orbital inclinations than any other launch vehicle.

Unlike NASA's larger ELVs, the Scout was assembled and the payload integrated and checked out in the horizontal position. The vehicle was raised to the vertical orientation prior to launch. The propulsion motors were arranged in tandem with transition sections between the stages to tie the structure together and to provide space for instrumentation. A standard fifth stage was available for highly elliptical and solar orbit missions.

Scout's first-stage motor was based on an earlier version of the Navy's Polaris missile motor; the second-stage motor was developed from the Army's Sergeant surface-to-surface missile; and the third- and fourth-stage motors were adapted by NASA's Langley Research Center from the Navy's Vanguard missile. The fourth-stage motor used on the G model could carry almost four times as much payload to low-Earth orbit as the original model in 1960—that is, 225 kilograms versus fifty-nine kilograms (Table 2–23).

Vought Corporation, a subsidiary of LTV Corporation, was the prime contractor for the Scout launch vehicle. The Langley Research Center managed the Scout program.

Space Shuttle

The reusable, multipurpose Space Shuttle was designed to replace the ELVs that NASA used to deliver commercial, scientific, and applications spacecraft into Earth's orbit. Because of its unique design, the Space Shuttle served as a launch vehicle, a platform for scientific laboratories, an orbiting service center for other satellites, and a return carrier for previously orbited spacecraft. Beginning with its inaugural flight in 1981 and through 1988, NASA flew twenty-seven Shuttle missions (Table 2–24). This section focuses on the Shuttle's use as a launch vehicle. Chapter 3 discusses its use as a platform for scientific laboratories and servicing functions.

The Space Shuttle system consisted of four primary elements: an orbiter spacecraft, two solid rocket boosters (SRBs), an external tank to house fuel and an oxidizer, and three main engines. Rockwell International built the orbiter and the main engines; Thiokol Corporation produced the SRB motors; and the external tank was built by Martin Marietta Corporation. The Johnson Space Center directed the orbiter and integration contracts, while the Marshall Space Flight Center managed the SRB, external tank, and main engine contracts.

The Shuttle could transport up to 29,500 kilograms of cargo into near-Earth orbit (185.2 to 1,111.2 kilometers). This payload was carried in a bay about four and a half meters in diameter and eighteen meters long. Major system requirements were that the orbiter and the two SRBs be reusable and that the orbiter have a maximum 160-hour turnaround time after landing from the previous mission. The orbiter vehicle carried personnel and payloads to orbit, provided a space base for performing their assigned tasks, and returned personnel and payloads to Earth. The orbiter provided a habitable environment for the crew and passengers, including scientists and engineers. Additional orbiter characteristics are addressed in Chapter 3.

The Shuttle was launched in an upright position, with thrust provided by the three main engines and the two SRBs. After about two minutes, at an altitude of about forty-four kilometers, the two boosters were spent and were separated from the orbiter. They fell into the ocean at predetermined points and were recovered for reuse.

The main engines continued firing for about eight minutes, cutting off at about 109 kilometers altitude just before the spacecraft was inserted into orbit. The external tank was separated, and it followed a ballistic trajectory back into a remote area of the ocean but was not recovered.

Two smaller liquid rocket engines made up the orbital maneuvering system (OMS). The OMS injected the orbiter into orbit, performed maneuvers while in orbit, and slowed the vehicle for reentry. After reentry, the unpowered orbiter glided to Earth and landed on a runway.

The Shuttle used two launch sites: the Kennedy Space Center in Florida and Vandenberg Air Force Base in California. Under optimum conditions, the orbiter landed at the site from which it was launched. However, as shown in the tables in Chapter 3 that describe the individual Shuttle missions, weather conditions frequently forced the Shuttle to land at Edwards Air Force Base in California, even though it had been launched from Kennedy.

Main Propulsion System

The main propulsion system (MPS) consisted of three Space Shuttle main engines (SSMEs), three SSME controllers, the external tank, the orbiter MPS propellant management subsystem and helium subsystem, four ascent thrust vector control units, and six SSME hydraulic servo-actuators. The MPS, assisted by the two SRBs during the initial phases of the ascent trajectory, provided the velocity increment from liftoff to a predetermined velocity increment before orbit insertion. The Shuttle jettisoned the two SRBs after their fuel had been expended, but the MPS continued to thrust until the predetermined velocity was achieved. At that time, main engine cutoff (MECO) was initiated, the external tank was jettisoned, and the OMS was ignited to provide the final velocity increment for orbital insertion. The magnitude of the velocity increment supplied by the OMS depended on payload weight, mission trajectory, and system limitations.

Along with the start of the OMS thrusting maneuver (which settled the MPS propellants), the remaining liquid oxygen propellant in the orbiter feed system and SSMEs was dumped through the nozzles of the engines. At the same time, the remaining liquid hydrogen propellant in the orbiter feed system and SSMEs was dumped overboard through the hydrogen fill and drain valves for six seconds. Then the hydrogen inboard fill and drain valve closed, and the hydrogen recirculation valve opened, continuing the dump. The hydrogen flowed through the engine hydrogen bleed valves to the orbiter hydrogen MPS line between the inboard and outboard hydrogen fill and drain valves, and the remaining hydrogen was dumped through the outboard fill and drain valve for approximately 120 seconds.

During on-orbit operations, the flight crew vacuum made the MPS inert by opening the liquid oxygen and liquid hydrogen fill and drain valves, which allowed the remaining propellants to be vented to space. Before entry into the Earth's atmosphere, the flight crew repressurized the MPS propellant lines with helium to prevent contaminants from being drawn into the lines during entry and to maintain internal positive pressure. MPS helium also purged the spacecraft's aft fuselage. The last activity involving the MPS occurred at the end of the landing rollout. At that time, the helium remaining in on-board helium storage tanks was released into the MPS to provide an inert atmosphere for safety.

Main Engine

The SSME represented a major advance in propulsion technology. Each engine had an operating life of seven and a half hours and fifty-five starts and the ability to throttle a thrust level that extended over a wide range (65 percent to 109 percent of rated power level). The SSME was the first large, liquid-fuel rocket engine designed to be reusable.

A cluster of three SSMEs housed in the orbiter's aft fuselage provided the main propulsion for the orbiter. Ignited on the ground prior to launch, the cluster of liquid hydrogen–liquid oxygen engines operated in parallel with the SRBs during the initial ascent. After the boosters separated, the main engines continued to operate. The nominal operating time was approximately eight and a half minutes. The SSMEs developed thrust by using high-energy propellants in a staged combustion cycle. The propellants were partially combusted in dual preburners to produce high-pressure hot gas to drive the turbopumps. Combustion was completed in the main combustion chamber. The cycle ensured maximum performance because it eliminated parasitic losses. The various thrust levels provided for high thrust during liftoff and the initial ascent phase but allowed thrust to be reduced to limit acceleration to three g's during the final ascent phase. The engines were gimbaled to provide pitch, yaw, and roll control during the orbiter boost phase.

Key components of each engine included four turbopumps (two low- and two high-pressure), two preburners, the main injector, the main combustion chamber, the nozzle, and the hot-gas manifold. The manifold was the structural backbone of the engine. It supported the two preburners, the high-pressure pumps, the main injector, the pneumatic control assembly, and the main combustion chamber with the nozzle. Table 2–25 summarizes SSME characteristics.

The SSME was the first rocket engine to use a built-in electronic digital controller. The controller accepted commands from the orbiter for engine start, shutdown, and change in throttle setting and also monitored engine operation. In the event of a failure, the controller automatically corrected the problem or shut down the engine safely.

Main Engine Margin Improvement Program. Improvements to the SSMEs for increased margin and durability began with a formal Phase II program in 1983. Phase II focused on turbomachinery to extend the time between high-pressure fuel turbopump (HPFT) overhauls by reducing the operating temperature in the HPFT and by incorporating margin improvements to the HPFT rotor dynamics (whirl), turbine blade, and HPFT bearings. Phase II certification was completed in 1985, and all the changes were incorporated into the SSMEs for the STS-26 mission.

In addition to the Phase II improvements, NASA made additional changes to the SSME to further extend the engine's margin and durability. The main changes were to the high-pressure turbomachinery, main combustion chamber, hydraulic actuators, and high-pressure turbine discharge temperature sensors. Changes were also made in the controller software to improve engine control. Minor high-pressure turbomachinery design changes resulted in margin improvements to the turbine blades, thereby extending the operating life of the turbopumps. These changes included applying surface texture to important parts of the fuel turbine blades to improve the material properties in the pressure of hydrogen and incorporating a damper into the high-pressure oxidizer turbine blades to reduce vibration.

Plating a welded outlet manifold with nickel increased the main combustion chamber's life. Margin improvements were also made to five hydraulic actuators to preclude a loss in redundancy on the launch pad. Improvements in quality were incorporated into the servo-component coil design, along with modifications to increase margin. To address a temperature sensor in-flight anomaly, the sensor was redesigned and extensively tested without problems.

To certify the improvements to the SSMEs and demonstrate their reliability through margin (or limit) testing, NASA initiated a ground test program in December 1986. Its primary purposes were to certify the improvements and demonstrate the engine's reliability and operating margin. From December 1986 to December 1987, 151 tests and 52,363 seconds of operation (equivalent to 100 Shuttle missions) were performed. These hot-fire ground tests were performed at the single-engine test stands at the Stennis Space Center in Mississippi and at the Rockwell International Rocketdyne Division's Santa Susana Field Laboratory in California.

NASA also conducted checkout and acceptance tests of the three main engines for the STS-26 mission. Those tests, also at Stennis, began in August 1987, and all three STS-26 engines were delivered to the Kennedy Space Center by January 1988.

Along with hardware improvements, NASA conducted several major reviews of requirements and procedures. These reviews addressed such topics as possible failure modes and effects, as well as the associated critical items list. Another review involved having a launch/abort reassessment team examine all launch-commit criteria, engine redlines, and software logic. NASA also performed a design certification review. Table 2–26 lists these improvements, as well as events that occurred earlier in the development of the SSME.

A related effort involved Marshall Space Flight Center engineers who, working with their counterparts at Kennedy, accomplished a comprehensive launch operations and maintenance review. This ensured that engine processing activities at the launch site were consistent with the latest operational requirements.

External Tank

The external tank contained the propellants (liquid hydrogen and liquid oxygen) for the SSMEs and supplied them under pressure to the three main engines in the orbiter during liftoff and ascent. Just prior to orbital insertion, the main engines cut off, and the external tank separated from the orbiter, descended through a ballistic trajectory over a predesignated area, broke up, and impacted in a remote ocean area. The tank was not recovered.

The largest and heaviest (when loaded) element of the Space Shuttle, the external tank had three major components: a forward liquid oxygen tank; an unpressurized intertank, which contained most of the electrical components; and an aft liquid hydrogen tank. Beginning with the STS-6 mission, NASA used a lightweight external tank (LWT). For each kilogram of weight reduced from the original external tank, the cargo-carrying capability of the Space Shuttle spacecraft increased one kilogram. The weight reduction was accomplished by eliminating portions of stringers (structural stiffeners running the length of the hydrogen tank), using fewer stiffener rings, and by modifying major frames in the hydrogen tank. Also, significant portions of the tank were milled differently to reduce thickness, and the weight of the external tank's aft SRB attachments was reduced by using a stronger, yet lighter and less expensive, titanium alloy. Earlier, the use of the LWT reduced the total weight by deleting the antigeyser line. The line paralleled the oxygen feed line and provided a circulation path for liquid oxygen to reduce the accumulation of gaseous oxygen in the feed line while the oxygen tank was being filled before launch. After NASA assessed propellant loading data from ground tests and the first four Space Shuttle missions, engineers removed the antigeyser line for STS-5 and subsequent missions. The total length and

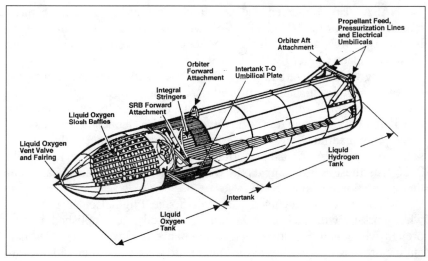

Figure 2–12. External Tank

diameter of the external tank remained unchanged (Figure 2–12). Table 2–27 summarizes the external tank characteristics, and Table 2–28 presents a chronology of external development.

As well as containing and delivering the propellant, the external tank served as the structural backbone of the Space Shuttle during launch operations. The external tank consisted of two primary tanks: a large hydrogen tank and a smaller oxygen tank, joined by an intertank to form one large propellant-storage container. Superlight ablator (SLA-561) and foam insulation sprayed on the forward part of the oxygen tank, the intertank, and the sides of the hydrogen tank protected the outer surfaces. The insulation reduced ice or frost formation during launch preparation, protecting the orbiter from free-falling ice during flight. This insulation also minimized heat leaks into the tank, avoided excessive boiling of the liquid propellants, and prevented liquification and solidification of the air next to the tank.

The external tank attached to the orbiter at one forward attachment point and two aft points. In the aft attachment area, umbilicals carried fluids, gases, electrical signals, and electrical power between the tank and the orbiter. Electrical signals and controls between the orbiter and the two SRBs also were routed through those umbilicals.

Liquid Oxygen Tank. The liquid oxygen tank was an aluminum monocoque structure composed of a fusion-welded assembly of preformed, chem-milled gores, panels, machined fittings, and ring chords. It operated in a pressure range of 1,035 to 1,138 mmHg. The tank contained antislosh and antivortex provisions to minimize liquid residuals and damp fluid motion. The tank fed into a 0.43-meter-diameter feedline that sent the liquid oxygen through the intertank, then outside the external tank to the aft righthand external tank/orbiter disconnect umbilical. The feedline permitted liquid oxygen to flow at approximately 1,268 kilograms per second, with the SSMEs operating at 104 percent of rated thrust, or permitted a maximum flow of 71,979 liters per minute. The liquid oxygen tank's double-wedge nose cone reduced drag and heating, contained the vehicle's ascent air data system, and served as a lightning rod.

Intertank. The intertank was not a tank in itself but provided a mechanical connection between the liquid oxygen and liquid hydrogen tanks. The primary functions of the intertank were to provide structural continuity to the propellant tanks, to serve as a protective compartment to house instruments, and to receive and distribute thrust loads from the SRBs. The intertank was a steel/aluminum semimonocoque cylindrical structure with flanges on each end for joining the liquid oxygen and liquid hydrogen tanks. It housed external tank instrumentation components and provided an umbilical plate that interfaced with the ground facility arm for purging the gas supply, hazardous gas detection, and hydrogen gas boiloff during ground operations. It consisted of mechanically joined skin, stringers, and machined panels of aluminum alloy. The intertank was vented during flight. It contained the forward SRB-external tank attach thrust beam and fittings that distributed the SRB loads to the liquid oxygen and liquid hydrogen tanks.

Liquid Hydrogen Tank. The liquid hydrogen tank was an aluminum semimonocoque structure of fusion-welded barrel sections, five major ring frames, and forward and aft ellipsoidal domes. Its operating pressure was 1,759 mmHg. The tank contained an antivortex baffle and siphon outlet to transmit the liquid hydrogen from the tank through a 0.43-meter line to the left aft umbilical. The liquid hydrogen feedline flow rate was 211.4 kilograms per second, with the SSMEs at 104 percent of rated thrust, or a maximum flow of 184,420 liters per minute. At the forward end of the liquid hydrogen tank was the external tank/orbiter forward attachment pod strut, and at its aft end were the two external tank/orbiter aft attachment ball fittings as well as the aft SRB-external tank stabilizing strut attachments.

External Tank Thermal Protection System. The external tank thermal protection system consisted of sprayed-on foam insulation and premolded ablator materials. The system also included the use of phenolic thermal insulators to preclude air liquefaction. Thermal isolators were required for liquid hydrogen tank attachments to preclude the liquefaction of air-exposed metallic attachments and to reduce heat flow into the liquid hydrogen. The thermal protection system weighed 2,192 kilograms.

External Tank Hardware. The external hardware, external tank/orbiter attachment fittings, umbilical fittings, and electrical and range safety system weighed 4,136.4 kilograms.

Each propellant tank had a vent and relief valve at its forward end. This dual-function valve could be opened by ground support equipment for the vent function during prelaunch and could open during flight when the ullage (empty space) pressure of the liquid hydrogen tank reached 1,966 mmHg or the ullage pressure of the liquid oxygen tank reached 1,293 mmHg.

The liquid oxygen tank contained a separate, pyrotechnically operated, propulsive tumble vent valve at its forward end. At separation, the liquid oxygen tumble vent valve was opened, providing impulse to assist in the separation maneuver and more positive control of the entry aerodynamics of the external tank.

There were eight propellant-depletion sensors, four each for fuel and oxidizer. The fuel-depletion sensors were located in the bottom of the fuel tank. The oxidizer sensors were mounted in the orbiter liquid oxygen feedline manifold downstream of the feedline disconnect. During SSME thrusting, the orbiter general purpose computers constantly computed the instantaneous mass of the vehicle because of the usage of the propellants. Normally, MECO was based on a predetermined velocity; however, if any two of the fuel or oxidizer sensors sensed a dry condition, the engines would be shut down.

The locations of the liquid oxygen sensors allowed the maximum amount of oxidizer to be consumed in the engines, while allowing sufficient time to shut down the engines before the oxidizer pumps ran dry. In addition, 500 kilograms of liquid hydrogen were loaded over and above that required by the six-to-one oxidizer/fuel engine mixture ratio. This assured that MECO from the depletion sensors was fuel rich; oxidizer-rich engine shutdowns could cause burning and severe erosion of engine components.

Four pressure transducers located at the top of the liquid oxygen and liquid hydrogen tanks monitored the ullage pressures. Each of the two aft external tank umbilical plates mated with a corresponding plate on the orbiter. The plates helped maintain alignment among the umbilicals. Physical strength at the umbilical plates was provided by bolting corresponding umbilical plates together. When the orbiter general purpose computers commanded external tank separation, the bolts were severed by pyrotechnic devices.

The external tank had five propellant umbilical valves that interfaced with orbiter umbilicals—two for the liquid oxygen tank and three for the liquid hydrogen tank. One of the liquid oxygen tank umbilical valves was for liquid oxygen, the other for gaseous oxygen. The liquid hydrogen tank umbilical had two valves for liquid and one for gas. The intermediate-diameter liquid hydrogen umbilical was a recirculation umbilical used only during the liquid hydrogen chill-down sequence during prelaunch.

The external tank also had two electrical umbilicals that carried electrical power from the orbiter to the tank and the two SRBs and provided information from the SRBs and external tank to the orbiter. A swing-arm-mounted cap to the fixed service structure covered the oxygen tank vent on top of the external tank during countdown and was retracted about two minutes before liftoff. The cap siphoned off oxygen vapor that threatened to form large ice on the external tank, thus protecting the orbiter's thermal protection system during launch.

External Tank Range Safety System. A range safety system, monitored by the flight crew, provided for dispersing tank propellants if necessary. It included a battery power source, a receiver/decoder, antennas, and ordnance.

Post-Challenger Modification. Prior to the launch of STS-26, NASA modified the external tank by strengthening the hydrogen pressurization line. In addition, freezer wrap was added to the hydrogen line. This permitted the visual detection of a hydrogen fire (Table 2–28).

Solid Rocket Boosters

The two SRBs provided the main thrust to lift the Space Shuttle off the pad and up to an altitude of about forty-four and a half kilometers. In addition, the two SRBs carried the entire weight of the external tank and orbiter and transmitted the weight load through their structure to the mobile launcher platform. The SRBs were ignited after the three SSMEs' thrust level was verified. The two SRBs provided 71.4 percent of the thrust at liftoff and during first-stage ascent. Seventy-five seconds after SRB separation, SRB apogee occurred at an altitude of approximately sixty-five kilometers. SRB impact occurred in the ocean approximately 226 kilometers downrange, to be recovered and returned for refurbishment and reuse.

The primary elements of each booster were the motor (including case, propellant, igniter, and nozzle), structure, separation systems, operational flight instrumentation, recovery avionics, pyrotechnics, deceleration system, thrust vector control system, and range safety destruct system (Figure 2–13). Each booster attached to the external tank at the SRB's aft frame with two lateral sway braces and a diagonal attachment. The forward end of each SRB joined the external tank at the forward end

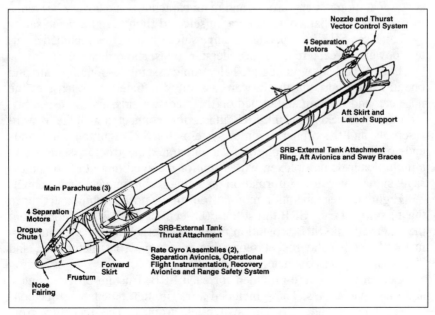

Figure 2–13. Solid Rocket Booster

of the SRB's forward skirt. On the launch pad, each booster also connected to the mobile launcher platform at the aft skirt with four bolts and nuts that were severed by small explosives at liftoff.

The SRBs were used as matched pairs. Each consisted of four solid rocket motor (SRM) segments. The pairs were matched by loading each of the four motor segments in pairs from the same batches of propellant ingredients to minimize any thrust imbalance. The exhaust nozzle in the aft segment of each motor, in conjunction with the orbiter engines, steered the Space Shuttle during the powered phase of launch. The segmented-casing design assured maximum flexibility in fabrication and ease of transportation and handling. Each segment was shipped to the launch site on a heavy-duty rail car with a specially built cover.

The propellant mixture in each SRB motor consisted of an ammonium perchlorate (oxidizer, 69.6 percent by weight), aluminum (fuel, 16 percent), iron oxide (a catalyst, 0.4 percent), a polymer (a binder that held the mixture together, 12.04 percent), and an epoxy curing agent (1.96 percent). The propellant was an eleven-point star-shaped perforation in the forward motor segment and a double-truncated-cone perforation in each of the aft segments and aft closure. This configuration provided high thrust at ignition and then reduced the thrust by approximately one-third fifty seconds after liftoff to prevent overstressing the vehicle during maximum dynamic pressure.

The cone-shaped aft skirt supported the four aft separation motors. The aft section contained avionics, a thrust vector control system that consisted of two auxiliary power units and hydraulic pumps, hydraulic systems, and a nozzle extension jettison system. The forward section of each booster contained avionics, a sequencer, forward separation motors, a nose cone separation system, drogue and main parachutes, a recovery beacon, a recovery light, a parachute camera on selected flights, and a range safety system. Each SRB incorporated a range safety system that included a battery power source, a receiver-decoder, antennas, and ordnance.

Each SRB had two integrated electronic assemblies, one forward and one aft. After burnout, the forward assembly initiated the release of the nose cap and frustum and turned on the recovery aids. The aft assembly, mounted in the external tank-SRB attach ring, connected with the forward assembly and the orbiter avionics systems for SRB ignition commands and nozzle thrust vector control. Each integrated electronic assembly had a multiplexer-demultiplexer, which sent or received more than one message, signal, or unit of information on a single communications channel.

Eight booster separation motors (four in the nose frustum and four in the aft skirt) of each SRB thrust for 1.02 seconds at SRB separation from the external tank. SRB separation from the external tank was electrically initiated. Each solid rocket separation motor was 0.8 meter long and 32.5 centimeters in diameter (Table 2–29).

Location aids were provided for each SRB, frustum-drogue chutes, and main parachutes. These included a transmitter, antenna, strobe/converter, battery, and saltwater switch electronics. The recovery crew retrieved the SRBs, frustum/drogue chutes, and main parachutes. The nozzles were plugged, the solid rocket motors were dewatered, and the crew towed the SRBs back to the launch site. Each booster was removed from the water, and its components disassembled and washed with fresh and de-ionized water to limit saltwater corrosion. The motor segments, igniter, and nozzle were shipped back to Thiokol for refurbishment. The SRB nose caps and nozzle extensions were not recovered.

Testing and production of the SRB were well under way in 1979. The booster performed well until the *Challenger* accident revealed flaws that had very likely existed for several missions but had resulted in little remedial action. The 1986 *Challenger* accident forced major modifications to the SRB and SRM.

***Post*-Challenger *Modifications*.** On June 13, 1986, President Reagan directed NASA to implement, as soon as possible, the recommendations of the Presidential Commission on the Space Shuttle *Challenger* Accident. During the downtime following the *Challenger* accident, NASA analyzed critical structural elements of the SRB, primarily focused in areas where anomalies had been noted during postflight inspection of recovered hardware.

Anomalies had been noted in the attach ring where the SRBs joined the external tank. Some of the fasteners showed distress where the ring attached to the SRB motor case. Tests attributed this to the high loads encountered during water impact. To correct the situation and ensure higher strength margins during ascent, the attach ring was redesigned to encircle the motor case completely (360 degrees). Previously, the attach ring formed a "C" and encircled the motor case 270 degrees.

In addition, NASA performed special structural tests on the aft skirt. During this test program, an anomaly occurred in a critical weld between the hold-down post and skin of the skirt. A redesign added reinforcement brackets and fittings in the aft ring of the skirt. These modifications added approximately 200 kilograms to the weight of each SRB.

Solid Rocket Motor Redesign. The Presidential Commission determined that the cause of the loss of the *Challenger* was "a failure in the joint between the two lower segments of the right solid rocket motor. The specific failure was the destruction of the seals that are intended to prevent hot gases from leaking through the joint during the propellant burn of the rocket motor."[18]

Consequently, NASA developed a plan for a redesigned solid rocket motor (RSRM). Safety in flight was the primary objective of the SRM redesign. Minimizing schedule impact by using existing hardware, to the extent practical, without compromising safety was another objective.

---

[18]*Report at a Glance,* report to the President by the Presidential Commission on the Space Shuttle *Challenger* Accident, Chapter IV, "The Cause of the Accident," Finding (no pg. number).

NASA established a joint redesign team with participants from the Marshall Space Flight Center, other NASA centers, Morton Thiokol, and outside NASA. The team developed an "SRM Redesign Project Plan" to formalize the methodology for SRM redesign and requalification. The plan provided an overview of the organizational responsibilities and relationships; the design objectives, criteria, and process; the verification approach and process; and a master schedule. Figure 2–14 shows the SRM Project Schedule as of August 1986. The companion "Development and Verification Plan" defined the test program and analyses required to verify the redesign and unchanged components of the SRM. The SRM was carefully and extensively redesigned. The RSRM received intense scrutiny and was subjected to a thorough certification process to verify that it worked properly and to qualify the motor for human spaceflight.

NASA assessed all aspects of the existing SRM and required design changes in the field joint, case-to-nozzle joint, nozzle, factory joint, propellant grain shape, ignition system, and ground support equipment. The propellant, liner, and castable inhibitor formulations did not require changes. Design criteria were established for each component to ensure a safe design with an adequate margin of safety. These criteria focused on loads, environments, performance, redundancy, margins of safety, and verification philosophy.

The team converted the criteria into specific design requirements during the Preliminary Requirements Reviews held in July and August 1986. NASA assessed the design developed from these requirements at the Preliminary Design Review held in September 1986 and baselined in October 1986. NASA approved the final design at the Critical Design

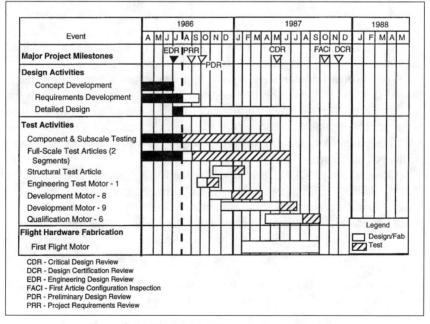

Figure 2–14. Solid Rocket Motor Redesign Schedule

Review held in October 1987. Manufacture of the RSRM test hardware and the first flight hardware began prior to the Preliminary Design Review and continued in parallel with the hardware certification program. The Design Certification Review considered the analyses and test results versus the program and design requirements to certify that the RSRM was ready to fly.

Specific Modifications. The SRM field-joint metal parts, internal case insulation, and seals were redesigned, and a weather protection system was added. The major change in the motor case was the new tang capture feature to provide a positive metal-to-metal interference fit around the circumference of the tang and clevis ends of the mating segments. The interference fit limited the deflection between the tang and clevis O-ring sealing surfaces caused by motor pressure and structural loads. The joints were designed so that the seals would not leak under twice the expected structural deflection and rate.

The new design, with the tang capture feature, the interference fit, and the use of custom shims between the outer surface of the tang and inner surface of the outer clevis leg, controlled the O-ring sealing gap dimension. The sealing gap and the O-ring seals were designed so that a positive compression (squeeze) was always on the O-rings. The minimum and maximum squeeze requirements included the effects of temperature, O-ring resiliency and compression set, and pressure. The redesign increased the clevis O-ring groove dimension so that the O-ring never filled more than 90 percent of the O-ring groove, and pressure actuation was enhanced.

The new field-joint design also included a new O-ring in the capture feature and an additional leak check port to ensure that the primary O-ring was positioned in the proper sealing direction at ignition. This new or third O-ring also served as a thermal barrier in case the sealed insulation was breached. The field-joint internal case insulation was modified to be sealed with a pressure-actuated flap called a j-seal, rather than with putty as in the STS 51-L (*Challenger*) configuration.

The redesign added longer field-joint-case mating pins, with a reconfigured retainer band, to improve the shear strength of the pins and increase the metal parts' joint margin of safety. The joint safety margins, both thermal and structural, were demonstrated over the full ranges of ambient temperature, storage compression, grease effect, assembly stresses, and other environments. The redesign incorporated external heaters with integral weather seals to maintain the joint and O-ring temperature at a minimum of 23.9 degrees Celsius. The weather seal also prevented water intrusion into the joint.

Original Versus Redesigned SRM Case-to-Nozzle Joint. The SRM case-to-nozzle joint, which experienced several instances of O-ring erosion in flight, was redesigned to satisfy the same requirements imposed on the case field joint. Similar to the field joint, case-to-nozzle joint modifications were made in the metal parts, internal insulation, and O-rings. The redesign added radial bolts with Stato-O-Seals to minimize the joint sealing gap opening. The internal insulation was modified to be sealed adhesively, and a third O-ring was included. The third O-ring served as a dam or wiper in front of the primary O-ring to prevent the polysulfide adhesive from being extruded in the primary O-ring groove. It also served as a thermal barrier in case the polysulfide adhesive was breached. The polysulfide adhesive replaced the putty used in the STS 51-L joint. Also, the redesign added an another leak check port to reduce the amount of trapped air in the joint during the nozzle installation process and to aid in the leak check procedure.

Nozzle. Redesigned internal joints of the nozzle metal parts incorporated redundant and verifiable O-rings at each joint. The modified nozzle steel fixed housing part permitted the incorporation of the 100 radial bolts that attached the fixed housing to the case's aft dome. The new nozzle nose inlet, cowl/boot, and aft exit cone assemblies used improved bonding techniques. Increasing the thickness of the aluminum nose inlet housing and improving the bonding process eliminated the distortion of the nose inlet assembly's metal-part-to-ablative-parts bond line. The changed tape-wrap angle of the carbon cloth fabric in the areas of the nose inlet and throat assembly parts improved the ablative insulation erosion tolerance. Some of these ply-angle changes had been in progress prior to STS 51-L. Additional structural support with increased thickness and contour changes to the cowl and outer boot ring increased their margins of safety. In addition, the outer boot ring ply configuration was altered.

Factory Joint. The redesign incorporated minor modifications in the case factory joints by increasing the insulation thickness and layup to increase the margin of safety on the internal insulation. Longer pins were also added, along with a reconfigured retainer band and new weather seal to improve factory joint performance and increase the margin of safety. In addition, the redesign changed the O-ring and O-ring groove size to be consistent with the field joint.

Propellant. The motor propellant forward transition region was recontoured to reduce the stress fields between the star and cylindrical portions of the propellant grain.

Ignition System. The redesign incorporated several minor modifications into the ignition system. The aft end of the igniter steel case, which contained the igniter nozzle insert, was thickened to eliminate a localized weakness. The igniter internal case insulation was tapered to improve the manufacturing process. Finally, although vacuum putty was still used at the joint of the igniter and case forward dome, it eliminated asbestos as one of its constituents.

Ground Support Equipment. Redesigned ground support equipment (1) minimized the case distortion during handling at the launch site, (2) improved the segment tang and clevis joint measurement system for more accurate reading of case diameters to facilitate stacking, (3) minimized the risk of O-ring damage during joint mating, and (4) improved leak testing of the igniter, case, and nozzle field joints. A ground support equipment assembly aid guided the segment tang into the clevis and rounded the two parts with each other. Other ground support equipment modifications included transportation monitoring equipment and the lifting beam.

Testing. Tests of the redesigned motor were carried out in a horizontal attitude, providing a more accurate simulation of actual conditions of the field joint that failed during the STS 51-L mission. In conjunction with the horizontal attitude for the RSRM full-scale testing, NASA incorporated externally applied loads. Morton Thiokol constructed a second horizontal test stand for certification of the redesigned SRM. The contractor used this new stand to simulate environmental stresses, loads, and temperatures experienced during an actual Space Shuttle launch and ascent. The new test stand also provided redundancy for the original stand.

The testing program included five full-scale firings of the RSRM prior to STS-26 to verify the RSRM performance. These included two development motor tests, two qualification motor tests, and a production verification motor test. The production verification motor test in August 1988 intentionally introduced severe artificial flaws into the test motor to make sure that the redundant safety features implemented during the redesign effort worked as planned. Laboratory and component tests were used to determine component properties and characteristics. Subscale motor tests simulated gas dynamics and thermal conditions for components and subsystem design. Simulator tests, consisting of motors using full-size flight-type segments, verified joint design under full flight loads, pressure, and temperature.

Full-scale tests verified analytical models and determined hardware assembly characteristics; joint deflection characteristics; joint performance under short duration, hot-gas tests, including joint flaws and flight loads; and redesigned hardware structural characteristics. Table 2–30 lists the events involved in the redesign of the SRB and SRM as well as earlier events in their development.[19]

Upper Stages

The upper stages boost payloads from the Space Shuttle's parking orbit or low-Earth orbit to geostationary-transfer orbit or geosynchronous orbit. They are also used on ELV missions to boost payloads from an early stage of the orbit maneuver into geostationary-transfer orbit or geosynchronous orbit. The development of the upper stages used by NASA began prior to 1979 and continued throughout the 1980s (Table 2–31).

The upper stages could be grouped into three categories, according to their weight delivery capacity:
- Low capacity: 453- to 1,360-kilogram capacity to geosynchronous orbit

---

[19]See Ezell, *NASA Historical Data Book, Volume III,* for earlier events in SRB development.

- Medium capacity: 1,360- to 3,175-kilogram capacity to geosynchronous orbit
- High capacity: 3,175- to 5,443-kilogram capacity to geosynchronous orbit

Inertial Upper Stages

DOD designed and developed the Inertial Upper Stage (IUS) medium-capacity system for integration with both the Space Shuttle and Titan launch vehicle. It was used to deliver spacecraft into a wide range of Earth orbits beyond the Space Shuttle's capability. When used with the Shuttle, the solid-propellant IUS and its payload were deployed from the orbiter in low-Earth orbit. The IUS was then ignited to boost its payload to a higher energy orbit. NASA used a two-stage configuration of the IUS primarily to achieve geosynchronous orbit and a three-stage version for planetary orbits.

The IUS was 5.18 meters long and 2.8 meters in diameter and weighed approximately 14,772 kilograms. It consisted of an aft skirt, an aft stage SRM with 9,707 kilograms of solid propellant generating 202,828.8 newtons of thrust, an interstage, a forward stage SRM with 2,727.3 kilograms of propellant generating 82,288 newtons of thrust and using an extendible exit cone, and an equipment support section. The equipment support section contained the avionics that provided guidance, navigation, telemetry, command and data management, reaction control, and electrical power. All mission-critical components of the avionics system and thrust vector actuators, reaction control thrusters, motor igniter, and pyrotechnic stage separation equipment were redundant to ensure better than 98-percent reliability (Figure 2–15).

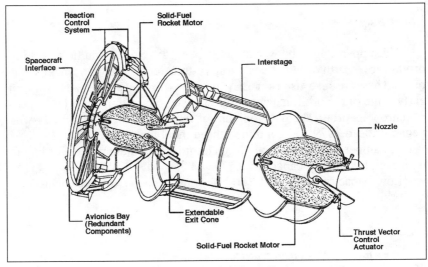

Figure 2–15. Inertial Upper Stage

The spacecraft was attached to the IUS at a maximum of eight attachment points. These points provided substantial load-carrying capability while minimizing thermal transfer. Several IUS interface connectors provided power and data transmission to the spacecraft. Access to these connectors could be provided on the spacecraft side of the interface plane or through the access door on the IUS equipment bay.

The IUS provided a multilayer insulation blanket of aluminized Kapton with polyester net spacers and an aluminized beta cloth outer layer across the IUS and spacecraft interface. All IUS thermal blankets vented toward and into the IUS cavity. All gases within the IUS cavity vented to the orbiter payload bay. There was no gas flow between the spacecraft and the IUS. The thermal blankets were grounded to the IUS structure to prevent electrostatic charge buildup.

Beginning with STS-26, the IUS incorporated a number of advanced features. It had the first completely redundant avionics system developed for an uncrewed space vehicle. This system could correct in-flight features within milliseconds. Other advanced features included a carbon composite nozzle throat that made possible the high-temperature, long-duration firing of the IUS motor and a redundant computer system in which the second computer could take over functions from the primary computer, if necessary.

Payload Assist Module

The Payload Assist Module (PAM), which was originally called the Spinning Stage Upper Stage, was developed by McDonnell Douglas at its own expense for launching smaller spacecraft to geostationary-transfer orbit. It was designed as a higher altitude booster of satellites deployed in near-Earth orbit but operationally destined for higher altitudes. The PAM-D could launch satellites weighing up to 1,247 kilograms. It was originally configured for satellites that used the Delta ELV but was used on both ELVs and the Space Shuttle. The PAM-DII (used on STS 61-B and STS 61-C) could launch satellites weighing up to 1,882 kilograms. A third PAM, the PAM-A, had been intended for satellites weighing up to 1,995 kilograms and was configured for missions using the Atlas-Centaur. NASA halted its development in 1982, pending definition of spacecraft needs. Commercial users acquired the PAM-D and PAM-DII directly from the manufacturer.

The PAM consisted of a deployable (expendable) stage and reusable airborne support equipment. The deployable stage consisted of a spin-stabilized SRM, a payload attach fitting to mate with the unmanned spacecraft, and the necessary timing, sequencing, power, and control assemblies.

The PAM's airborne support equipment consisted of the reusable hardware elements required to mount, support, control, monitor, protect, and operate the PAM's expendable hardware and untended spacecraft from liftoff to deployment from the Space Shuttle or ELV. It also provided these functions for the safing and return of the stage and spacecraft in case of an aborted mission. The airborne support equipment was designed to be as self-contained as possible. The major airborne support equipment elements included the cradle for structural mounting and support, the spin table and drive system, the avionics system to control and monitor the airborne support equipment and the PAM vehicle, and the thermal control system.

The PAM stages were supported through the spin table at the base of the motor and through restraints at the PAF. The forward restraints were retracted before deployment. The sunshield of the PAM-D and DII provided thermal protection of the PAM/untended spacecraft when the Space Shuttle orbiter payload bay doors were open on orbit.

Transfer Orbit Stage

The development of the Transfer Orbit Stage (TOS) began in April 1983 when NASA signed a Space System Development Agreement with Orbital Sciences Corporation (OSC) to develop a new upper stage. Under the agreement, OSC provided technical direction, systems engineering, mission integration, and program management of the design, production, and testing of the TOS. NASA, with participation by the Johnson and Kennedy Space Centers, provided technical assistance during TOS development and agreed to provide technical monitoring and advice during TOS development and operations to assure its acceptability for use with major national launch systems, including the STS and Titan vehicles. NASA also established a TOS Program Office at the Marshall Space Flight Center. OSC provided all funding for the development and manufacturing of TOS (Figure 2–16).

In June 1985, Marshall awarded a 16-month contract to OSC for a laser initial navigation system (LINS) developed for the TOS. Marshall would use the LINS for guidance system research, testing, and other purposes related to the TOS program.

Production of the TOS began in mid-1986. It was scheduled to be used on the Advanced Communications Technology Satellite (ACTS) and the Planetary Observer series of scientific exploration spacecraft, beginning with the Mars Observer mission in the early 1990s.

The TOS could place 2,490 to 6,080 kilograms payloads into geostationary-transfer orbit from the STS and up to 5,227 kilograms from the Titan III and IV and could also deliver spacecraft to planetary and other high-energy trajectories. The TOS allowed smaller satellites to be placed into geostationary-transfer orbit in groups of

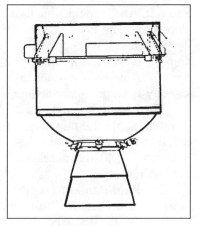

Figure 2–16.
Transfer Orbit Stage

two or three. Two payloads of the Atlas class (1,136 kilograms) or three payloads of the Delta class (636 kilograms) could be launched on a single TOS mission. Besides delivery of commercial communications satellites, its primary market, the TOS would be used for NASA and DOD missions.

The TOS system consisted of flight vehicle hardware and software and associated airborne and ground support equipment required for buildup. Table 3–32 lists its characteristics. Performance capabilities of the TOS included:
- Earth escape transfer capability
- Geosynchronous transfer orbit capability
- Orbit inclination change capability
- Low-altitude transfer capability
- Intermediate transfer orbit capability
- De-orbit maneuver
- Satellite repair and retrieval

Apogee and Maneuvering System

The liquid bipropellant Apogee and Maneuvering System (AMS) was designed to be used both with and independently of the TOS. The AMS would boost the spacecraft into a circular orbit and allow on-orbit maneuvering. Martin Marietta Denver Aerospace worked to develop the AMS with Rockwell International's Rocketdyne Division, providing the AMS RS-51 bipropellant rocket engine, and Honeywell, Inc., supplied the TOS/AMS LINS avionics system.

When it became operational, the TOS/AMS combination would deliver up to approximately 2,950 kilograms into geosynchronous orbit from the orbiter's parking orbit into final geosynchronous orbit. The TOS/AMS would have a delivery capability 30 percent greater than the IUS and would reduce stage and STS user costs. The main propulsion, reaction control, avionics, and airborne support equipment systems would be essentially the same as those used on the TOS. In particular, the avionics would be based on a redundant, fault-tolerant LINS.

Operating alone, the AMS would be able to place communications satellites weighing up to approximately 2,500 kilograms into geostationary-transfer orbit after deployment in the standard Space Shuttle parking orbit. Other missions would include low-orbit maneuvering between the Shuttle and the planned space station, delivery of payloads to Sun-synchronous and polar orbits, and military on-demand maneuvering capability. The AMS was planned to be available for launch in early 1989 and would provide an alternative to the PAM-DII.

The avionics, reaction control system, and airborne support equipment designs of the AMS would use most of the standard TOS components. Main propulsion would be provided by the 2,650-pound thrust Rocketdyne RS-51 engine. This engine was restartable and operable over extended periods. A low-thrust engine option that provided 400 pounds of thrust would also be available for the AMS.

Centaur Upper Stage

NASA studied and began production in the early 1980s of a modified Centaur upper stage for use with the STS for planetary and heavier geosynchronous mission applications. The proposed modifications would increase the size of the propellant tanks to add about 50 percent more propellant capacity and make the stage compatible with the Space Shuttle. This wide-body version would use the same propulsion system and about 85 percent of the existing Centaur's avionics systems. Contracts were negotiated with General Dynamics, Honeywell, Pratt & Whitney, and Teledyne for the design, development, and procurement of Centaur upper stages for the Galileo and International Solar Polar missions that were scheduled for 1986.

However, following the *Challenger* accident, NASA determined that even with modifications, the Centaur could not comply with necessary safety requirements for use on the Shuttle. The Centaur upper stage initiative was then dropped.

Advanced Programs

Advanced programs focused on future space transportation activities, including improving space transportation operations through the introduction of more advanced technologies and processes, and on servicing and protecting U.S. space assets. The following sections describe NASA's major advanced program initiatives. Several of the efforts progressed from advanced program status to operational status during this decade.

Orbital Transfer Vehicle

NASA's Advanced Planning/Programs Division of the Office of Space Transportation identified the need for an Orbital Transfer Vehicle (OTV) in the early 1980s, when it became obvious that a way was needed to transport payloads from the Space Shuttle's low-Earth orbit to a higher orbit and to retrieve and return payloads to the Shuttle or future space station. The Marshall Space Flight Center was designated as the lead center for the development effort, and the Lewis Research Center led the propulsion system studies. An untended OTV was proposed for a first flight in the early 1990s.

NASA believed that the use of aerobraking was necessary to make the OTV affordable. Studies beginning in 1981 conducted at Marshall by definition phase contractors Boeing Aerospace Company and General Electric Reentry Systems determined that aerodynamic braking was an efficient fuel-saving technique for the OTV, perhaps doubling payload capacity. This technique would use the Earth's atmosphere as a braking mechanism for return trips, possibly supplemented by the use of a ballute, an inflatable drag device. When the transfer vehicle passed through the atmosphere, the friction of the air against the vehicle would provide enough drag to slow the vehicle. Otherwise, a rocket engine firing would be required to brake the vehicle. Aeroassist braking would save one burn, and the extra fuel could be used to transport a larger payload to a high orbit. The aeroassisted braking could result in about a twofold increase in the amount of payload that could be ferried to high altitudes.

Boeing's studies emphasized low lifting-body designs—"low lift-to-drag ratio"—designs with a relatively low capability of lift to enable them to fly, but ones that weigh less. General Electric Reentry Systems focused on moderate lift-to-drag ratio designs—relatively moderate lift capability and somewhat heavier weight.

In 1981, NASA designated the Lewis Research Center the lead center for OTV propulsion technology. This program supported technology for three advanced engine concepts that were developed by Aerojet TechSystems, Pratt & Whitney, and Rocketdyne to satisfy a NASA-supplied set of goals. The proposed engines would be used to transfer loads—both personnel and cargo—between low-Earth orbit and geosynchronous orbit, and beyond. In addition, because OTVs would face requirements ranging from high-acceleration round-trip transfers for resupply to very low-acceleration one-way transfers of large, flexible structures, NASA investigated variable thrust propulsion systems, which would provide high performance over a broad throttling range.

In 1983, NASA chose the same three contractors to begin a program leading to the design, development, test, and engineering of the OTV. These contracts expired in 1986. NASA sponsored another competitive procurement to continue the OTV propulsion program. Funding was reduced, and only Rocketdyne and Aerojet continued the advanced engine technology development. Component testing began in 1988, and further investigations into aerobraking continued into the 1990s.

The OTV would be used primarily to place NASA, DOD, and commercial satellites and space platforms into geosynchronous orbit. The OTV could also deliver large payloads into other orbits and boost planetary exploration spacecraft into high-velocity orbits approaching their mission trajectory. The vehicle was expected to use liquid oxygen–liquid hydrogen propellants.

The OTV's reusable design provided for twenty flights before it had to be refurbished or replaced. Because of its reusability, the OTV would significantly reduce payload transportation costs.

At the same time, that Lewis was leading propulsion studies, Marshall initiated studies in 1984 to define OTV concepts and chose Boeing Aerospace and Martin Marietta to conduct the conceptual studies. The studies examined the possibilities of both a space-based and an Earth-based OTV. Both would initially be uncrewed upper stages. The ultimate goal, however, was to develop a crewed vehicle capable of ferrying a crew capsule to geosynchronous orbit. The vehicle would then return the crew and capsule for other missions. The development of a crew capsule for the OTV was planned for the 1990s.

The Space Shuttle would carry the Earth-based OTV into space. It would be launched from the Shuttle's payload bay or from an aft cargo carrier attached to the aft end of the Shuttle's external tank. The OTV would transfer payloads from a low orbit to a higher one. It would also retrieve payloads in high orbits and return them to the Shuttle. The OTV would then return to Earth in the Shuttle's payload bay. The OTV would separate from the Shuttle's external tank at about the same time that the payload was deployed from the orbiter's cargo bay. The two components would then join together and begin to travel to a higher orbit. This Earth-based OTV offered the advantage of performing vehicle maintenance and refueling on the ground with the help of gravity, ground facilities, and workers who do not have to wear spacesuits.

A space-based OTV would be based at the future space station. It would move payloads into higher orbit from the space station and then return to its home there. It would be refueled and maintained at the space station. Studies showed cost savings for space-based OTVs. This type of OTV could be assembled in orbit rather than on the ground so it could be larger than a ground-based unit and capable of carrying more payload. Initial studies of an OTV that would be based at the space station were completed in 1985.

A single-stage OTV could boost payloads of up to 7,272 kilograms to high-Earth or geosynchronous orbit. A multistage OTV could provide up to 36,363 kilograms to lunar orbit with 6,818.2 kilograms returned to low-Earth orbit. After completing its delivery or servicing mission, the OTV would use its rocket engines to start a descent. Skimming through the thin upper atmosphere (above sixty kilometers), the OTV's aerobrake would slow the OTV without consuming extra propellant. Then, because of orbital dynamics, the OTV would navigate back to a low-Earth orbit. When the OTV reached the desired orbital altitude, its rocket engines would again fire, circularizing its orbit until it was retrieved by the Space Shuttle or an orbital maneuvering vehicle (OMV) dispatched from the space station.

NASA Administrator James M. Beggs stated in June 1985 that the OTV would complement the proposed OMV. The OTV would transport payloads from low-Earth orbit to destinations much higher than the OMV could reach. The majority of the payloads transported by the OTV would be delivered to geostationary orbit. Beggs envisioned that most OTVs would be based at the space station, where they would be maintained, fueled, and joined to payloads. In time, the OTV would also be used to transport people to geostationary orbit.

Orbital Maneuvering Vehicle

The OMV (Figure 2–17) was designed to aid satellite servicing and retrieval. This uncrewed vehicle could be characterized as a "space tug," which would move satellites and other orbiting objects from place to

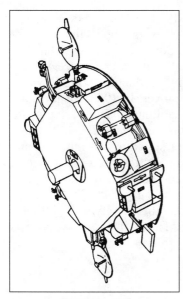

Figure 2–17.
Orbital Maneuvering Vehicle

place above the Earth. A reusable, remotely operated unmanned propulsive vehicle to increase the range of the STS, the OMV was designed to be used primarily for spacecraft delivery, retrieval, boost, deboost, and close proximity visual observation beyond the operating range of the Space Shuttle. The vehicle would extend the reach of the Shuttle up to approximately 2,400 kilometers.

Concept definition studies were completed in 1983, and development began toward a flight demonstration of the ability to refuel propellant tanks of an orbiting satellite. In 1984, an in-flight demonstration of hydrazine fuel transfer took place successfully on STS 41-G. System definition studies were completed in 1985, and in June 1986, TRW was selected by NASA for negotiations leading to the award of a contract to develop the OMV. The Preliminary Requirements Review took place in 1987, and the Preliminary Design Review was held in 1988, with the Marshall Space Flight Center managing the effort.

NASA planned for the OMV to be available for its first mission in 1993, when it would be remotely controlled from Earth. In the early years of use, NASA envisioned that the OMV would be deployed from the Space Shuttle for each short-duration mission and returned to Earth for servicing. Later, the vehicle would be left parked in orbit for extended periods, for use with both the Shuttle and the space station. However, the OMV was the victim of budget cuts, and the contract with TRW was canceled in June 1990.

Tethered Satellite System

The Tethered Satellite System (TSS) program was a cooperative effort between the government of Italy and NASA to provide the capability to perform science in areas of space outside the reach of the Space Shuttle. The TSS would enable scientists to conduct experiments in the upper atmosphere and ionosphere while tethered to the Space Shuttle as its operating base. The system consisted of a satellite anchored to the Space Shuttle by a tether up to 100 kilometers long. (Tethers are long, superstrong tow lines joining orbiting objects together.)

The advanced development stage of the program was completed in 1983, and management for the TSS moved to the Space Transportation and Capability Development Division. In 1984, a study and laboratory program was initiated to define and evaluate several applications of tethers in space. Possible applications included power generation, orbit raising in the absence of propellants, artificial gravity, and space vehicle constellations. In 1986, the Critical Design and Manufacturing Reviews were conducted on the satellite and the deployer. In 1988, manufacture and qualification of the flight subsystems continued. The twelve-meter deployer boom, reel motor, and on-board computer were all qualified and delivered. Also, manufacture of the deployer structure was initiated, and the tether control mechanisms were functionally tested. A test program was completed for the satellite structural and engineering models. The flight satellite structure was due for delivery in early 1989. The development of the scientific instruments continued, with delivery of flight satellite instruments scheduled for early 1989. The first TSS mission was scheduled for 1991.

Advanced Launch System

The Advanced Launch System, a joint NASA-DOD effort, was a systems definition and technology advanced development program aimed at defining a new family of launchers for use after 2000, including a new heavy-lift vehicle. President Reagan signed a report to Congress in January 1988 that officially created the program. Within this DOD-funded program, NASA managed the liquid engine system and advanced development efforts.

Next Manned Launch Vehicle

In 1988, attention was focused on examining various next-generation manned launch vehicle concepts. Three possible directions were considered: Space Shuttle evolution, a personnel launch system, and an advanced manned launch system. The evolution concept referred to the option of improving the current Shuttle design through the incorporation of upgraded technologies and capabilities. The personnel launch system would be a people carrier and have no capability to launch payloads into space. The advanced manned launch system represented an innovative crewed transportation system. Preliminary studies on all three possibilities progressed during 1988.

Shuttle-C

Shuttle-C (cargo) was a concept for a large, uncrewed launch vehicle that would make maximum use of existing Space Shuttle systems with a cargo canister in place of the orbiter. This proposed cargo-carrying launch vehicle would be able to lift 45,454.5 to 68,181.8 kilograms to low-Earth orbit. This payload capacity is two to three times greater than the Space Shuttle payload capability.

In October 1987, NASA selected three contractors to perform the first of a two-phase systems definition study for Shuttle-C. The efforts focused on vehicle configuration details, including the cargo element's length and diameter, the number of liquid-fueled main engines, and an operations concept evaluation that included ground and flight support systems. A major purpose of the study was to determine whether Shuttle-C would be cost effective in supporting the space station. Using Shuttle-C could free the Space Shuttle for STS-unique missions, such as solar system exploration, astronomy, life sciences, space station crew rotation, and logistics and materials processing experiments. Shuttle-C also would be used to launch planetary missions and serve as a test bed for new Shuttle boosters.

The results of the Shuttle-C efforts were to be coordinated with other ongoing advanced launch systems studies to enable a joint steering group, composed of DOD and NASA senior managers. The purpose of the steering group was to formulate a national heavy-lift vehicle strategy that best accommodated both near-term requirements and longer term objectives for reducing space transportation operational costs.

Advanced Upper Stages

Advanced missions in the future would require even greater capabilities to move from low- to high-Earth orbit and beyond. During 1988, activity in the advanced upper stages area focused on the space transfer vehicle (STV) and the possibility of upgrading the existing Centaur upper stage. The STV concept involved a cryogenic hydrogen-oxygen vehicle that could transport payloads weighing from 909.1 to 8,636 kilograms from low-Earth orbit to geosynchronous orbit or the lunar surface, as well as for unmanned planetary missions. The STV concept could potentially lead to a vehicle capable of supporting human exploration missions to the Moon or Mars.

Advanced Solid Rocket Motor

The Advanced Solid Rocket Motor (ASRM) was an STS improvement intended to replace the RSRM that was used on STS-26. The ASRM would be based on a better design than the former rocket motor, contain more reliable safety margins, and use automated manufacturing techniques. The ASRM would also enhance Space Shuttle performance by offering a potential increase of payload mass to orbit from 5454.5 kilograms to 9090.9 kilograms for the Shuttle. In addition, a new study on liquid rocket boosters was conducted that examined the feasibility of replacing SRMs with liquid engines.

In March 1988, NASA submitted the "Space Shuttle Advanced Solid Rocket Motor Acquisition Plan" to Congress. This plan reviewed procurement strategy for the ASRM and discussed implementation plans and schedules. Facilities in Mississippi would be used for production and testing of the new rocket motor. In August 1988, NASA issued an request for proposals to design, develop, test, and evaluate the ASRM. Contract award was anticipated for early 1989, and the first flight using the new motor was targeted for 1994.

LAUNCH SYSTEMS 59

Table 2–1. Appropriated Budget by Launch Vehicle and Launch-Related Component (in thousands of dollars)

Vehicle/Year	1979	Supp. Appr.	1980	Supp. Appr.	1981	1982	1983
Atlas E/F	—		a		—	—	—
Atlas Centaur	b		c		5,600	—	—
Delta	b		d		47,900	30,400	42,800
Scout	b		e		900	800	—
Space Shuttle Main Engine (SSME) Design, Development, Test, and Evaluation (DDTE)	b	f	g	h	145,700	127,000	262,000
SSME Production	b	f	g	h	121,500	105,000	—
Solid Rocket Booster (SRB)	b	i	j	h	14,000	17,000	k
External Tank	b	l	m	h	48,000	25,000	n
STS Upper Stages (STS Operations Capability Development)	b	—	o	—	29,000	75,000 p	q
Upper Stages Operations (STS Operations)	b	—	o		30,900	40,000 p	—
Orbital Maneuvering Vehicle	Program did not begin until 1983 when it was incorporated into Advanced Programs.						r
Tethered Satellite System	Program did not begin until 1982 when it was incorporated into Advanced Programs.					r	r
Advanced Programs	b	—	s	—	t	8,800	11,900

Table 2-1 continued

Vehicle/Year	1984	1985	Supp. Appr.	1986	1987	1988
Atlas E/F	u	v	—	v	v	—
Atlas Centaur	u	v	—	v	v	—
Delta	u	v	—	v	v	28,000
Scout	u	v	—	v	v	—
SSME	w	x	—	y	z	aa
SRB (Propulsion System)	w	x	—	y	z	aa
Solid Rocket Booster (Flight Hardware)	w	x	—	y	z	
External Tank (Propulsion System)	w	x	—	y	z	z
External Tank (Flight Hardware)	w			y	z	aa
Upper Stages	143,200 bb	92,400	40,000	122,000	202,100	159,700
Orbital Maneuvering Vehicle	r	r	—	10,000	45,000	55,000
Tethered Satellite System	3,300	18,200	—	10,000	10,600	7,300
Advanced Programs	15,000	20,500	—	21,000	16,600	30,900

a Undistributed. Only total 1980 R&D appropriation specified: $4,091,086,000. (Authorization for Atlas F = $2,000,000.)
b Undistributed. Total 1979 R&D appropriation = $3,477,200,000.
c Undistributed. Only total 1980 R&D appropriation specified: $4,091,086,000. (Authorization for Atlas Centaur = $18,300,000.)
d Undistributed. Only total 1980 R&D appropriation specified: $4,091,086,000. (Authorization for Delta = $43,100,000.)
e Undistributed. Only total 1980 R&D appropriation specified: $4,091,086,000. (Authorization for Scout = $7,300,000
f Supplemental appropriation specified for overall R&D activities = $185,000,000. (Authorization for SSME = $48,000,000.)
g Undistributed. Only total 1980 R&D appropriation specified: $4,091,086,000. (Authorization for SSME DDT&E = $140,600,000; production = $109,000,000.)
h Undistributed. Total Space Shuttle supplemental appropriation = $285,000,000. No specific Shuttle elements listed.

Table 2–1 continued

i   Supplemental appropriation specified for overall R&D activities = $185,000,000. (Authorization for SRB = $36,700,000.)
j   Undistributed. Only total 1980 R&D appropriation specified: $4,091,086,000. (Authorization for SRB = $57,500,000.)
k   No budget item listed. Supporting committee documentation includes SRB in Space Shuttle Production category with no amount specified. Total Production appropriation = $1,636,600,000.
l   Supplemental appropriation specified for overall R&D activities = $185,000,000. (Authorization for external tank = $27,100,000.)
m   Undistributed. Only total 1980 R&D appropriation specified: $4,091,086,000. (Authorization for external tank = $68,400,000.)
n   No budget item listed. Supporting committee documentation includes external tank in Space Shuttle Production category with no amount specified. Total Production appropriation = $1,636,600,000.
o   No specific funding.
p   Included in narrative for Public Law 97–101, December 23, 1981, 97th Cong.
q   Includes $140,000,000 for Centaur upper stage development (from Appropriations Conference Report to accompany H.R. 6956). Total Space Flight Operations appropriation = $1,796,000,000.
r   Included in Advanced Planning/Programs.
s   Undistributed. Total 1980 R&D appropriation = $4,091,086,000.
t   Undistributed. Included in R&D appropriation of $4,396,200 (modified by General Provision, Sec. 412, to $4,340,788).
u   No budget submission, authorization, or appropriation for specific expendable launch vehicles (ELVs). Total undistributed ELV submission = $50,000,000; authorization = $50,000,000; and appropriation = $50,000,000. ELV appropriation removed from R&D and placed in Space Flight, Control & Data Communications (SFC&DC) (Office of Space Transportation Systems) appropriation. NASA Budget Estimate for FY 1984 shows $50,000,000 for Delta ($0 for Scout) but specific appropriation for Delta not confirmed by congressional committee documentation.
v   FY 1985–1987—no appropriation for ELVs. All ELV costs would be completely funded on a reimbursable basis.
w   No specific appropriation for SSME, external tank, or SRB. Appropriation for Space Shuttle activities of $1,545,000,000 moved from R&D to SFC&DC. Amount of $427,400,000 remained in R&D for upper stages, Spacelab, engineering and technology base, planetary operations and support equipment, Advanced Programs, Tethered Satellite System, and Teleoperator Maneuvering System. NASA Budget Estimate documents indicate estimated amount of $280,700,000 for SSME, $108,400,000 for SRB, and $83,100,000 for external tank under Propulsion Systems/Shuttle Production and Capability Development category. According to NASA Budget Estimate documents, the Shuttle Production and Capability Development /Propulsion Systems "provides for the production of the Space Shuttle's main engines and the development of the capability to support operational requirements established for the orbiter, main engine, solid rocket booster, and external tank." Congressional documents also state that the category includes continuing "capability development tasks for the orbiter, main engine, external tank, and SRB, . . ." and "the development of the filament wound case (FWC) SRB." Some launch system-related appropriated funding is included in the Flight Hardware/Shuttle Operations category (also in SFC&DC) undistributed, included in Shuttle Operations appropriation = $1,520,600,000. NASA Budget Estimate documents indicate estimated amount of $336,200,000 for external tank and $353,200,000 for SRB under Flight Hardware/Shuttle Operations category.

Table 2–1 continued

x Production and residual development tasks for the orbiter, SSME, external tank, and SRB fall under Space Production and Operational Capability, Propulsion Systems. SRBs and external tank procurement (production) falls under Space Transportation Operations, Flight Hardware. No breakdown is provided for individual Space Shuttle propulsion components. The 1985 appropriation for Propulsion Systems = $599,000,000; Flight Hardware appropriation = $758,000,000.

y No breakdown for individual Space Shuttle propulsion components. The 1986 appropriation for Propulsion Systems = $454,000,000; no Flight Hardware appropriation in 1986.

z No breakdown for individual Space Shuttle propulsion components. The 1987 appropriation for Propulsion Systems = $338,400,000; appropriation for Flight Hardware = $646,200,000.

aa No breakdown for individual Space Shuttle propulsion components. The 1988 appropriation for Propulsion Systems = $249,300,000; appropriation for Flight Hardware = $923,100,000.

bb Includes funding for modification of the Centaur for use in the Shuttle.

Source: *NASA Chronological History Fiscal Years 1979–1983 Budget Submissions.*

Table 2–2. Atlas E/F Funding History *(in thousands of dollars)*

Year (Fiscal)	Submission	Authorization	Appropriation	Programmed (Actual)
1980	2,000 *a*	2,000	*b*	1,200
1981	No direct funds authorized or appropriated; no proposed use of Atlas E/F after 1980 by NASA			—
1982	No budget line item			—
1983	Reimbursable only			—
1984	No budget line item for specific ELVs *c*			—
1985 *d*	There were no direct appropriated fund requirements for the ELV program. DOD and NOAA continued to use the Delta, Scout, Atlas, and Atlas—Centaur ELVs on a fully reimbursable basis. Atlas E/F not in use by NASA			—
1986 *e*				—
1987 *f*				—
1988				—

a Atlas F only.
b Undistributed. Included in 1980 R&D appropriation of $4,091,086,000.
c No budget submission, authorization, or appropriation for specific ELVs. Total undistributed ELV submission = $50,000,000; authorization = $50,000,000; and appropriation = $50,000,000. ELV appropriation removed from R&D and placed in Space Flight, Control & Data Communications (SFC&DC) (Office of Space Transportation Systems) appropriation.
d No budget line item for ELVs. Support for ELVs paid for as part of Space Transportation Operations Program. Budget data for Space Transportation Operations Program found in Chapter 3 budget tables.
e No budget line item for ELVs. Support for ELVs paid for as part of Space Transportation Operations Program. Budget data for Space Transportation Operations Program found in Chapter 3 budget tables.
f Included in Flight Hardware category. Budget data for Flight Hardware found in Chapter 3 budget tables.

64   NASA HISTORICAL DATA BOOK

Table 2–3. Atlas-Centaur Funding History (in thousands of dollars)

Year (Fiscal)	Submission	Authorization	Appropriation	Programmed (Actual)
1979	21,500	a	b	17,320
1980	18,300	18,300	c	18,000
1981	5,600	5,600	5,600	5,600 d
1982	Reimbursable only			—
1983	Reimbursable only			—
1984	No budget line item for specific ELVs e			—
1985 f	There were no direct appropriated fund requirements for the,			—
1986 g	ELV program. DOD and NOAA continued to use the Delta			—
1987 h	Scout, Atlas, and Atlas-Centaur ELVs on a fully reimbursable basis.			—
1988	Atlas-Centaur not in use by NASA			

a   Not distributed by vehicle—total 1979 ELV authorization = $74,000,000.
b   Not distributed by vehicle—1979 R&D appropriation = $3,477,200,000.
c   Undistributed. Included in R&D appropriation of $4,091,086,000.
d   Based on anticipated closeout of the NASA program by the end of 1981.
e   No budget submission, authorization, or appropriation for specific ELVs. Total undistributed ELV submission = $50,000,000; authorization = $50,000,000; and appropriation = $50,000,000. ELV appropriation removed from R&D and placed in Space Flight, Control & Data Communications (SFC&DC) (Office of Space Transportation Systems) appropriation.
f   No budget line item for ELVs. Support for ELVs paid for as part of Space Transportation Operations Program. Budget data for Space Transportation Operations Program found in Chapter 3 budget tables.
g   No budget line item for ELVs. Support for ELVs paid for as part of Space Transportation Operations Program. Budget data for Space Transportation Operations Program found in Chapter 3 budget tables.
h   Included in Flight Hardware category. Budget data for Flight Hardware found in Chapter 3 budget tables.

LAUNCH SYSTEMS 65

Table 2–4. Delta Funding History (in thousands of dollars)

Year (Fiscal)	Submission	Authorization	Appropriation	Programmed (Actual)
1979	38,600	a	b	45,680
1980	43,100	43,100	c	43,100
1981	47,900	47,900	47,900	47,900
1982	30,400	30,400	30,400	30,400
1983	42,800	42,800	42,800	83,000
1984		No budget line item for specific ELVs d		50,000 e
1985 f		No budget line item for ELVs		—
1986 g		No budget line item for ELVs		—
1987 h		No budget line item for ELVs		—
1988	28,000 i	60,000 i	28,000 i	28,000 i

a Not distributed by vehicle—total 1979 ELV authorization = $74,000,000.
b Not distributed by vehicle—1979 R&D appropriation = $3,477,200,000.
c Undistributed. Included in R&D appropriation of $4,091,086,000.
d No budget submission, authorization, or appropriation for specific ELVs. Total undistributed ELV submission = $50,000,000; authorization = $50,000,000; and appropriation = $50,000,000. ELV appropriation removed from R&D and placed in SFC&DC (Office of Space Transportation Systems) appropriation. Congressional supporting documentation indicates that $50,000,000 is for "continued procurement of the Delta ELVs in FY 1984."
e NASA budget summary data do not specifically indicate that programmed amount was for the Delta. However, the narrative that accompanies congressional committee reports describes programs that use the Delta as the launch vehicle.
f No budget line item for ELVs. Support for ELVs paid for as part of Space Transportation Operations Program. It was anticipated that the NASA ELV program would be completely funded on a reimbursable basis in 1985.
g No budget line item for ELVs. Support for ELVs paid for as part of Space Transportation Operations Program. It was anticipated that the NASA ELV program would be completely funded on a reimbursable basis in 1986.
h Included in Flight Hardware category. It was anticipated that the NASA ELV program would be completely funded on a reimbursable basis in 1987.
i Vehicle not specified in budget figures but indicated in supporting congressional committee documentation, which specifies two Delta II vehicles for 1990 and 1991 launches.

Table 2–5. Scout Funding History (in thousands of dollars)

Year (Fiscal)	Submission	Authorization	Appropriation	Programmed (Actual)
1979	16,400	a	b	10,600
1980	7,300	7,300	c	5,100
1981	2,200	2,200	900 d	900
1982	800	800	800	800
1983	No budget line item (Scout not in use by NASA after 1982)			—

a Not distributed by vehicle—total 1979 ELV authorization = $74,000,000.
b Not distributed by vehicle—1979 R&D appropriation = $3,477,200,000.
c Undistributed. Included in R&D appropriation of $4,091,086,000.
d Basic appropriation of $2,200,000. Effect of General Provision, Sec. 412 (Public Law 96–526), reduced funding level to $900,000.
e No budget submission, authorization, or appropriation for specific ELVs. Total undistributed ELV submission = $50,000,000; authorization = $50,000,000; and appropriation = $50,000,000. ELV appropriation removed from R&D and placed in Space Flight, Control & Data Communications (SFC&DC) (Office of Space Transportation Systems) appropriation.
f No budget line item for ELVs. Support for ELVs paid for as part of Space Transportation Operations Program. It was anticipated that the NASA ELV program would be completely funded on a reimbursable basis in 1985.
g No budget line item for ELVs. Support for ELVs paid for as part of Space Transportation Operations Program. It was anticipated that the NASA ELV program would be completely funded on a reimbursable basis in 1986.
h Included in Flight Hardware category. It was anticipated that the NASA ELV program would be completely funded on a reimbursable basis in 1987.

Table 2–6. Space Shuttle Main Engine Funding History (in thousands of dollars)

Year (Fiscal)	Submission	Authorization	Appropriation	Programmed (Actual)
1979 DDT&E	176,700	176,700	a	172,700
Production	18,000	b	a	264,500
Suppl. Appropriation	c	48,000 d	e	
1980 DDT&E	140,600	140,600	f	140,600
Production	109,900	109,900	f	123,600
Suppl. Appropriation g				
1981 DDT&E	145,700	145,700	145,700	134,000
Production	121,500	121,500	121,500	779,000
1982 DDT&E	127,000	127,000	127,000	h
Production	105,000	105,000	105,000	163,300
1983	262,000	262,000	262,000	355,700
1984	i	i	i	418,100
1985	j	i	i	419,000
1986	k	k	k	394,400
1987	l	l	l	432,700
1988	m	m	m	395,900

a Not distributed by element/vehicle—1979 R&D appropriation = $3,477,200,000.
b No SSME Production category broken out. Total Production amount = $458,000,000.
c No breakout of Supplemental Appropriation submission; included in general R&D supplemental appropriation submission.
d Breakdown of supplemental authorization not provided in budget request or public law. Breakdown provided in supporting documentation for authorization only.
e Supplemental appropriation specified for overall R&D activities = $185,000,000.
f Undistributed. Included in R&D appropriation of $4,091,086,000.
g Supplemental appropriation for Space Shuttle in response to amended NASA budget submission of $300,000,000. No authorization activity. Supplemental appropriation of $285,000,000 approved with no distribution to individual components.
h Programmed amount for SSME DDT&E in 1982 not indicated.

Table 2–6 continued

i   No specific authorization for SSME, external tank, or SRB. According to congressional reports, the Space Transportation and Capability Development program supported the production of the SSME, SRB, and external tank, in addition to providing for critical spares (as well as other items). The total authorization for this category = $2,009,400,000. Category also included continuing "capability development tasks for the orbiter, main engine, external tank, and SRB, . . ." and "the development of the filament wound case (FWC) SRB." Appropriation for Space Shuttle activities of $1,545,000,000 moved from R&D to SFC&DC. Some Space Shuttle funding was included in the Flight Hardware category: submission = $848,400,000; authorization undistributed, included in Shuttle Operations authorization = $1,495,600,000; and appropriation (moved to SFC&DC) undistributed, included in Shuttle Operations appropriation = $1,520,600,000. Amount of $427,400,000 remained in R&D for other activities.

j   SSME production and residual development tasks for the orbiter, SSME, external tank, and SRB fell under Space Production and Operational Capability, Propulsion Systems. SRBs and external tank procurement (production) fell under Space Transportation Operations, Flight Hardware. No breakdown for individual Space Shuttle propulsion components. The 1985 amounts for Propulsion Systems were: submission = $599,000,000; authorization = $599,000,000; and appropriation = $599,000,000. Flight Hardware submission = $758,000,000; authorization = $758,000,000; and appropriation = $758,000,000.

k   No breakdown for individual Space Shuttle propulsion components. The 1986 amounts for Propulsion Systems were: submission = $454,000,000; authorization = $454,000,000; and appropriation = $454,000,000. No Flight Hardware budget category in 1986.

l   No breakdown for individual Space Shuttle propulsion components. The 1987 amounts for Propulsion Systems were: submission = $338,400,000; authorization = $646,200,000. Flight Hardware submission = $646,200,000; authorization = $879,100,000; and appropriation = $646,200,000.

m   No breakdown for individual Space Shuttle propulsion components. The 1988 amounts for Propulsion Systems were: submission = $552,100,000; authorization = $552,100,000; and appropriation = $249,300,000. Flight Hardware submission = $923,100,000; authorization = $923,100,000; and appropriation = $923,100,000.

LAUNCH SYSTEMS

Table 2–7. *Solid Rocket Boosters Funding History (in thousands of dollars)*

Year (Fiscal)	Submission	Authorization	Appropriation	Programmed (Actual)
1979	63,500	63,500	a	115,400
Suppl. Appropriation	b	36,700 c	d	
1980	57,500	57,500	e	65,200
Suppl. Appropriation f	—	—	—	
1981	14,000	14,000	14,000	50,500
1982 Propulsion Systems g	17,000	17,000	17,000	22,000
Flight Hardware g				156,200
1983 Propulsion Systems g	h	h	h	102,300
Flight Hardware g				309,200
1984 Propulsion Systems	i	i	i	140,500
Flight Hardware	i	i	i	341,200
1985 Propulsion Systems	j	j	j	105,100
Flight Hardware	j	j	j	298,600
1986 Propulsion Systems	k	k	k	328,500
Flight Hardware	k	k	k	335,000
1987 Propulsion Systems	l	l	l	322,100
Flight Hardware	l	l	l	144,300
1988 Propulsion Systems	m	m	m	161,200
Flight Hardware	m	m	m	200,500

a Not distributed by element/vehicle—1979 R&D appropriation = $3,477,200,000.
b No breakout of Supplemental Appropriation submission; included in general R&D request of $185,000,000.
c Breakdown of supplemental authorization not provided in budget request or public law. Breakdown provided in supporting documentation for authorization only.
d Supplemental Appropriation for general R&D activities.
e Undistributed. Included in 1980 R&D appropriation of $4,091,086,000.

Table 2–7 continued

f   Supplemental appropriation for Space Shuttle in response to amended NASA budget submission of $300,000,000. No authorization activity. Supplemental appropriation of $285,000,000 approved with no distribution to individual components.

g   Propulsion Systems and Flight Hardware budget categories were not used in NASA's budget prior to 1984. However, programmed amounts used these categories to be consistent with categories used in estimates for the future years.

h   No budget item listed. Supporting committee documentation included SRB in Space Shuttle Production category with no amount specified. Total Production amount: submission = $1,585,500,000; authorization = $1,670,500,000; and appropriation = $1,636,600,000.

i   No specific authorization for SSME, external tank, or SRB. According to congressional reports, the Space Transportation and Capability Development program supported the production of the SSME, SRB, and external tank, in addition to providing for critical spares (as well as other items). The total authorization for this category = $2,009,400,000. Category also included continuing "capability development tasks for the orbiter, main engine, external tank, and SRB, . . ." and "the development of the filament wound case (FWC) SRB." Appropriation moved from R&D to SFC&DC = $1,545,000,000. Some Space Shuttle funding was included in the Flight Hardware category (see above for definition): submission = $848,400,000; authorization undistributed, included in Shuttle Operations authorization of $1,495,600,000; and appropriation (moved to SFC&DC) undistributed, included in Shuttle Operations appropriation = $1,520,600,000. Appropriation moved from R&D to SFC&DC = $1,545,000,000.

j   SSME production and residual development tasks for the orbiter, SSME, external tank, and SRB fell under Space Production and Operational Capability, Propulsion Systems. SRB and external tank procurement (production) fell under Space Transportation Operations, Flight Hardware. No breakdown for individual Space Shuttle propulsion components. The 1985 amount for Propulsion Systems was: submission = $599,000,000; authorization = $599,000,000; and appropriation = $599,000,000. Authorization for submission = $758,000,000. Procurement of external tank, solid rocket motor, and SRB hardware included in Space Transportation Operations Program, Flight Hardware amount of $758,000,000; appropriation = $758,000,000. SSME production and residual development tasks for the orbiter, SSME, external tank, and SRB fall under Space Production and Operational Capability, Propulsion Systems. SRB and external tank procurement (production) fell under Space Transportation Operations, Flight Hardware. Flight Hardware submission = $758,000,000; authorization = $758,000,000; and appropriation = $758,000,000.

k   No breakdown for individual Space Shuttle propulsion components. The 1986 amounts for Propulsion Systems were: submission = $454,000,000; authorization = $454,000,000; and appropriation = $454,000,000. No Flight Hardware budget category in 1986.

l   No breakdown for individual Space Shuttle propulsion components. The 1987 amounts for Propulsion Systems were: submission = $338,400,000; authorization = $338,400,000; and appropriation = $338,400,000. Flight Hardware submission = $646,200,000; authorization = $879,100,000; and appropriation = $646,200,000.

m   No breakdown for individual Space Shuttle propulsion components. The 1988 amounts for Propulsion Systems were: submission = $552,100,000; authorization = $552,100,000; and appropriation = $249,300,000. Funds deleted from Propulsion Systems; $302,800,000 appropriated moved to Launch and Mission Support category. Flight Hardware submission = $923,100,000; authorization = $923,100,000; and appropriation = $923,100,000.

LAUNCH SYSTEMS   71

Table 2–8. *External Tank Funding History (in thousands of dollars)*

Year (Fiscal)	Submission	Authorization	Appropriation	Programmed (Actual)
1979	80,500	80,500	a	104,800
Suppl. Appropriation	b	27,100	c	
1980	68,400	68,400	d	79,400
Suppl. Appropriation e	—	—	—	
1981	48,000	48,000	48,000	63,500
1982 Propulsion Systems f	25,000	25,000	25,000	45,700
Flight Hardware f				176,200
1983 Propulsion Systems f	g	g	g	97,600
Flight Hardware f	g	g	g	269,400
1984 Propulsion Systems	h	h	h	74,400
Flight Hardware	h	h	h	242,700
1985 Propulsion Systems	i	i	i	60,500
Flight Hardware	i	i	i	267,000
1986 Propulsion Systems	j	j	j	63,200
Flight Hardware	j	j	j	285,100
1987 Propulsion Systems	k	k	k	51,700
Flight Hardware	k	k	k	251,400
1988 Propulsion Systems	l	l	l	36,000
Flight Hardware	l	l	l	286,600

a Undistributed. No amount specified for external tank appropriation. Included in total R&D appropriation of $3,477,200,000.
b No breakout of supplemental appropriation submission; included in R&D submission of $185,000,000.
c Supplemental Appropriation of $185,000,000 specified for general R&D activities.
d Undistributed. Included in R&D appropriation of $4,091,096,000.
e Supplemental appropriation for Space Shuttle in response to amended NASA budget submission of $300,000,000. No authorization activity. Supplemental appropriation of $285,000,000 approved with no distribution to individual components.

Table 2–8 continued

f   Propulsion Systems and Flight Hardware budget categories were not used by NASA prior to FY 1984. However, programmed amounts used these categories in FY 1982 and FY 1983 to be consistent with categories used in estimates for future years.

g   No budget item listed. Supporting committee documentation included external tank in Space Shuttle Production with no amount specified. Total Production amount: submission = $1,585,500,000; authorization = $1,670,500,000; and appropriation = $1,636,500,000.

h   No specific authorization for SSME, external tank, or SRB. According to congressional reports, the Space Transportation and Capability Development program supported the production of the SSME, SRB, and external tank, in addition to providing for critical spares (see above for definition): submission = $848,400,000; authorization = $2,009,400,000. Some Space Shuttle funding was included in the Flight Hardware category (see above for definition): submission = $848,400,000; authorization undistributed, included in Shuttle Operations authorization = $1,495,600,000; and appropriation (moved to SFC&DC) undistributed, included in Shuttle Operations appropriation = $1,520,600,000.

i   SSME production and residual development tasks for the orbiter, SSME, external tank, and SRB fell under Space Production and Operational Capability, Propulsion Systems. SRB and external tank procurement (production) fell under Space Transportation Operations, Flight Hardware. No breakdown for individual Space Shuttle propulsion components. The 1985 amounts for Propulsion Systems were: submission = $599,000,000; authorization = $599,000,000; and appropriation = $599,000,000. Authorization for submission = $758,000,000. Procurement of external tank, solid rocket motor, and SRB hardware included in Space Transportation Operations Program, Flight Hardware amount of $758,000,000; appropriation = $758,000,000. SSME production and residual development tasks for the orbiter, SSME, external tank, and SRB fell under Space Production and Operational Capability, Propulsion Systems. SRB and external tank procurement (production) fell under Space Transportation Operations, Flight Hardware. Flight Hardware submission = $758,000,000; authorization = $758,000,000; and appropriation = $758,000,000.

j   No breakdown for individual Space Shuttle propulsion components. The 1986 amounts for Propulsion Systems were: submission = $454,000,000; authorization = $454,000,000; and appropriation = $454,000,000. No Flight Hardware budget category in 1986.

k   No breakdown for individual Space Shuttle propulsion components. The 1987 amounts for Propulsion Systems were: submission = $338,400,000; authorization = $338,400,000; and appropriation = $338,400,000. Flight Hardware submission = $646,200,000; authorization = $879,100,000; appropriation = $646,200,000.

l   No breakdown for individual Space Shuttle propulsion components. The 1988 amounts for Propulsion Systems were: submission = $552,100,000; authorization = $552,100,000; and appropriation = $249,300,000. Funds deleted from Propulsion Systems; $302,800,000 moved to Launch and Mission Support category. Flight Hardware submission = $923,100,000; authorization = $923,100,000; and appropriation = $923,100,000.

Table 2–9. Upper Stages Funding History (in thousands of dollars)

Year (Fiscal)	Submission	Authorization	Appropriation	Programmed (Actual)
1979 STS Upper Stages	a	a	a	19,300
Upper Stage Operations				6,300
1980 STS Upper Stages	b	b	b	18,300
Upper Stage Operations	b	b	b	18,700
1981 STS Upper Stages	c	c	c	38,300
Upper Stage Operations	c	c	c	30,900
1982 d	—	—	—	106,700
1983 e	—	—	—	167,000
1984 f	143,200	143,200	143,200	143,200
1985 g	92,400	92,400	92,400	137,400
Suppl. Appropriation		40,000 h		
1986	122,000	122,000	122,000	122,000
1987	202,100 i	200,100 j	202,100 k	156,100
1988	159,700	159,700	159,700	

a No specific funding. Submission for Space Transportation System Operations Capability Development = $110,500,000; authorization for Space Transportation System Operations Capability Development by Senate committee = $110,500,000 (no final authorization); and appropriation for Space Transportation System Operations Capability Development was undistributed. Total R&D appropriation = $3,477,200,000. Submission for Space Transportation System Operations = $33,400,000. Authorization for Space Transportation System Operations by Senate committee = $33,400,000 (no final authorization). Appropriation undistributed.

b No specific funding for Upper Stages.

c Upper Stages were included in the Space Flight Operations Space Transportation Systems Operational Capability budget line item. House Committee documentation indicated that NASA submission, as well as congressional authorization, for upper stage activities was $29,000,000; Public Law 96–526 report referred to both STS Upper Stages and STS Operations Upper Stages: appropriation = $29,000,000 for STS Upper Stages and $30,900,000 for STS Operations Upper Stages.

d No NASA submission, final authorization, or appropriation for Upper Stages indicated.

e Upper Stages included in Space Flight Operations, but no amount specified. Total Space Flight Operations: submission = $1,707,000,000; authorization = $1,699,000,000; and appropriation = $1,796,000,000.

Table 2–9 continued

f   Included modification of the Centaur for use in the Shuttle.
g   Included development of Transfer Orbit Stage for use in launching the Mars geoscience/climatology orbiter in 1990. Also included joint development program between NASA and DOD for use of the Centaur as an STS upper stage. Procurement would be initiated in FY 1985 for two Centaur G vehicles to support the Venus Radar Mapper mission planned for 1988 and the TDRS-E mission.
h   Supplemental Appropriation added $40,000,000 to initial appropriation for Upper Stages for total of $132,400,000
i   Amended budget submission increased amount from $85,100,000 to $202,100,000.
j   Figure reflects authorization act, which was vetoed.
k   Figure reflects Appropriation Conference Committee action, which was subsequently included in the Omnibus Appropriation Act of 1987 (Public Law 99–591).

Table 2–10. *Orbital Maneuvering Vehicle Funding History (in thousands of dollars)*

Year (Fiscal)	Submission	Authorization	Appropriation	Programmed (Actual)
1983		Included in Advanced Programs		
1984		Included in Advanced Programs		
1985		Included in Advanced Programs		
1986	25,000	13,000	10,000	5,000
1987	45,000 *a*	50,000 *b*	45,000 *c*	45,000
1988	80,000	75,000	55,000	

a Reflects revised budget submission, which decreased amount from $70,000,000 to $45,000,000
b Figure reflects authorization act, which was vetoed.
c Figure reflects Appropriation Conference Committee action, which was subsequently included in the Omnibus Appropriation Act of 1987 (Public Law 99–591).

Table 2–11. Tethered Satellite System Funding History (in thousands of dollars)

Year (Fiscal)	Submission	Authorization	Appropriation	Programmed (Actual)
1979		Included in Advanced Programs		
1980		Included in Advanced Programs		
1981		Included in Advanced Programs		
1982		Included in Advanced Programs		
1983		Included in Advanced Programs		
1984	3,300	3,300	3,300	3,300
1985	18,200	18,200	18,200	15,800
1986	21,000	14,000	21,000	15,000
1987	10,600 a	11,600 b	10,600 c	10,600
1988	7,300	7,300	7,300	

a Figure reflects $1,000,000 reduction from initial budget submission.
b Figure reflects authorization act, which was vetoed.
c Figure reflects Appropriation Conference Committee action, which was subsequently included in the Omnibus Appropriation Act of 1987 (Public Law 99–591).

Table 2–12. *Advanced Programs/Planning Funding History (in thousands of dollars)*

Year (Fiscal)	Submission	Authorization	Appropriation	Programmed (Actual)
1979	5,000	a	b	7,000
1980	13,000	c	d	13,000
1981	8,800	13,800 e	f	11,800
1982	8,800	12,800 g	8,800	9,700
1983	11,900	h	11,900	12,600
1984	15,000	25,000 i	15,000	21,500
1985	14,500	14,500	20,500	20,500
1986	21,000	21,000	21,000	19,400
1987	16,600	16,600 j	16,600 k	33,600
1988	24,900	24,900	30,900	

a   Undistributed. Included in Space Flight Operations Program authorization of $315,900,000.
b   Undistributed. Included in R&D appropriation of $3,477,200,000.
c   Undistributed. Included in Space Flight Operations Program authorization of $463,300,000.
d   Undistributed. Included in R&D appropriation of $4,091,086,000.
e   Increased authorization recommended by House Committee to support enhanced Phase B definition studies and technical development for the power extension package (PEP) and the 25-kilowatt (kW) power module.
f   Undistributed. Included in R&D appropriation of $4,396,200,000 (modified by General Provision, Sec. 412, to $4,340,788).
g   House recommended additional authorization of $5,000,000 for PEP, 25-kW power module, space platforms, space operations definition studies, and advanced technical development. Conference Committee reduced additional authorization to $12,800,000.
h   Undistributed. Included in Space Flight Operations authorization of $1,699,000,000.
i   House authorized additional $10,000,000 for space station studies and space platform. Senate authorized additional $5,000,000 for space station studies. Conference Committee authorized additional $10,000,000 for space station studies.
j   Figure reflects authorization act, which was vetoed.
k   Figure reflects Appropriation Conference Committee action, which was subsequently included in the Omnibus Appropriation Act of 1987 (Public Law 99-591).

Table 2–13. ELV Success Rate by Year and Launch Vehicle for NASA Launches

Year (Fiscal)	Atlas-Centaur	Atlas E/F	Delta	Scout	Total
1979	2/2	1/1	3/3	3/3	9/9
1980	3/3	0/1	1/1	—	1/1
1981	4/4	1/1	5/5	1/1	11/11
1982	2/2	—	7/7	—	9/9
1983	1/1	1/1	8/8	1/1	11/11
1984	0/1	1/1	4/4	1/1	6/7
1985	3/3	—	—	2/2	5/5
1986	1/1	1/1	1/2	1/1	4/5
1987	0/1	—	2/2	1/1	3/4
1988	—	1/1	1/1	4/4	6/6
Total	16/18 (88.9%)	6/7 (85.7%)	34/35 (97.1%)	14/14 (100%)	70/74 (94.6%)

Table 2–14. *NASA Atlas E/F Vehicle Launches*

Atlas-E/F Vehicle	Date	Mission	Atlas Successful [a]
Atlas F	June 27, 1979	NOAA-6	Yes
Atlas F	May 29, 1980	NOAA-B	No. Launch vehicle malfunctioned; failed to place satellite into proper orbit.
Atlas F	June 23, 1981	NOAA-7	Yes
Atlas E	March 28, 1983	NOAA-8	Yes
Atlas E	Dec. 12, 1984	NOAA-9	Yes
Atlas E	Sept. 17, 1986	NOAA-10	Yes
Atlas E	Sept. 24, 1988	NOAA-11	Yes

[a] One failure out of seven attempts (85.7% success rate).

Table 2–15. Atlas E/F Characteristics

	1-1/2 Stages (Booster & Sustainer)	Apogee Kick Motor	Fairing
Length	21.3 meters (m)	—	7.0 m
Overall Length	Up to 28.3 m including fairing		
Diameter	3.05 m		2.1 m
Gross Weight (Liftoff)	121,000 kilograms (kg)	47.7 kg (weight of motor)	735 kg assembly case after depletion of fuel)
Fuel Weight	112,900 kg	666 kg	
Engine Type/Name	MA-3 system consisting of LR 89-NA-5 booster, LR 105-NA-5 sustainer, LR 101-NA-7 vernier engines	TE-M-364-15	
Number of Engines	2 booster engines, 1 sustainer engine, & 2 vernier engines (VE)	1	
Propellant	LOX & RJ-1-1	Solid	
Burn Time (Avg.)	120-sec booster, 309-sec. sustainer	45 sec.	
Liftoff Thrust	1,743,000 newtons		
Avg. Thrust per Engine	1,470,000 newtons (boosters); (267,000 newtons (sustainer); 3,000 newtons (each VE)	650,800 newtons	
Max. Payload	2,090 kg in 185-km orbit from polar launch with dual TE-364 4 engines; 1,500 kg in 185-km orbit from polar launch with single TE 374-4 engine		

Table 2–15 continued

	1-1/2 Stages (Booster & Sustainer)	Apogee Kick Motor	Fairing
Prime Contractor	General Dynamics		
Contractors	Rocketdyne	Thiokol	
How Utilized	To launch meteorological satellites		
Remarks	The Atlas E/F series was originally deployed as ICBMs. By the late 1970s, the remaining Atlas E/Fs were converted for space launch. During 1979–1988, they were used only to launch meteorological satellites. On particular missions, the fairings were lengthened to 7.4 m to accommodate additional equipment—for instance, search-and-rescue equipment on NOAA missions.		

Table 2–16. NASA Atlas-Centaur Vehicle Launches

Atlas-Centaur Vehicle Serial Number	Date	Mission	Atlas-Centaur Successful [a]
AC-47	May 4, 1979	FltSatCom 2	Yes
AC-53	Sept. 20, 1979	HEAO 3	Yes
AC-49	Jan. 17, 1980	FltSatCom 3	Yes
AC-52	Oct. 30, 1980	FltSatCom 4	Yes
AC-54	Dec. 6, 1980	Intelsat V-A F-2	Yes
AC-42	Feb. 21, 1981	Comstar 4	Yes
AC-56	May 23, 1981	Intelsat V-B F-1	Yes
AC-59	Aug. 6, 1981	FltSatCom 5	Yes
AC-55	Dec. 15, 1981	Intelsat V F-3	Yes
AC-58	Mar. 4, 1982	Intelsat V-D F-4	Yes
AC-60	Sept. 28, 1982	Intelsat V-E F-5	Yes
AC-61	May 19, 1983	Intelsat V-F F-6	Yes
AC-62	June 9, 1984	Intelsat V-G F-9	No. Vehicle failed to place satellite in useful orbit.
AC-36	Mar. 22, 1985	Intelsat V-A F-10	Yes
AC-64	June 29, 1985	Intelsat V-A F-11	Yes
AC-65	Sept. 28, 1985	Intelsat V-A F-12	Yes
AC-66	Dec. 4, 1986	FltSatCom 7	Yes
AC-67	Mar. 26, 1987	FltSatCom 6	No. Telemetry lost shortly after launch; destruct signal sent at 70.7 seconds into flight. An electrical transient, caused by lightning strike on launch vehicle, was most probable cause of loss.

[a] Two failures out of eighteen attempts (88.9% success rate).

Table 2–17. Atlas-Centaur Characteristics

	Atlas Booster & Sustainer SLV-3D	Centaur Stage D-1A
Length	21.1 meters (m)	9.1 m without fairing; 18.6 m (with payload fairing)
Overall Length	40.8 m including fairing	
Diameter	3.05 m	3.05 m
Engine Type/Name	MA-5 system consisting of 2 boosters 1 sustainer, and 2 vernier engines	RL-10
Prime Contractor Contractors	General Dynamics Rocketdyne	Pratt & Whitney Aircraft
Number of Engines	5 (2 booster engines, 1 thrust sustainer engine, 2 vernier engines)	2 thrust engines and 14 small hydrogen peroxide thrusters
Liftoff Thrust (Avg.)	1,931,000 newtons (at sea level) using two 828,088-newton-thrust booster engines, one 267,000-newton-thrust sustainer engine, and two vernier engines developing 3,006 newtons each	133,440 newtons (vacuum) using two 67,000-newton-thrust RL-10 engines and 14 small hydrogen peroxide thrusters
Burn Time	174-sec. booster, 226-sec. sustainer	450 sec.
Propellant	LOX as the oxidizer and RP-1	LOX and LH$_2$
Max. Payload	6,100 kilograms (kg) in 185-km orbit; 2,360 kg in geosynchronous transfer orbit; 900 kg to Venus or Mars	
Launch Weight	128,934 kg	17,676 kg
How Utilized	Primarily to launch communications satellites	
Remarks	Unlike earlier Atlas-Centaur combinations, the SLV-3D and later models were integrated electronically with the Centaur D-1A upper stage. The Intelsat V-A F-10, Intelsat V-A F-11, Intelsat V-A F-12, FltSatCom 5, and FltSatCom 6 missions used the Atlas G configuration. The Atlas stage on the "G" configuration is 2.06 m longer than the SLV-3D, and its engine provided 33,600 newtons more thrust than the SLV-3D engines.	

Table 2–18. Chronology of Delta Vehicle Launches

Delta Vehicle Type	Date	Mission	Delta Successful [a]
2914/148	Jan. 30, 1979	SCATHA	Yes
2914/149	Aug. 9, 1979	Westar-C	Yes
3914/150	Dec. 6, 1979	RCA-C	Yes
3910/151	Feb. 14, 1980	Solar Max Mission	Yes
3914/152	Sept. 9, 1980	GOES 4	Yes
3910-PAM/153	Nov. 15, 1980	SBS-A (first use of PAM)	Yes
3914/154	May 22, 1981	GOES 5	Yes
3913/155	Aug. 3, 1981	Dynamic Explorer DE-A/B	Yes
3910-PAM/156	Sept. 24, 1981	SBS-B	Yes
2310/157 [b]	Oct. 6, 1981	SME/Uosat	Yes
3910-PAM/158	Nov. 20, 1981	RCA-D	Yes
3910-PAM/159	Jan. 15, 1982	RCA-C	Yes
3910-PAM/160	Feb. 25, 1982	Westar IV	Yes
32910-PAM/161	Apr. 10, 1982	Insat-1A	Yes
3915/162	June 8, 1982	Westar-V	Yes
3920/163	July 16, 1982	Landsat-D	Yes
3920-PAM/164	Aug. 26, 1982	(Telesat-F) Anik-D-1	Yes
3924/165	Oct. 27, 1982	RCA-E	Yes
3910/166	Jan. 25, 1983	IRAS/PIX II	Yes
3924/167	Apr. 11, 1983	RCA-F	Yes
3914/168	Apr. 28, 1983	GOES F	Yes
3914/169	May 26, 1983	EXOSAT	Yes
3920-PAM/170	June 28, 1983	Galaxy-A	Yes
3920-PAM/171	July 28, 1983	Telstar-3A	Yes

Table 2–18 continued

Delta Vehicle Type	Date	Mission	Delta Successful [a]
3924/172	Sept. 8, 1983	Satcom-IIR (RCA-G)	Yes
3920-PAM/173	Sept. 22, 1983	Galaxy-B	Yes
3920/174	Mar. 1, 1984	Landsat-D/Uosat	Yes
3924/175	Aug. 16, 1984	AMPTE	Yes
3920-PAM/176	Sept. 21, 1984	Galaxy-C	Yes
3914/177	Nov. 13, 1984	NATO-IIID	Yes
3914/178	May 3, 1986	GOES G	No. Vehicle failed.
3920/180	Sept. 5, 1986	SDI	Yes
3924/179	Feb. 26, 1987	GOES H	Yes
3920-PAM/182	Mar. 20, 1987	Palapa-B2P	Yes
3910/181	Feb. 8, 1988	SDI	Yes

[a] One failure out of thirty-five attempts (97.1% success rate).
[b] Three strap-on engines.

Table 2–19. Delta 2914 Characteristics

	Strap-on	Stage I	Stage II	Stage III
Length		21.3 m	6.4 m	1.4 m
Overall Length	35.5 m including spacecraft shroud			
Diameter	Overall basic diameter of 2.4 m			
Engine Type/Name	TX-354-5 Castor II	RS-27 extended long tank Thor	8-foot-diameter TR-201	TE-364-4
No. of Engines	9	1 main and 2 vernier	1	1
Thrust (per Engine) (Avg.)	233,856 newtons	911,840 newtons	45,800 newtons	66,586 newtons
Liftoff Thrust	1,765,315 newtons (includes 6 of 9 strap-ons, which are ignited at liftoff)			
Burn Time	37 sec.	209 sec.	335 sec.	44 sec.
Propellant	Solid	RP-1/LOX	N_2O_4 & aerozine-50	Solid
Fuel Weight	18,084 kg each strap-on	80,264 kg	4,593 kg	1,039 kg
Liftoff Weight	40,320 kg	84,330 kg	6,125 kg	1,120 kg
Prime Contractor	McDonnell Douglas			
Contractors	Thiokol	Rocketdyne	TRW	Thiokol
How Utilized	Medium-weight payloads			
Remarks	Only three Deltas in the 2900 series were used between 1979 and 1988. Two were 2914s and one was a 2310, which had only three strap-on motors and two stages.			

Table 2–20. Delta 3910/3914 Characteristics

	Strap-on	Stage I	Stage II	Stage III
Overall Length	35.5 m including spacecraft shroud			
Length	11.3 m	21.3 m	6.0 m a	1.8 m a
Diameter	Overall 2.4 m max.			
Engine Type/Name	Castor IV/TX-526-2	RS-27 modified long tank Thor booster	TR-201	Thiokol TE-364-3 or TE-364-4
No. of Engines	9	1 main and 2 vernier	1	1
Burn Time (Avg.)	57 sec.	224 sec.	320 sec.	44 sec.
Specific Impulse (Avg.)	229.9 sec.	262.4 sec.	319 sec.	283 sec.
Thrust	377,165 newtons	911,887 newtons	43,815 newtons	TE 364-3 engine: 42,169 newtons TE 364-4 engine: 66,586 newtons
Propellant	Solid TP-H-8038	LOX and RP-1 (hydrazine) or LOX and RJ-1 (liquid hydrocarbon) b	N_2O_4 and aerozine-50	Solid
Fuel Weight	9,373 kg	80,264 kg	4,593 kg	1,039 kg
Gross Weight	10,840 kg each	85,076 kg	6,115 kg	1,158 kg
Prime Contractor	McDonnell Douglas			
Contractors	Thiokol	Rocketdyne	TRW	Thiokol
How Utilized	Medium-weight payloads			
Remarks	With the exceptions noted below in notes a and b, the 3910 was identical to the 3914 but had only two stages. For launches from the Eastern Space and Missile Center, six strap-on motors were ignited at liftoff and jettisoned approximately nine seconds after ignition of the second set of three strap-on motors. The remaining three motors were jettisoned at approximately 126 seconds after liftoff. For the Western Space and Missile Center, the six ground-ignited motors were jettisoned at a later time for range safety considerations.			

a The length of the second stage on the 3910 equaled the sum of the lengths of the second and third stages on the 3914. The lengths of the individual stages did not include the length of the spacecraft shroud.
b The 3910 used LOX and RP-1 propellant; the 3914 used LOX and RJ-1 propellant.

Table 2–21. Delta 3920/3924 Characteristics

	Strap-on	Stage I	Stage II	Stage III
Length		21.3 m	6m	1.8m
Overall Length	35.5 m including spacecraft shroud			
Diameter	Overall 2.4 m max.			
Engine Type/Name	Castor IV TX-526-2 solid boosters	RS-27 modified long tank Thor booster	Improved Transtage Injector Program	TE-364-4
No. of Engines	9	1 main & 2 vernier	1	1
Specific Impulse (Avg.)	229.9 sec.	262.4 sec.	319 sec.	283.6 sec.
Thrust (Avg.)	377,165 newtons	911,007 newtons	44,000 newtons	66,586 newtons
Burn Time	57 sec.	224 sec.	320 sec.	44 sec.
Propellant	Solid TP-H-8038	RP-1 and LOX	Aerozine-50 and N_2O_4 oxide	Solid
Fuel Weight	9,373 kg	79,380 kg	4,593 kg	1,039 kg
Max Payload	3,045 kg in 185-km orbit with due east launch; 1,275 kg in geosynchronous transfer orbit with due east launch; 2,135 kg in circular Sun-synchronous orbit with polar launch; 2,180 kg in 185-km orbit with polar launch			
Gross Weight	10,840 kg	85,076 kg	6,920 kg	1,122 kg
Prime Contractor	McDonnell Douglas			
Contractors	Thiokol	Rocketdyne	Aerojet	Thiokol
How Utilized	Mid-size communication and meteorological satellite			
Remarks	The length of the second stage of the 3920 equalled the combined lengths of the second and third stages of the 3924. Lengths of individual stages did not include length of the spacecraft shroud.			

Table 2–22. NASA Scout Launches

Scout Vehicle	Date	Mission	Scout Successful
S-202	Feb. 18, 1979	SAGE	Yes
S-198	June 2, 1979	UK-6	Yes
S-203	Oct. 30, 1979	Magsat	Yes
S-192	May 14, 1981	NOVA-I	Yes
S-205	June 27, 1983	Hilat	Yes
S-208	Oct. 11, 1984	NOVA-III	Yes
S-209	Aug. 2, 1985	SOOS-1	Yes
S-207	Dec. 12, 1985	AFITV	Yes
S-199	Nov. 13, 1986	AF Polar BEAR	Yes
S-204	Sept. 16, 1987	SOOS-2	Yes
S-206	Mar. 25, 1988	San Marco-DL	Yes
S-211	Apr. 25, 1988	SOOS-3	Yes
S-213	June 15, 1988	NOVA-II	Yes
S-214	Aug. 25, 1988	SOOS-4	Yes

All of attempted launches were successful.

Table 2–23. Scout Characteristics (G-1)

	First Stage	Second Stage	Third Stage	Fourth Stage
Length	9.94 m	6.56 m	3.28 m	1.97 m
Overall Length	22.86 m including transition and payload sections			
Weight	14,255 kg	4,424 kg	1,395 kg	302 kg
Diameter	1.01 m max.			
Engine Type/Name	Algol IIIA	Castor IIA	Antares IIIA a	Altair IIIA/ Star 31
Thrust (Avg.)	481,000 newtons	281,000 newtons	83,100 newtons	25,593 newtons
Fuel	Solid	Solid	Solid	Solid
Fuel Weight	12,684 kg	3,762 kg	1,286 kg	275 kg
Launch Weight	14,215 kg	4,433 kg	1,394 kg	301 kg
Burn Time (Avg.)	90 sec.	46 sec.	48.4 sec.	30 sec.
Payload Capacity	227.2 kg payload to a 480-km Earth orbit			
Prime Contractor	Vought Corp. (LTV Corp.)			
Contractors	United Technologies	Thiokol	Thiokol	Thiokol
How Utilized	Smaller payloads			
Remarks	An optional fifth stage used the Alcyone IA engine, with a thrust of approximately 26,230 newtons, a burn time of 8.42 sec., and a total weight of 98.2 kg.			

a Missions prior to Magsat (SAGE and UK-6) used the Antares II third stage engine.

Table 2–24. STS-Launched Missions

Vehicle	Mission	Deployed Payload	Date
Columbia	STS-1	First test flight, no deployable payload	Apr. 12–14, 1981
Columbia	STS-2	Second test flight, no deployable payload	Nov. 12–14, 1981
Columbia	STS-3	Third test flight, no deployable payload	Mar. 22–30, 1982
Columbia	STS-4	Fourth and final test flight; DOD payload 82-1	June 27–July 4, 1982
Columbia	STS-5	SBS-C/PAM-D, Anik C-3/PAM-D (Telesat-E) (Canada)	Nov. 11–16, 1982
Challenger	STS-6	Tracking and Data Relay Satellite (TDRS)-1/IUS	Apr. 4–9, 1983
Challenger	STS-7	Telesat 7 (Anik C-2)/PAM-D (Canada)/PAM-D, Palapa B-1 (Indonesia)/PAM-D	June 18–24, 1983
Challenger	STS-8	INSAT-1B/PAM-D (India)	Aug. 30–Sept. 5, 1983
Columbia	STS-9	Spacelab-1 (no satellites deployed)	Nov. 28–Dec. 8, 1983
Challenger	STS 41-B	Palapa-B2/PAM-D (Indonesia), Westar VI/PAM-D	Feb. 3–11, 1984
Challenger	STS 41-C	Long Duration Exposure Facility (LDEF-1)	Apr. 6–13, 1984
Discovery	STS 41-D	Syncom IV-2 (Leasat 2)/Unique Upper Stage*, Telstar 3-C/PAM-D, SBS-D/PAM-D	Aug. 30–Sept. 5, 1984
Challenger	STS 41-G	Earth Radiation Budget Satellite (ERBS)	Oct. 5–13, 1984
Discovery	STS 51-A	Syncom IV-1 (Leasat 1)/Unique Upper Stage*, Anik (Telesat-H)/PAM-D	Nov. 8–16, 1984
Discovery	STS 51-C	DOD classified payload/IUS	Jan. 24–27, 1984
Discovery	STS 51-D	Anik C-1 (Telesat-I)/PAM-D, Syncom IV (Leasat 3)/Unique Upper Stage*	Apr. 12–19, 1985
Challenger	STS 51-B	Spacelab 3, NUSAT, GLOMR (failed to deploy)	Apr. 29–May 6, 1985
Discovery	STS 51-G	Morelos-A/PAM-D (Mexico), Arabsat-A/PAM-D, Telstar 3-D/PAM-D, Spartan-1/MPESS	June 17–24, 1985

Table 2–24 continued

Vehicle	Mission	Deployed Payload	Date
Challenger	STS 51-F	Spacelab 2 (no satellites deployed)	July 29–Aug. 6, 1985
Discovery	STS 51-I	ASC-1/PAM-D, Aussat-1/PAM-D (Australia), Syncom IV (Leasat-4)/Unique Upper Stage*	Aug. 27–Sept. 3, 1985
Atlantis	STS 51-J	DOD Mission	Oct. 3–7, 1985
Challenger	STS 61-A	GLOMR GAS (DOD classified mission)	Oct. 30–Nov. 6, 1985
Atlantis	STS 61-B	Morelos-B/PAM-D (Mexico), Aussat-2/PAM-D (Australia), Satcom Ku-2/PAM-DII (RCA)	Nov. 26–Dec. 3, 1985
Columbia	STS 61-C	Satcom Ku-1/PAM-DII (RCA)	Jan. 12–18, 1986
Challenger	STS 51-L	TDRS-B/IUS and Spartan 203 (carried but not deployed because of the destruction of *Challenger*)	Jan. 28, 1986
Discovery	STS-26	TDRS-3/IUS	Sept. 29–Oct. 3, 1988
Atlantis	STS-27	DOD payload	Dec. 2–6, 1988

* Unique Upper Stage—Minuteman missile third stage used as a solid propellant perigee motor.

Table 2–25. Space Shuttle Main Engine Characteristics

Number of Engines	Three on each Shuttle
Thrust	2,000,000 newtons each
Operating Life	7.5 hours and 55 starts
Range of Thrust Level	65%–109% of rated power level
Propellant	LOX/LH$_2$
Nominal Burn Time	522 sec.
Prime Contractor	Rockwell International

Table 2–26. *Main Engine Development and Selected Events*

Date	Event
June 1980	Surpassed original goal of achieving 80,000 seconds of engine test time before the first orbital flight.
Feb. 20, 1981	Flight readiness firing (20-second firing of all three SSMEs), *Columbia* (OV-102) at Kennedy Space Center.
Feb. 28, 1982	Completed main propulsion test program, National Space Technology Laboratories (NSTL), Mississippi.
1983	Phase II program began for improvements to SSMEs for increased margin and durability.
Dec. 1983	Completed certification of main engines at 109 percent of present rated power level to full power level. Certification process included 400 tests of more than 40,000 seconds of static firing operation.
June 26, 1984	Launch of STS 41-D postponed indefinitely because of shutdown of SSMEs 3 and 2 at T-4 seconds caused by slow opening SSME 3 main fuel valve. SSME 1 never received a start command.
Aug. 30, 1984	STS 41-D conducted successfully.
July 12, 1985	STS 51-F launch scrubbed at T-3 seconds and shutdown of SSMEs because of loss of redundancy (channel A) on SSME 2 chamber coolant valve.
July 29, 1985	STS 51-F conducted successfully.
July 16, 1986	250-second test conducted successfully at NSTL. The test was the first in a series to verify a modification designed to extend the operational service life of turbine blades on the engine's high-pressure oxidizer turbopump.
Aug. 13, 1986	NASA announced selection of Pratt & Whitney for alternate turbopump development contract, which would provide extended life capability and enhance safety margins.
Dec. 1986	Ground test program initiated.
Dec. 1986-Dec. 1987	151 tests and 52,363 seconds of operation (equivalent to 100 Shuttle missions) were performed at NSTL (Mississippi) and Rockwell International's Rocketdyne Division (California).
Aug. 1987–Jan. 1988	Acceptance tests at Stennis Space Center (formerly NSTL).
Sept. 1987	Beginning of acceptance testing of main engines to be used on STS-26 at NSTL. A number of improvements were made on the engines as a result of an extensive, ongoing test program.
Jan. 6, 1988	Engine 2016 arrived at Kennedy.
Jan. 10, 1988	Engine 2106 installed in number one position on *Discovery*.
Jan. 15, 1988	Engine 2022 arrived at Kennedy.
Jan. 21, 1988	Engine 2028 arrived at Kennedy.
Jan. 24, 1988	Engine 2022 installed in number-two position and Engine 2028 installed in number-three position on *Discovery*.
Aug. 10, 1988	Conducted a 22-second flight readiness firing of *Discovery*'s main engine. Verified that the entire Shuttle system was ready for flight

Table 2–27. Space Shuttle External Tank Characteristics

Propellants	LH_2, LOX
Length	46.8 m
Diameter	8.4 m
Weight of Propellant	700,000 kg
Gross Liftoff Weight	750,980 kg
Inert Weight of Lightweight Tank	30, 096 kg
Liquid Oxygen Max. Weight	617,774 kg
Liquid Oxygen Tank Volume	542,583 liters
Liquid Oxygen Tank Diameter	8.4 m
Liquid Oxygen Tank Length	15 m
Liquid Oxygen Tank Weight	5,454.5 kg empty
Liquid Hydrogen Max. Weight	103, 257 kg
Liquid Hydrogen Tank Diameter	8.4 m
Liquid Hydrogen Tank Length	29.46 m
Liquid Hydrogen Tank Volume	1,458,228 liters
Liquid Hydrogen Tank Weight (Empty)	13,181.8 kg
Intertank Length	6.9 m
Intertank Diameter	8.4 m
Intertank Weight	5,500 kg
Prime Contractor	Martin Marietta Aerospace

Table 2–28. External Tank Development and Selected Events*

Date	Event
Mar. 19, 1979	First external tank leaves Marshall Space Flight Center for Kennedy Space Center.
June 25, 1979	First external tank ready for flight.
Feb. 28, 1980	Successful completion of full duration test of MPTA-098.
June 30, 1980	NASA awards contract for external tank to Martin Marietta Corp.
Oct. 8, 1980	"All Systems Test" conducted.
Nov. 3, 1980	First external tank mated to SRBs for STS-1.
Nov. 11, 1980	External tank and SRBs mated to orbiter for STS-1.
Dec. 2, 1980	Assembly of first lightweight tank begins.
Jan. 17, 1981	Static firing at NSTL. External tank test without anti-geyser line to verify feasibility of eventually removing it from later external tank versions.
Jan. 22 and 24, 1981	External tank liquid hydrogen load of *Columbia* at Kennedy.
Apr. 12, 1981	First tank flown successfully.
Nov. 12, 1981	Second tank flown successfully.
Oct. 1981	Third tank delivered to Kennedy.
1981	Major welding and structural assembly completed on the first production version of a lightweight tank.
Apr. 4, 1983	First lightweight tank (LWTR 1) flown on STS-6 mission; design changes reduced weight of external tank by 4,000 kg, permitting heavier payload.
Aug. 1, 1988	Wet Countdown Demonstration Test held; external tank loaded with liquid oxygen and liquid hydrogen.

* Relatively few events were associated with the development of the external tank, and there were no events over a five-year period from April 1983 to August 1988. The external tank performed successfully on the STS missions during this period and required little attention.

Table 2–29. Space Shuttle Solid Rocket Booster Characteristics

Length	45.5 m
Diameter	3.7 m
Outside Diameter of Nozzle and Thrust Vector Control System	12.4 feet
Weight at Launch (Each)	589,680 kg
Propellant Weight (Each)	500,000 kg
Inert Weight (Each)	87,273 kg
Propellant Mixture	Ammonium perchlorate, aluminum, iron oxide, a polymer, an epoxy curing agent
Thrust (Sea Level) of Each Booster in Vacuum	14,409,740 newtons at launch
Separation Motors	Four motors in the nose frustum and four motors in the aft skirt
Length	0.8 m
Diameter	32.5 cm
Thrust of Separation Motors	98,078 newtons each
Inert Weight	87373.6 kg
Burn Time (Nominal)	123 sec.
Prime Contractors	SRB motors: Morton Thiokol Corp. SRB assembly, checkout, and refurbishment for all non–solid rocket motor components and for SRB integration: Booster Production Co.
NASA Lead Center	Marshall Space Flight Center
Remarks	Structural modifications following the Challenger accident added approximately 204 kg to the weight of each SRB.

Table 2–30. Chronology of Selected Solid Rocket Booster Development Events

Date	Event
Jan. 30, 1979	Began orbiter/external tank/SRB burnout mated vertical ground vibration test at Marshall Space Flight Center.
Feb. 17, 1979	Fourth SRB firing at Thiokol, Utah.
June 15, 1979	First SRB qualification firing, Thiokol, 122 seconds; nozzle extension severed at end of run as in actual mission; full cycle gimbal.
July 23, 1979	*Enterprise* (OV-101), external tank, and SRBs transported on mobile launcher platform from Launch Complex 39-A to Vehicle Assembly Building at Kennedy Space Center.
Aug. 1979	Second SRB qualification firing, Thiokol.
Feb. 14, 1980	Final qualification firing SRB, Thiokol.
Aug. 4, 1980	*Columbia* mated with SRBs and external tank for STS-2.
Nov. 3, 1980	External tank mated to SRBs in Vehicle Assembly Building, Kennedy, for STS-1.
Nov. 5, 1980	External tank mated to SRBs at Kennedy.
Nov. 26, 1980	Mating of *Columbia* (OV-102) to external tank and SRBs in Vehicle Assembly Building for STS-1, Kennedy.
Apr. 20, 1981	SRB stacking began on mobile launcher platform for STS-2, Kennedy.
July 30, 1981	Start mating of external tank to SRBs on mobile launcher platform for STS-2, Kennedy.
Apr. 12, 1981, Nov. 12, 1981	STS-1 and STS-2 flights verified reusability of SRBs; some redesign of aft skirts indicated.
Sept. 9, 1981	*Columbia* mated with SRBs and external tank in preparation for STS-5.
Nov. 23, 1981	Start SRB stacking on mobile launcher platform for STS-3, Kennedy.
Dec. 19, 1981	Start mating of external tank to SRBs on mobile launcher platform for STS-3, Kennedy.
Apr. 16, 1982	Complete mating of SRBs and external tank for STS-4 in Vehicle Assembly Building, Kennedy.
Apr. 4, 1983	New lightweight SRB case first flown on STS-6.
Aug. 30, 1983	First high-performance solid-fueled rocket motor flown on STS-8.
Aug. 2, 1984	*Discovery* (OV-103) transported from Orbiter Processing Facility to Vehicle Assembly Building for remate with original 41-D SRBs and external tank, Kennedy.
Dec. 19, 1985	STS 61-C, seventh flight of *Columbia* (OV-102), launch scrubbed at T-13 seconds because of righthand SRB auxiliary power unit turbine system B overspeed.

LAUNCH SYSTEMS 99

Table 2–30 continued

Date	Event
Jan. 6, 1986	STS 61-C launch scrubbed at T-31 seconds because of a launch facility liquid oxygen replenish valve problem.
Jan. 7, 1986	STS 61-C launch scrubbed at T-9 seconds because of adverse weather conditions.
Jan. 12, 1986	STS 61-C launch conducted successfully.
Jan. 28, 1986	STS 51-L launched. Destruction of *Challenger* and all crew aboard.
Mar. 25, 1986	Formation of Solid Rocket Motor Redesign Team to requalify the SRB motor.
July–Aug. 1986	Preliminary Requirements Reviews held.
Aug. 22, 1986	NASA announced the beginning of a series of tests designed to verify the ignition pressure dynamics of the Space Shuttle solid rocket motor (SRM) field joint. The series was conducted over the next year at Thiokol's facility and at Marshall.
Sept. 5, 1986	Study contracts awarded to five aerospace firms for conceptual designs of an alternative or Block II Space Shuttle SRM.
Sept. 1986	Preliminary Design Review held to assess design requirements.
Oct. 2, 1986	NASA announced the decision to test-fire the redesigned SRM in a horizontal attitude to best simulate the critical conditions on the field joint that failed during the 51-L mission.
Oct. 9, 1986	Transfer of *Atlantis* (OV-104) mated, minus SSMEs, from the Vehicle Assembly Building to Launch Complex 39-B for weather protection fit checks, payload bay operations, SRB flight readiness test, terminal countdown demonstration test, and emergency egress simulation, Kennedy.
Oct. 16, 1986	NASA announced it would proceed with constructing a second horizontal test stand for redesign and certification of the Space Shuttle SRM at the Thiokol facility. The new test stand was designed to simulate, more closely than the existing SRM stand, the stresses on the SRM during an actual Shuttle launch and ascent.
Oct. 1986	Design requirements baselined.
1987	Primary design changes made to the SRM field joints, nozzle-to-case joints, case insulation, and seals.
Jan. 1987	Results of studies relating to innovative change to the existing (pre-*Challenger*) SRM joint design and design of new concepts for improved SRM performance were reported to Congress.
Mar. 1987	SRM Acquisition Strategy and Plan submitted to Congress. Plan indicated that NASA proposed to initiate Phase B (Definition) studies for an Advanced Solid Rocket Motor (ASRM); SRM redesign team evaluated design alternatives that would minimize the redesign time but ensure adequate safety margins. The team conducted analyses and tests of the redesign baseline.

Table 2–30 continued

Date	Event
May 22, 1987	First in a series of test firings conducted at Thiokol's facility. Objectives of the test, called the Nozzle Joint Environment Simulator test, included diversified motor system operations, such as evaluating and characterizing the SRM nozzle-to-case joint, obtaining information on the joint deflection data, and validating the test article in its original design.
May 27, 1987	An engineering test motor (ETM) was test-fired at Thiokol's facility in Utah as part of the Shuttle motor redesign program. The extensively instrumented ETM-1A was successfully fired for 120 seconds, a full-duration test.
Aug. 29, 1987	First full-duration test firing of the redesigned SRM at Thiokol. Designated DM-8, the 2-minute test evaluated the performance of the major features of the redesigned motor and completed several tests of case and nozzle-to-case joints with intentionally flawed insulation and O-rings.
Aug. 1987	Five solid propulsion contractors were awarded contracts for 9-month preliminary design and definition studies of both a monolithic and segmented ASRM that would permit performance increases of up to 5,443 kg of payload.
Oct. 1987	Critical Design Review—final design was approved.
Dec. 19, 1987	Second full-duration test firing of the redesigned Space Shuttle SRM at Thiokol.
Mar. 1, 1988	Redesigned SRM segments began arriving at Kennedy.
Mar. 28, 1988	Began stacking of *Discovery*'s SRM segments beginning with left aft booster.
Apr. 1988	Full-duration test firing of redesigned solid rocket motor (RSRM) at Thiokol's facility in Utah.
May 5, 1988	Began stacking lefthand booster segments.
May 28, 1988	Complete stacking of *Discovery*'s SRBs.
June 1988	Full-duration test firing of RSRM at Thiokol's facility in Utah.
June 10, 1988	SRBs and external tank are mated for STS-26; interface test between boosters and external tank conducted to verify connection.
July 1988	Solid Propulsion Integrity Program conducted most highly instrumented SRM nozzle test up to that time.
Aug. 1988	Full-duration test firing of RSRM. For this test, production verification motor-1 was extensively flawed to demonstrate the fail-safe characteristics of the redesign.
Aug. 1988	RFP issued for design, development, test, and evaluation of a Space Shuttle ASRM to replace the current RSRM in the mid-1990s. Contract award was anticipated for the spring of 1989.
Sept. 29, 1988	STS-26 returns Shuttle to operational status.

Table 2-31. Upper Stage Development

Date	Event
1979	DOD's detailed design of the two-stage Inertial Upper Stage (IUS) configuration completed.
1979	Detailed design of the NASA three-stage IUS configuration initiated.
1979	Designs completed and qualification program initiated for the PAM.
1979	Most PAM flight hardware manufactured and ready for assembly.
1979	NASA ordered PAM-As for Comsat's Intelsat V communications satellite missions.
Nov. 15, 1980	First flight of PAM-D on the Delta launched the SBS 1 spacecraft.
Sept. 1981	SBS 2 used PAM-D on Delta vehicle.
Nov. 1981	RCA-Satcom 3-R launched using PAM-D.
1982	PAM-A qualification and production halted, pending definition of spacecraft needs and launch schedules.
1982	PAM-D completed qualification and verification tests.
1982	PAM-D flew six commercial flights as third stage of Delta ELV.
May 1982	Transfer Orbit Stage (TOS) conceptual studies initiated.
Oct. 30, 1982	First IUS flown on DOD mission.
Dec. 1982	NASA/Orbital Sciences Corp. TOS Memorandum of Understanding.
Apr. 4, 1983	First IUS launched from Space Shuttle on STS-6, carrying the Tracking and Data Relay Satellite (TDRS). Second stage failed to place satellite in final geosynchronous orbit. Additional maneuvers placed TDRS 1 in its required functional orbit. NASA-Air Force team determined the IUS problem was in the gimbal mechanism of the second stage.
1983	PAM-D launched nine communications satellites, three from the Space Shuttle's cargo bay and six from ELVs.
Apr. 1983	NASA and Orbital Sciences signed a joint agreement for commercial development of the TOS.
May 1983	TOS design studies initiated.
Oct. 1983	TOS full-scale development initiated.
Feb. 3, 1984	PAM failed to boost Westar 6 and Palapa B-2 to proper orbit on STS 41-B mission.
May 1984	TOS Preliminary Design Review.
June 1984	Laser initial navigation system development begun by Honeywell for use in upper stages.
Dec. 1984	Orbital Sciences and Martin Marietta sign a development contract for TOS/Apogee Maneuvering Stage.

Table 2–31 continued

Date	Event
Mar. 1985	TOS Critical Design Review.
June 1985	Contract awarded to Orbital Sciences for the laser initial navigation system.
Aug. 1985	TOS Production Readiness Review; factory rollout of first TOS upper stage.
Nov. 26, 1985	PAM DII used on STS 61-B.
Jan. 12, 1986	PAM DII used on STS 61-C.
Feb. 1986	Boeing Aerospace selected to provide upper stage for TDRS-E and -F.
Mar. 1986	Mars Observer TOS contract selection.
June 19, 1986	Termination of Centaur upper stage development.
Nov. 26, 1986	NASA announced the selection of the IUS as the baseline option for three planetary missions: Galileo, Magellan, and Ulysses.
Nov. 26, 1986	The TOS was selected to place the Mars Observer spacecraft into the proper interplanetary trajectory.
1989	Martin Marietta and Boeing chosen to conduct studies on space transfer concepts, successor to the Orbital Transfer Vehicle.
July 14, 1993	Advanced Communications Technology Satellite (ACTS) launched from Shuttle with TOS for transfer to higher orbit.

Table 2–32. Transfer Orbit Stage Characteristics

Length	3.3 m
Weight With Full Propellant Load	10,886 kg
Airborne Support Equipment Weight	1,450 kg
Payload to Geotransfer Orbit	6,080 kg from Shuttle
Payload to Planetary and High-Energy Orbits	5,227 kg from Titan III and IV
Propulsion System	Orbis 21 solid rocket motor and attitude control system
Capacity	1,360 kg to 3,175 kg capacity

CHAPTER THREE
SPACE TRANSPORTATION/ HUMAN SPACEFLIGHT

CHAPTER THREE
SPACE TRANSPORTATION/ HUMAN SPACEFLIGHT

Introduction

In April 1981, after a hiatus of six years, American astronauts returned to space when they left the launch pad aboard the Space Shuttle orbiter *Columbia*. This chapter describes the major technology used by the Space Shuttle; each Space Shuttle mission through 1988, their payloads, and the operations surrounding the missions; the events surrounding the 1986 *Challenger* accident and the changes that occurred as a result of the accident; and the development of the Space Station program through 1988, one of NASA's major initiatives of the decade. It also describes the budget for human spaceflight at NASA and the management of human spaceflight activities.

The Last Decade Reviewed (1969–1978)

The successful culmination of three major spaceflight programs and steady progress in the Space Shuttle program highlighted NASA's second decade. The Apollo program concluded with its lunar landings; Skylab demonstrated the possibility of a space-based platform that could support human life over an extended period of time; and the Apollo-Soyuz Test Project showed that international cooperation in the space program was possible in the face of political differences. Steady progress in the human spaceflight program encouraged NASA to commit major resources to the Shuttle program.

The successful Apollo lunar expeditions caught the imagination of the American public. The first lunar landing took place on July 20, 1969, and was followed by the lunar landings of Apollo 12, 14, 15, 16, and 17. (Apollo 13 experienced a major anomaly, and the mission was aborted before a lunar landing could take place.) However, by the later missions, enthusiasm over the scientific and technological advances gave way to budget concerns, which ended the program with Apollo 17.

Skylab was the first American experimental space station to be built and could be considered a predecessor of the space station efforts of the 1980s. Skylab was an orbital workshop constructed from a Saturn IVB stage. It was launched in May 1973 and visited by three crews over the next nine months, each remaining at the orbiting laboratory for increasingly extended periods of time. The mission confirmed that humans could productively function in a space environment. It also provided solar observations, Earth resource studies, and tests of space manufacturing techniques.

The 1975 Apollo-Soyuz Test Project involved the docking of an American Apollo vehicle and a Soviet Soyuz vehicle. Joined by a docking module, the two crews conducted joint activities on their docked vehicles for two days before separating. Even though many hoped that this program would be the first of ongoing cooperative ventures between the two superpowers, the political situation prevented further efforts during this decade.

Although a six-year period interrupted human spaceflights between the 1975 Apollo-Soyuz mission and the first Shuttle flight in 1981, development of the new Space Shuttle moved slowly but steadily toward its inaugural launch in 1981. The major component of the Space Transportation System (STS), the Shuttle would perform a variety of tasks in orbit, including conducting scientific and technological experiments as well as serving as NASA's primary launch vehicle. NASA received presidential approval to proceed with the program in August 1972, and Rockwell International, the prime Shuttle contractor, rolled out *Enterprise,* the first test orbiter, in September 1976, setting off a series of system and flight tests. The production of *Columbia,* the first orbiter that would actually circle the Earth, already under way, continued during this time. Even though qualifying *Columbia* for spaceflight took longer than anticipated, as the decade closed, NASA was eagerly awaiting its first orbital flight test scheduled for the spring of 1981.

Overview of Space Transportation/Human Spaceflight (1979–1988)

The inauguration of Space Shuttle flights dominated the decade from 1979 through 1988. Twenty-seven Shuttle flights took place, and twenty-six of them were successful. However, from January 28, 1986, the memory of STS 51-L dominated the thoughts of many Americans and effectively overshadowed NASA's considerable achievements. The loss of life and, in particular, the loss of individuals who were not career astronauts haunted both the public and the agency. The agency conducted a far-reaching examination of the accident and used the findings of the independent Rogers Commission and the NASA STS 51-L Data and Design Analysis Task Force to implement a series of recommendations that improved the human spaceflight program from both a technical and management perspective. Two successful Shuttle missions followed at the end of the decade, demonstrating that NASA was able to recover from its worst accident ever.

The first twenty-four Shuttle missions and the two following the *Challenger* accident deployed an assortment of government and commercial satellites and performed an array of scientific and engineering experiments. The three Spacelab missions highlighted NASA's investigations aboard the Shuttle, studying everything from plant life and monkey nutrition to x-ray emissions from clusters of galaxies.

The 1980s also included a push toward the development of a permanently occupied space station. Announced by President Ronald Reagan in his 1984 State of the Union address, which directed NASA to have a permanently manned space station in place within ten years, NASA invested considerable time and money toward bringing it about. The European Space Agency (ESA), Canada, and Japan signed on as major participants in both the financial and technical areas of the Space Station program, and by the end of 1988, Space Station Freedom had completed the Definition and Preliminary Design Phase of the project and had moved into the Design and Development Phase.

Management of the Space Transportation/Human Spaceflight Program

The organizational elements of the space transportation program have been addressed in Chapter 2, "Launch Systems." Briefly, Code M, at different times called the Office of Space Transportation, Office of Space Transportation Systems (Acquisition), and Office of Space Flight, managed space transportation activities for the decade from 1979 through 1988. From November 1979 to August 1982, Code M split off the operations function of the spaceflight program into Code O, Office of Space Operations. Also, in 1984, the Office of Space Station, Code S, superseded the Code M Space Station Task Force, in response to President Reagan's directive to develop and build an occupied space station within the next ten years. Space Station program management is addressed later in this section.

The Space Shuttle program was the major segment of NASA's National Space Transportation System (NSTS), managed by the Office of Space Flight at NASA Headquarters. (The Space Shuttle Program Office was renamed the National Space Transportation System Program Office in March 1983.) The office was headed by an associate administrator who reported directly to the NASA administrator and was charged with providing executive leadership, overall direction, and effective accomplishment of the Space Shuttle and associated programs, including expendable launch vehicles.

The associate administrator for spaceflight exercised institutional management authority over the activities of the NASA field organizations whose primary functions were related to the NSTS program. These were the Johnson Space Center in Houston, the Kennedy Space Center at Cape Canaveral, Florida, the Marshall Space Flight Center in Huntsville, Alabama, and the Stennis Space Center (formerly National Space Technology Laboratories) in Bay St. Louis, Mississippi. Organizational elements of the NSTS office were located at NASA Headquarters, Johnson, Kennedy, Marshall, and at the Vandenberg Launch Site in California.

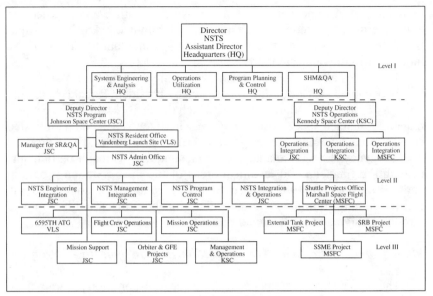

Figure 3–1. NSTS Organization

The organization of the NSTS was divided into four levels (Figure 3–1). The NSTS director served as the Level I manager and was responsible for the overall program requirements, budgets, and schedules. The NSTS deputy directors were Level II managers and were responsible for the management and integration of all program elements, including integrated flight and ground system requirements, schedules, and budgets. NSTS project managers located at Johnson, Kennedy, and Marshall were classified as Level III managers and were responsible for managing the design, qualification, and manufacturing of Space Shuttle components, as well as all launch and landing operations. NSTS design authority personnel and contractors were Level IV managers (not shown in Figure 3–1) and were responsible for the design, development, manufacturing, test, and qualification of Shuttle systems.

Initially, the NSTS was based at Johnson Space Center, which was designated as the lead center for the Space Shuttle program. Johnson had management responsibility for program control and overall systems engineering and systems integration. Johnson was also responsible for the development, production, and delivery of the Shuttle orbiter and managed the contract of the orbiter manufacturer.

Kennedy Space Center was responsible for the design of the launch and recovery facilities. Kennedy served as the launch and landing site for the Shuttle development flights and for most operational missions. Marshall Space Flight Center was responsible for the development, production, and delivery of the Space Shuttle main engines, solid rocket boosters, and external tank.

Robert F. Thompson served as manager of the Space Shuttle Program Office until 1981, when Glynn S. Lunney assumed the position of NSTS program manager. He had been with NASA since 1959 and involved in the Shuttle program since 1975. Lunney held the position of manager until his retirement in April 1985. He was replaced by Arnold D. Aldrich in July 1985, a twenty-six-year NASA veteran and head of the Space Shuttle Projects Office at Johnson Space Center. Aldrich's appointment was part of a general streamlining of the NSTS that took effect in August of that year, which reflected the maturation of the Shuttle program. In that realignment, the Level II NSTS organization at Johnson was renamed the NSTS Office and assimilated the Projects Office, consolidating all program elements under Aldrich's direction. Richard H. Kohrs, who had been acting program manager, and Lt. Col. Thomas W. Redmond, U.S. Air Force, were named deputy managers.

Aldrich took charge of the integration of all Space Shuttle program elements, including flight software, orbiter, external tank, solid rocket boosters, main engines, payloads, payload carriers, and Shuttle facilities. His responsibilities also included directing the planning for NSTS operations and managing orbiter and government-furnished equipment projects.

Post-Challenger Restructuring

The *Challenger* accident brought about major changes in the management and operation of the NSTS. The Rogers Commission concluded that flaws in the management structure and in communication at all levels were elements that needed to be addressed and rectified. Two of the recommendations (Recommendations II and V, respectively) addressed the management structure and program communication. In line with these recommendations, NASA announced in November 1986 a new Space Shuttle management structure for the NSTS. These changes aimed at clarifying the focal points of authority and responsibility in the Space Shuttle program and to establish clear lines of communication in the information-transfer and decision-making processes.

Associate Administrator for Space Flight Admiral Richard Truly issued a detailed description of the restructured NSTS organization and operation in a memorandum released on November 5, 1986. As part of the restructuring, the position of director, NSTS, was established, with Arnold Aldrich, who had been manager, NSTS, at the Johnson Space Center since July 1985, assuming that position in Washington, D.C. He had full responsibility and authority for the operation and conduct of the NSTS program. This included total program control, with full responsibility for budget, schedule, and balancing program content. He was responsible for overall program requirements and performance and had the approval authority for top-level program requirements, critical hardware waivers, and budget authorization adjustments that exceeded a predetermined level. He reported directly to the associate administrator for spaceflight and had two deputies, one for the program and one for operations.

NASA appointed Richard H. Kohrs, who had been deputy manager, NSTS, at the Johnson Space Center, to the position of deputy director, NSTS program. He was responsible for the day-to-day management and execution of the Space Shuttle program, including detailed program planning, direction, scheduling, and STS systems configuration management. Other responsibilities encompassed systems engineering and integration for the STS vehicle, ground facilities, and cargoes. The NSTS Engineering Integration Office, reporting to the deputy director, NSTS program, was established and directly participated with each NSTS project element (main engine, solid rocket booster, external tank, orbiter, and launch and landing system). Kohrs was located at Johnson, but he reported directly to the NSTS director.

Five organizational elements under the deputy director, NSTS program, were charged with accomplishing the management responsibilities of the program. The first four was located at Johnson, and the last was at the Marshall Space Flight Center.

- NSTS Engineering Integration
- NSTS Management Integration
- NSTS Program Control
- NSTS Integration and Operations
- Shuttle Projects Office

The Shuttle Projects Office had overall management and coordination responsibility for the Marshall elements involved in the Shuttle program: the solid rocket boosters, external tank, and main engines.

NASA named Captain Robert L. Crippen to the position of deputy director, NSTS operations, reporting directly to the NSTS director and responsible for all operational aspects of STS missions. This included such functions as final vehicle preparation, mission execution, and return of the vehicle for processing for its next flight. In addition, the deputy director, NSTS operations, presented the Flight Readiness Review, which was chaired by the associate administrator for spaceflight, managed the final launch decision process, and chaired the Mission Management Team.

Three operations integration offices located at Johnson, the Kennedy Space Center, and Marshall carried out the duties of the NSTS deputy director. In addition to the duties of the director and deputy directors described above, Admiral Truly's memorandum addressed the role of the centers and project managers in the programmatic chain and budget procedures and control. In the programmatic chain, the managers of the project elements located at the various field centers reported to the deputy director, NSTS program. Depending on the individual center organization, this chain was either direct (such as the Orbiter Project Office at Johnson) or via an intermediate office (such as the Shuttle Projects Office at Marshall).

The NSTS program budget continued to be submitted through the center directors to the director, NSTS, who had total funding authority for the program. The deputy directors, NSTS program and NSTS operations, each provided an assessment of the budget submittal to the director, NSTS, as an integral part of the decision process.

The restructuring also revitalized the Office of Space Flight Management Council. The council consisted of the associate administrator for spaceflight and the directors of Marshall, Kennedy, Johnson, and the NSTS. This group met regularly to review Space Shuttle program progress and to provide an independent and objective assessment of the status of the overall program.

Management relationships in the centralized NSTS organization were configured into four basic management levels, which were designed to reduce the potential for conflict between the program organizations and the NASA institutional organizations.

Office of Safety, Reliability, and Quality Assurance

Although not part of the Office of Space Flight, the Office of Safety, Reliability, and Quality Assurance (Code Q) resulted from the findings of the Rogers Commission, which recommended that NASA establish such an office with direct authority throughout the agency. NASA established this office in July 1986, with George A. Rodney, formerly of Martin Marietta, named as its first associate administrator (Figure 3–2). The objectives of the office were to ensure that a NASA Safety, Reliability, and Quality Assurance program monitored equipment status, design validation problem analysis, and system acceptability in agencywide plans and programs.

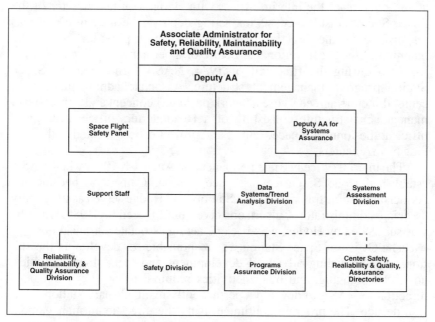

Figure 3–2. Safety, Reliability, and Quality Assurance Office Organization

The responsibilities of the associate administrator included the oversight of safety, reliability, and quality assurance functions related to all NASA activities and programs. In addition, he was responsible for the direction of reporting and documentation of problems, problem resolution, and trends associated with safety.

Management of the Space Station Program

NASA first officially committed to a space station on May 20, 1982, when it established the Space Station Task Force under the direction of John D. Hodge, assistant for space station planning and analysis, Office of the Associate Deputy Administrator in the Office of Space Flight (Code M). Hodge reported to Philip E. Culbertson, associate deputy administrator, and drew from space station-related activities of each of the NASA program offices and field centers.

The task force was responsible for the development of the programmatic aspects of a space station as they evolved, including mission analysis, requirements definition, and program management. It initiated industry participation with Phase A (conceptual analysis) studies that focused on user requirements and their implications for design. The task force developed the space station concept that formed the basis for President Reagan's decision to commit to a space station.

The task force remained in existence until April 6, 1984, when, in response to Reagan's January 1984 State of the Union address, NASA established an interim Space Station Program Office. Culbertson, in addition to his duties as associate deputy administrator, assumed the role of acting director of the interim office, with Hodge (former director of the Space Station Task Force) as his acting deputy. The interim office was responsible for the direction of the Space Station program and for the planning of the organizational structure of a permanent program office.

Also during the first half of 1984, NASA formulated the Space Station program management structure. Associate administrators and center directors agreed to use a "work package" concept and a three-level management structure consisting of a Headquarters office, a program office at the Johnson Space Center, and project offices located at the various NASA centers.

The interim office became permanent on August 1, 1984, when NASA established Code S, Office of Space Station. Culbertson became the Associate Administrator for Space Station, and Hodge served as the deputy associate administrator. Culbertson served until December 1985, when he was succeeded by Hodge, who became acting associate administrator.

The Office of Space Station was responsible for developing the station and conducting advanced development and technology activities, advanced planning, and other activities required to carry out Reagan's direction to NASA to develop a permanently manned space station within a decade. The program continued using the three-tiered management structure developed earlier in the year. The Headquarters Level A office encompassed the Office of the Associate Administrator for the Office of Space Station and provided overall policy and program direction for the Space Station program. The Level B Space Station Program Office at Johnson in Houston reported to the Headquarters office. Space Station Level C project offices at other NASA centers also were responsible to the Office of Space Station through the Johnson program office. Johnson had been named lead center for the Space Station program in February 1984. The associate administrator was supported by a chief scientist, policy and plans and program support offices, and business management, engineering, utilization and performance requirements, and operations divisions.

On June 30, 1986, Andrew J. Stofan, who had been director of NASA's Lewis Research Center in Cleveland, was appointed Associate Administrator for Space Station. Along with this appointment, NASA Administrator James C. Fletcher announced several management structural actions that were designed to strengthen technical and management capabilities in preparation for moving into the development phase of the Space Station program.

The decision to create the new structure resulted from recommendations made by a committee headed by former Apollo program manager General Samuel C. Phillips. General Phillips had conducted a review of space station management as part of a long-range assessment of NASA's overall capabilities and requirements, including relationships between the various space centers and NASA Headquarters. His report reflected discussions with representatives from all the NASA centers and the contractors involved in the definition and preliminary design of the space station, as well as officials from other offices within NASA. His report recommended the formation of a program office, which was implemented in October 1986 when NASA Administrator Fletcher named Thomas L. Moser director of the Space Station Program Office, reporting to Associate Administrator Stofan.

Fletcher stated that the new space station management structure was consistent with recommendations of the Rogers Commission, which investigated the Space Shuttle *Challenger* accident. The commission had recommended that NASA reconsider management structures, lines of communication, and decision-making processes to ensure the flow of important information to proper decision levels. As part of the reconfiguration of the management structure, the Johnson Space Center was no longer designated as Level B. Instead, a Level A' was substituted, located in the Washington metropolitan area, assuming the same functions Johnson previously held (Figure 3–3).

Fletcher said the program would use the services of a top-level, non-hardware support contractor. In addition to the systems engineering role, the program office would contain a strong operations function to ensure that the program adequately addressed the intensive needs of a permanent facility in space.

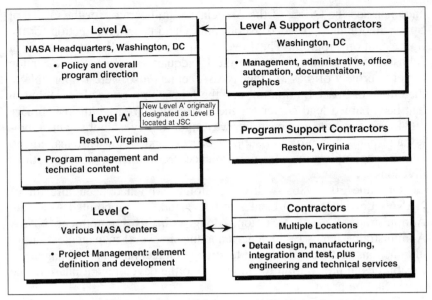

Figure 3–3. Space Station Program Management Approach

NASA established a systems integration field office in Houston as part of the program office organization. Project managers at the Goddard Space Flight Center, Johnson, Kennedy, Lewis, and Marshall reported functionally to the associate administrator. They coordinated with their respective center directors to keep them informed of significant program matters.

NASA assigned John Hodge the job of streamlining and clarifying NASA's procurement and management approach for the Space Station program and issuing instructions related to work package assignments, the procurement of hardware and services, and the selection of contractors for the development phase of the program. In addition, NASA tasked Hodge with developing a program overview document that would define the role automation and robotics would play in the Space Station program and with conducting further studies in the areas of international involvement, long-term operations, user accommodations, and servicing.

At the same time, Fletcher authorized NASA to procure a Technical and Management Information System (TMIS), a computer-based information network. It would link NASA and contractor facilities together and provide engineering services, such as computer-aided design, as well as management support on items such as schedules, budgets, labor, and facilities. TMIS was implemented in 1988.

The Space Station Program Office was responsible for the overall technical direction and content of the Space Station program, including systems engineering and analysis, configuration management, and the integration of all elements into an operating system that was responsive to customer needs. NASA approved a further reorganization of the Office of Space Station in December 1986.

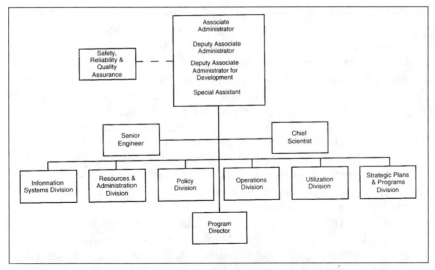

Figure 3–4. Office of Space Station Organization (December 1986)

In addition to the associate administrator and two deputies, the approved Space Station program organization included a chief scientist, a senior engineer, and six division directors responsible for resources and administration, policy, utilization, operations, strategic plans and programs, and information systems. There was also a position of special assistant to the associate administrator (Figure 3–4).

Andrew Stofan continued in the position of associate administrator. Franklin D. Martin continued as the deputy associate administrator for space station. Previously director of space and Earth sciences at the Goddard Space Flight Center, Martin had been named to the post in September 1986.

Thomas L. Moser became the deputy associate administrator for development in October 1986, a new position established by the reorganization. In this position, Moser also served as the program director for the Office of Space Station, directing the Washington area office that was responsible for overall technical direction and content of the Space Station program, including systems engineering and analysis, program planning and control, configuration management, and the integration of all the elements into an operating system. The creation of the program director position was the central element of program restructuring in response to recommendations of the committee headed by General Phillips. The Phillips Committee conducted an extensive examination of the Space Station organization.

As a result of this restructuring, NASA centers performed a major portion of the systems integration through Space Station field offices that were established at Goddard, Johnson, Kennedy, Lewis, and Marshall. The space station project manager at each of the five centers headed the field office and reported directly to the program manager in Washington.

A program support contractor assisted the program office and field offices in systems engineering, analysis, and integration activities.

Also as part of this reorganization, NASA named Daniel H. Herman senior engineer, a new staff position. The senior engineer advised the associate administrator on the policy, schedule, cost, and user implications of technical decisions. Previously, Herman was director of the engineering division, whose functions and responsibilities were absorbed by Moser's organization, and was on the original Space Station Task Force, which defined the basic architecture of the space station system.

David C. Black continued to serve as chief scientist for the space station. Black, chief scientist of the Space Research Directorate at the Ames Research Center, had served as chief scientist for the space station since the post was created in 1984.

Paul G. Anderson acted as the director of the Resources and Administration Division, which combined the former business management and program support organizations. Anderson previously served as comptroller at the Lewis Research Center.

Margaret Finarelli, director of the Policy Division, had functional responsibility for the former policy and plans organization. This element of the reorganization reflected the strong policy coordination role required of the Space Station Program Office in working with other elements of NASA, the international partners, and other external organizations. Prior to this assignment, Finarelli was chief of the International Planning and Programs Office in the International Affairs Division at NASA Headquarters.

Richard E. Halpern became the director of the Utilization Division, which had responsibility for developing user requirements for the space station, including science and applications, technology development, commercial users, and the assurance that those requirements could be efficiently and economically accommodated on the space station. Halpern was the director of the Microgravity Science and Applications Division in the Office of Space Science and Applications prior to accepting this position.

The Operations Division had the responsibility for developing an overall philosophy and management approach for space station system operations, including user support, prelaunch and postlanding activities, logistics support, and financial management. Granville Paules served as acting director of the Operations Division.

Under the new organization, NASA formed two new divisions, Strategic Plans and Programs and Information Systems. The Information Systems Division provided a management focus for the total end-to-end information system complex for Space Station.

Alphonso V. Diaz assumed the position of director of strategic plans and programs and had responsibility for ensuring that the evolution of the space station infrastructure was well planned and coordinated with other NASA offices and external elements. As part of its responsibility, this division managed and acted as the single focus for space station automation and robotics activities and program-focused technology and advanced development work.

The Strategic Plans and Program Division under Mr. Diaz became responsible for determining requirements and managing the Transition Definition program at Level A. The division maintained the Space Station Evolution Technical and Management Plan, which detailed evolution planning for the long-term use of the space station. The Level A' Space Station Program Office in Reston, Virginia, managed the program, including provision for the "hooks and scars," which were design features for the addition or update of computer software (hooks) or hardware (scars). The Langley Evolution Definition Office chaired the agencywide Evolution Working Group, which provided interagency communication and coordination of station evolution, planning, and interfaces with the baseline Work Packages (Level C). (Work Packages are addressed later in this chapter.)

William P. Raney, who had served as director of the Utilization and Performance Requirements Division, served as special assistant to the associate administrator. Stofan served as Associate Administrator for Space Station until his retirement from NASA in April 1988, when he was replaced by James B. Odom.

Money for Human Spaceflight

As with money for launch systems, Congress funded human spaceflight entirely from the Research and Development (R&D) appropriation through FY 1983. Beginning with FY 1984, the majority of funds for human spaceflight came from the Space Flight, Control, and Data Communications (SFC&DC) appropriation. Only funds for the Space Station and Spacelab programs remained with R&D. In FY 1985, Space Station became a program office with its own budget. Spacelab remained in the Office of Space Flight.

As seen in Table 3–1, appropriated funding levels for human spaceflight for most years met NASA's budget requests as submitted to Congress. The last column in the table shows the actual amounts that were programmed for the major budget items.

Program funding generally increased during 1979–1988 (Table 3–2). However, the reader must note that these figures are all current year money—that is, the dollar amounts do not take into account the reduced buying power caused by inflation. In addition, the items that are included in a major budget category change from one year to the next, depending on the current goals and resources of the agency and of Congress. Thus, it is difficult to compare dollar amounts because the products or services that those dollars are intended to buy may differ from year to year.

Tables 3–3 through 3–10 show funding levels for individual programs within the human spaceflight category.

Space Station

NASA's initial estimate of the U.S. investment in the Space Station program was $8 billion in 1984 dollars. By March 1988, this estimate had grown to $14.5 billion, even though, in 1987, the National Research Council had priced the Space Station program at $31.8 billion.[1]

President Reagan strongly endorsed the program and persuaded an ambivalent Congress of its importance. Program funding reflected both his persuasive powers and the uncertainty in which members of Congress looked at the space station, who took the view that it had little real scientific or technological purpose. The congressional Office of Technology Assessment reported that Congress should not commit to building a space station until space goals were more clear and that the potential uses of the proposed station did not justify the $8 billion price tag.

Congress passed the FY 1985 appropriation of $155.5 million for starting the design and development work on the space station based on NASA's initial $8 billion figure. The FY 1986 appropriation reduced the Administration's request from $230 million to $205 million.

President Reagan's FY 1987 budget asked for $410 million for the Space Station program, doubling the station funds from the previous year. Congress approved this increase in August 1986, which would move space station into the development phase toward planned operation by the mid-1990s. However, Congress placed limitations on the appropriation; it stipulated that NASA funds could not be spent to reorganize the program without congressional approval. In addition, $150 million was to be held back until NASA met several design and assembly requirements set by the House Appropriations Committee. About $260 million of the $410 million were to be spent for Phase B activities, and the other $150 million was reserved for initial hardware development. NASA must comply with the following conditions: a minimum of thirty-seven and a half kilowatts of power for initial operating capability, rather than the twenty-five kilowatts envisioned by NASA; a fully equipped materials processing laboratory by the sixth Space Shuttle flight and before crew habitat was launched; early launch of scientific payloads; and deployment of U.S. core elements before foreign station elements.[2]

During the next month, NASA Administrator James Fletcher stated that the $8 billion estimated for the Space Station program was now seen to be insufficient and that the station must either receive additional funds or be scaled down. The Reagan Administration submitted a request in

---

[1]National Research Council, *Report of the Committee on the Space Station of the National Research Council* (Washington, DC: National Academy Press, September 1987).

[2]Report to accompany Department of Housing and Urban Development–Independent Agencies Appropriations Budget, 1987, House of Representatives.

January 1987 for $767 million for the Space Station program. However, after much debate, which raised the possibility of freezing the entire program, Congress appropriated only $425 million, but again, conditions were attached. In the FY 1988 Continuing Resolution that funded the program, Congress ordered NASA to provide a rescoping plan for the space station. In addition, only $200 million of the $425 million was to be available before June 1, 1988, while the rescoping was under discussion. By the time the rescoping plan had gone to Congress, the cost of the Station was up to $14.5 billion. Further talks in Congress later during the year proposed reducing funding for FY 1989 to an even lower level.

The Space Transportation System

This section focuses on the structure and operation of the equipment and systems used in the Space Transportation System (STS) and describes the mission and flight operations. The overview provides a brief chronology of the system's development. The next section looks at the orbiter as the prime component of STS. (The launch-related elements—that is, the external tank, solid rocket boosters, main engines, and the propulsion system in general—have been addressed previously in Chapter 2, "Launch Systems.") The last part of this section addresses STS mission operations and support.

A vast quantity of data exists on the Space Shuttle, and this document presents only a subset of the available material. It is hoped that the primary subject areas have been treated adequately and that the reader will get a useful overview of this complex system. It is highly recommended that readers who wish to acquire more detailed information consult the *NSTS Shuttle Reference Manual* (1988).[3]

Overview

The history of NASA's STS began early in the 1970s when President Richard Nixon proposed the development of a reusable space transportation system. The *NASA Historical Data Book, Volume III, 1969–1978*, presents an excellent account of events that took place during those early days of the program.[4]

By 1979, all major STS elements were proceeding in test and manufacture, and major ground test programs were approaching completion. NASA completed the design certification review of the overall Space Shuttle configuration in April 1979. Development testing throughout the

---

[3] *NSTS Shuttle Reference Manual* (1988), available both through the NASA History Office and on-line through the NASA Kennedy Space Center Home Page.

[4] Linda Neuman Ezell, *NASA Historical Data Book, Volume III: Programs and Projects, 1969–1978* (Washington, DC: NASA SP-4012, 1988).

program was substantially complete, and the program was qualifying flight-configured systems.

The orbiter's structural test article was under subcontract for structural testing and would ultimately be converted to become the second orbital vehicle, *Challenger*. The development of *Columbia* was proceeding more slowly than anticipated, with much work remaining to be completed before the first flight, then scheduled for late 1980. The main engine had accumulated more than 50,000 seconds of test time toward its goal of 80,000 seconds before the first orbital flight, and the first external tank that would be used during flight had been delivered as well as three test tanks. Three flight tanks were also being manufactured for flight in the orbital flight test program. By the end of 1979, Morton Thiokol, the solid rocket booster contractor, had completed four development firings of the solid rocket boosters, and the qualification firing program had started. Two qualification motor firings had been made, and one more was scheduled before the first flight. Most of the rocket segments for the first flight boosters had been delivered to Kennedy Space Center.

All launch and landing facilities at Kennedy were complete and in place for the first orbital flight. Ground support equipment and the computerized launch-processing installations were almost complete, and software validation was progressing. All hardware for the launch processing system had been delivered, simulation support was continuing for the development of checkout procedures, and checkout software was being developed and validated.

By the end of 1979, nine commercial and foreign users had reserved space on Space Shuttle flights. Together with NASA's own payloads and firm commitments from the Department of Defense (DOD) and other U.S. government agencies, the first few years of STS operations were fully booked.

During 1980, testing and manufacture of all major system continued, and by the end of 1980, major ground-test programs neared completion. The first flight-configuration Space Shuttle stood on the launch pad. Additional testing of the vehicle was under way; qualification testing of flight-configured elements continued toward a rescheduled launch in the spring of 1981.

In December 1980, *Columbia* was in final processing at the Kennedy Space Center. The main engines had surpassed their goal of 80,000 seconds of engine test time, with more than 90,000 seconds completed. Technicians had mated the orbiter with the solid rocket boosters and external tank in November and rolled it out onto the launch pad in December. Contractors had delivered the final flight hardware, which was in use for vehicle checkout. Hardware and thermal protection system certifications were nearly complete. Further manufacture and testing of the external tanks and solid rocket boosters had also been completed.

The Kennedy launch site facilities were completed during 1980 in anticipation of the first launch. The computerized launch processing system had been used extensively for Space Shuttle testing and facility activation. The high-energy fuel systems had been checked out, and the integrated test of the Shuttle was complete.

The mission control center and Shuttle mission simulator facilities at the Johnson Space Center were ready to support the first Shuttle flight. Both the flight crew and ground flight controllers had used these facilities extensively for training and procedure development and verification. Seven full-duration (fifty-four-hour) integrated simulations had been successfully conducted, with numerous ascent, orbit, entry, and landing runs completed. The mission flight rules and launch-commit criteria had also been completed.

Follow-on orbiter production was in progress, leading to the four-orbiter fleet for the STS's future needs. The structural test article was being modified to a flight-configured orbiter, *Challenger*. Secondary and primary structural installations were under way, and thermal protection installations had begun for vehicle delivery in June 1982.

The Space Shuttle program made its orbital debut with its first two flights in 1981. All major mission objectives were met on both flights. Details of these missions and other STS missions through 1988 appear in later sections of this chapter.

The following pages describe the orbiter's structure, major systems, and operations, including crew training. Because this volume concentrates on the period from 1979 through 1988, the wording reflects configurations and activities as they existed during that decade. However, most of the Space Shuttle's physical characteristics and operations have continued beyond 1988 and are still valid.

Orbiter Structure

NASA designed the Space Shuttle orbiter as a space transport vehicle that could be reused for approximately 100 missions. The orbiter was about the same length and weight as a commercial DC-9 airplane. Its structure consisted of the forward fuselage (upper and lower forward fuselage and the crew module, which could accommodate up to seven crew members in normal operations and up to ten during emergency operations), the wings, the mid-fuselage, the payload bay doors, the aft fuselage, and the vertical stabilizer. Its appearance, however, differed markedly from a conventional airplane. High-performance double-delta (or triangular) wings and a large cargo bay gave the Shuttle its squat appearance (Figure 3–5 and Table 3–11).

A cluster of three Space Shuttle Main Engines (SSMEs) in the aft fuselage provided the main propulsion for the orbiter vehicle. The external tank carried fuel for the orbiter's main engines. Both the solid rocket boosters and the external tank were jettisoned prior to orbital insertion. In orbit, the orbital maneuvering system (OMS), contained in two pods on the aft fuselage, maneuvered the orbiter. The OMS provided the thrust for orbit insertion, orbit circularization, orbit transfer, rendezvous, deorbit, abort-to-orbit, and abort-once-around and could provide up to 453.6 kilograms of

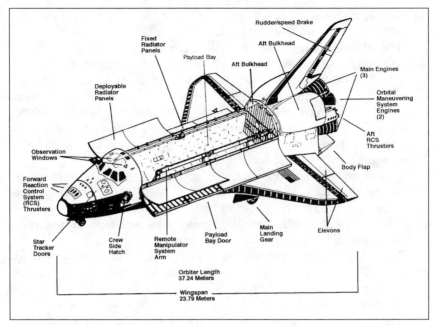

Figure 3–5. Space Shuttle Orbiter

propellant to the aft reaction control system (RCS). The RCS, contained in the two OMS pods and in a module in the nose section of the forward fuselage, provided attitude control in space and during reentry and was used during rendezvous and docking maneuvers. When it completed its orbital activities, the orbiter landed horizontally, as a glider, at a speed of about ninety-five meters per second and at a glide angle of between eighteen and twenty-two degrees.

The liquid hydrogen–liquid oxygen engine was a reusable high-performance rocket engine capable of various thrust levels. Ignited on the ground prior to launch, the cluster of three main engines operated in parallel with the solid rocket boosters during the initial ascent. After the boosters separated, the main engines continued to operate for approximately eight and a half minutes. The SSMEs developed thrust by using high-energy propellants in a staged combustion cycle. The propellants were partially combusted in dual preburners to produce high-pressure hot gas to drive the turbopumps. Combustion was completed in the main combustion chamber. The SSME could be throttled over a thrust range of 65 to 109 percent, which provided for a high thrust level during liftoff and the initial ascent phase but allowed thrust to be reduced to limit acceleration to three g's during the final ascent phase.

The orbiter was constructed primarily of aluminum and was protected from reentry heat by a thermal protection system. Rigid silica tiles or some other heat-resistant material shielded every part of the Space Shuttle's external shell. Tiles covering the upper and forward fuselage sections and the tops of the wings could absorb heat as high as

650 degrees Centigrade. Tiles on the underside absorbed temperatures up to 1,260 degrees Centigrade. Areas that had to withstand temperatures greater than 1,260 degrees Centigrade, such as the nose and leading edges of the wings on reentry, were covered with black panels made of reinforced carbon-carbon.

A five-computer network configured in a redundant operating group (four operate at all times and one is a backup) monitored all Space Shuttle subsystems. They simultaneously processed data from every area of the Shuttle, each interacting with the others and comparing data.

During ascent, acceleration was limited to less than three g's. During reentry, acceleration was less than two and a half g's. By comparison, Apollo crews had to withstand as much as eight g's during reentry into the Earth's atmosphere. The Space Shuttle's relatively comfortable ride allowed crew other than specially trained astronauts to travel on the Shuttle. While in orbit, crew members inhabited a "shirtsleeve" environment—no spacesuits or breathing apparatus were required. The microgravity atmosphere remained virtually the only non-Earth-like condition that crew members had to encounter.

NASA named the first four orbiter spacecraft after famous exploration sailing ships:

- *Columbia* (OV-102), the first operational orbiter, was named after a sailing frigate launched in 1836, one of the first Navy ships to circumnavigate the globe. *Columbia* also was the name of the Apollo 11 command module that carried Neil Armstrong, Michael Collins, and Edwin "Buzz" Aldrin on the first lunar landing mission in July 1969. *Columbia* was delivered to Rockwell's Palmdale assembly facility for modifications on January 30, 1984, and was returned to the Kennedy Space Center on July 14, 1985, for return to flight.

- *Challenger* (OV-099) was also the name of a Navy ship, one that explored the Atlantic and Pacific Oceans from 1872 to 1876. The name also was used in the Apollo program for the Apollo 17 lunar module. *Challenger* was delivered to Kennedy on July 5, 1982.

- *Discovery* (OV-103) was named after two ships. One was the vessel in which Henry Hudson in 1610–11 attempted to search for a northwest passage between the Atlantic and Pacific Oceans and instead discovered the Hudson Bay. The other was the ship in which Captain Cook discovered the Hawaiian Islands and explored southern Alaska and western Canada. *Discovery* was delivered to Kennedy on November 9, 1983.

- *Atlantis* (OV-104) was named after a two-masted ketch operated for the Woods Hole Oceanographic Institute from 1930 to 1966 that traveled more than half a million miles conducting ocean research. *Atlantis* was delivered to Kennedy on April 3, 1985.

A fifth orbiter, *Endeavour* (OV-105), was named by Mississippi school children in a contest held by NASA. It was the ship of Lieutenant James Cook in 1769–71, on a voyage to Tahiti to observe the planet Venus passing between the Earth and the Sun. This orbiter was delivered to NASA by Rockwell International in 1991.

Major Systems

Avionics Systems

The Space Shuttle avionics system controlled, or assisted in controlling, most of the Shuttle systems. Its functions included automatic determination of the vehicle's status and operational readiness; implementation sequencing and control for the solid rocket boosters and external tank during launch and ascent; performance monitoring; digital data processing; communications and tracking; payload and system management; guidance, navigation, and control; and electrical power distribution for the orbiter, external tank, and solid rocket boosters.

Thermal Protection System

A passive thermal protection system helped maintain the temperature of the orbiter spacecraft, systems, and components within their temperature limits primarily during the entry phase of the mission. It consisted of various materials applied externally to the outer structural skin of the orbiter.

Orbiter Purge, Vent, and Drain System

The purge, vent, and drain system on the orbiter provided unpressurized compartments with gas purge for thermal conditioning and prevented the accumulation of hazardous gases, vented the unpressurized compartments during ascent and entry, drained trapped fluids (water and hydraulic fluid), and conditioned window cavities to maintain visibility.

Orbiter Communications System

The Space Shuttle orbiter communications system transferred (1) telemetry information about orbiter operating conditions and configurations, systems, and payloads; (2) commands to the orbiter systems to make them perform some function or configuration change; (3) documentation from the ground that was printed on the orbiter's teleprinter or text and graphics system; and (4) voice communications among the flight crew members and between the fight crew and ground. This information was transferred through hardline and radio frequency links.

Direct communication took place through Air Force Satellite Control Facility remote tracking station sites, also known as the Spaceflight Tracking and Data Network ground stations for NASA missions or space-ground link system ground stations for military missions. Direct signals from the ground to the orbiter were referred to as uplinks, and signals from the orbiter to the ground were called downlinks.

Tracking and Data Relay Satellite (TDRS) communication took place through the White Sands Ground Terminal. These indirect signals from TDRS to the orbiter were called forward links, and the signal from the orbiter to the TDRS was called the return link. Communication with a detached payload from the orbiter was also referred to as a forward link, and the signal from the payload to the orbiter was the return link. Refer to Chapter 4, "Tracking and Data Acquisition Systems," in Volume VI of the *NASA Historical Databook* for a more detailed description of Shuttle tracking and communications systems.

Data Processing System

The data processing system, through the use of various hardware components and its self-contained computer programming (software), provided the vehicle with computerized monitoring and control. This system supported the guidance, navigation, and control of the vehicle, including calculations of trajectories, SSME thrusting data, and vehicle attitude control data; processed vehicle data for the flight crew and for transmission to the ground and allowed ground control of some vehicle systems via transmitted commands; checked data transmission errors and crew control input errors; supported the annunciation of vehicle system failures and out-of-tolerance system conditions; supported payloads with flight crew/software interface for activation, deployment, deactivation, and retrieval; processed rendezvous, tracking, and data transmissions between payloads and the ground; and monitored and controlled vehicle subsystems.

Guidance, Navigation, and Control

Guidance, navigation, and control software commanded the guidance, navigation, and control system to effect vehicle control and to provide the sensor and controller data needed to compute these commands. The process involved three steps: (1) guidance equipment and software computed the orbiter location required to satisfy mission requirements; (2) navigation tracked the vehicle's actual location; and (3) flight control transported the orbiter to the required location. A redundant set of four orbiter general purpose computers (GPCs) formed the primary avionics software system; a fifth GPC was used as the backup flight system.

The guidance, navigation, and control system operated in two modes: auto and manual (control stick steering). In the automatic mode, the primary avionics software system essentially allowed the GPCs to fly the vehicle; the flight crew simply selected the various operational sequences. In the manual mode, the flight crew could control the vehicle using hand controls, such as the rotational hand controller, translational hand controller, speed brake/thrust controller, and rudder pedals. In this mode, flight crew commands still passed through and were issued by the GPCs. There were no direct mechanical links between the flight crew and the orbiter's various propulsion systems or aerodynamic surfaces; the orbiter was an entirely digitally controlled, fly-by-wire vehicle.

Dedicated Display System

The dedicated displays provided the flight crew with information required to fly the vehicle manually or to monitor automatic flight control system performance. The dedicated displays were the attitude director indicators, horizontal situation indicators, alpha Mach indicators, altitude/vertical velocity indicators, a surface position indicator, RCS activity lights, a g-meter, and a heads-up display.

Main Propulsion System

The Space Shuttle's main propulsion system is addressed in Chapter 2, "Launch Systems."

Crew Escape System

The in-flight crew escape system was provided for use only when the orbiter would be in controlled gliding flight and unable to reach a runway. This condition would normally lead to ditching. The crew escape system provided the flight crew with an alternative to water ditching or to landing on terrain other than a landing site. The probability of the flight crew surviving a ditching was very slim.

The hardware changes required to the orbiters following the STS 51-L (*Challenger*) accident enabled the flight crew to equalize the pressurized crew compartment with the outside pressure via the depressurization valve opened by pyrotechnics in the crew compartment aft bulkhead that a crew member would manually activate in the mid-deck of the crew compartment. The crew could also pyrotechnically jettison the crew ingress/egress side hatch manually in the mid-deck of the crew compartment and bail out from the mid-deck through the ingress/egress side hatch opening after manually deploying the escape pole through, outside, and down from the side hatch opening.

Emergency Egress Slide. The emergency egress slide replaced the emergency egress side hatch bar. It provided the orbiter flight crew members with a rapid and safe emergency egress through the orbiter mid-deck ingress/egress side hatch after a normal opening of the side hatch or after jettisoning of the side hatch at the nominal end-of-mission landing site or at a remote or emergency landing site. The emergency egress slide supported return-to-launch-site, transatlantic-landing, abort-once-around, and normal end-of-mission landings.

Secondary Emergency Egress. The lefthand flight deck overhead window provided the flight crew with a secondary emergency egress route.

Side Hatch Jettison. The mid-deck ingress/egress side hatch was modified to provide the capability of pyrotechnically jettisoning the side hatch for emergency egress on the ground. In addition, a crew compartment pressure equalization valve provided at the crew compartment aft bulkhead was also pyrotechnically activated to equalize cabin/outside pressure before the jettisoning of the side hatch.

Crew Equipment

Food System and Dining. The mid-deck of the orbiter was equipped with facilities for food stowage, preparation, and dining for each crew member. Three one-hour meal periods were scheduled for each day of the mission. This hour included time for eating and cleanup. Breakfast, lunch, and dinner were scheduled as close to the usual hours as possible. Dinner was scheduled at least two to three hours before crew members began preparations for their sleep period.

Shuttle Orbiter Medical System. The Shuttle orbiter medical system provided medical care in flight for minor illnesses and injuries. It also provided support for stabilizing severely injured or ill crew members until they were returned to Earth. The medical system consisted of the medications and bandage kit and the emergency medical kit.

Operational Bioinstrumentation System. The operational bioinstrumentation system provided an amplified electrocardiograph analog signal from either of two designated flight crew members to the orbiter avionics system, where it was converted to digital tape and transmitted to the ground in real time or stored on tape for dump at a later time. On-orbit use was limited to contingency situations.

Radiation Equipment. The harmful biological effects of radiation must be minimized through mission planning based on calculated predictions and monitoring of dosage exposures. Preflight requirements included a projection of mission radiation dosage, an assessment of the probability of solar flares during the mission, and a radiation exposure history of flight crew members. In-flight requirements included the carrying of passive dosimeters by the flight crew members and, in the event of solar flares or other radiation contingencies, the readout and reporting of the active dosimeters.

Crew Apparel. During launch and entry, crew members wore the crew altitude protection system consisting of a helmet, a communications cap, a pressure garment, an anti-exposure, anti-gravity suit, gloves, and boots. During launch and reentry, the crew wore escape equipment over the crew altitude protection system, consisting of an emergency oxygen system; parachute harness, parachute pack with automatic opener, pilot chute, drogue chute, and main canopy; a life raft; two liters of drinking water; flotation devices; and survival vest pockets containing a radio/beacon, signal mirror, shroud cutter, pen gun flare kit, sea dye marker, smoke flare, and beacon.

Sleeping Provisions. Sleeping provisions consisted of sleeping bags, sleep restraints, or rigid sleep stations. During a mission with one shift, all crew members slept simultaneously and at least one crew member would wear a communication headset to ensure the reception of ground calls and orbiter caution and warning alarms.

Personal Hygiene Provisions. Personal hygiene and grooming provisions were furnished for both male and female flight crew members. A water dispensing system provided water.

Housekeeping. In addition to time scheduled for sleep periods and meals, each crew member had housekeeping tasks that required from five to fifteen minutes at intervals throughout the day. These included cleaning the waste management compartment, the dining area and equipment, floors and walls (as required), the cabin air filters, trash collection and disposal, and change-out of the crew compartment carbon dioxide (lithium hydroxide) absorber canisters.

Sighting Aids. Sighting aids included all items used to aid the flight crew within and outside the crew compartment. They included the crewman optical alignment sight, binoculars, adjustable mirrors, spotlights, and eyeglasses.

Microcassette Recorder. The microcassette recorder was used primarily for voice recording of data but could also be used to play prerecorded tapes.

Photographic Equipment. The flight crew used three camera systems—16mm, 35mm, and 70mm—to document activities inside and outside the orbiter.

Wicket Tabs. Wicket tabs helped the crew members activate controls when vision was degraded. The tabs provided the crew members with tactile cues to the location of controls to be activated as well as a memory aid to their function, sequence of activation, and other pertinent information. Controls that were difficult to see during the ascent and entry flight phases had wicket tabs.

Reach Aid. The reach aid, sometimes known as the "swizzle stick," was a short adjustable bar with a multipurpose end effector that was used to actuate controls that were out of the reach of seated crew members. It could be used during any phase of flight, but was not recommended for use during ascent because of the attenuation and switch-cueing difficulties resulting from acceleration forces.

Restraints and Mobility Aids. Restraints and mobility aids enabled the flight crew to perform all tasks safely and efficiently during ingress, egress, and orbital flight. Restraints consisted of foot loop restraints, the airlock foot restraint platform, and the work/dining table as well as temporary stowage bags, Velcro, tape, snaps, cable restraints, clips, bungees, and tethers. Mobility aids and devices consisted of handholds for ingress and egress to and from crew seats in the launch and landing configuration, handholds in the primary interdeck access opening for ingress and egress in the launch and landing configuration, a platform in the mid-deck for ingress and egress to and from the mid-deck when the orbiter is in the launch configuration, and an interdeck access ladder to enter the flight deck from the mid-deck in the launch configuration and go from the flight deck to the mid-deck in the launch and landing configuration.

Crew Equipment Stowage. Crew equipment aboard the orbiter was stowed in lockers with two sizes of insertable trays. The trays could be adapted to accommodate a wide variety of soft goods, loose equipment, and food. The lockers were interchangeable and attached to the orbiter with crew fittings. The lockers could be removed or installed in flight by the crew members.

Exercise Equipment. The only exercise equipment on the Shuttle was a treadmill.

Sound Level Meter. The sound level meter determined on-orbit acoustical noise levels in the cabin. Depending on the requirements for each flight, the flight crew took meter readings at specified crew compartment and equipment locations. The data obtained by the flight crew were logged and/or voice recorded.

Air Sampling System. The air sampling system consisted of air bottles that were stowed in a modular locker. They were removed for sampling and restowed for entry.

On-Board Instrumentation. Orbiter operational instrumentation collected, routed, and processed information from transducers and sensors on the orbiter and its payloads. This system also interacted with the solid rocket boosters, external tank, and ground support equipment. More than 2,000 data points were monitored, and the data were routed to operational instrumentation multiplexers/demultiplexers. The instrumentation system consisted of transducers, signal conditioners, two pulse code modulation master units, encoding equipment, two operational recorders, one payload recorder, master timing equipment, and on-board checkout equipment.

Payload Accommodations

The Space Shuttle had three basic payload accommodation categories: dedicated, standard, and mid-deck accommodations:

- **Dedicated payloads** took up the entire cargo-carrying capacity and services of the orbiter, such as the Spacelab and some DOD payloads.

- **Standard payloads**—usually geosynchronous communications satellites—were the primary type of cargo carried by the Space Shuttle. Normally, the payload bay could accommodate up to four standard payloads per flight. Power, command, and data services for standard payloads were provided by the avionics system through a standard mixed cargo harness.

- **Mid-deck payloads**—small, usually self-contained packages—were stored in compartments on the mid-deck. These were often manufacturing-in-space or small life sciences experiments.

Structural attach points for payloads were located at 9.9-centimeter intervals along the tops of the two orbiter mid-fuselage main longerons. Some payloads were not attached directly to the orbiter but to payload carriers that were attached to the orbiter. The inertial upper stage, Spacelab and Spacelab pallet, and any specialized cradle for holding a payload were typical carriers.

Small payloads mounted in the payload bay required a smaller range of accommodations. These payloads received a reduced level of electric power, command, and data services, and their thermal conditions were those in the payload bay thermal environment. Small payloads could be mounted in either a side-mounted or an across-the-bay configuration.

The Space Shuttle could also accommodate small payloads in the mid-deck of the crew compartment. This location was ideal for payloads that required a pressurized crew cabin environment or needed to be operated directly by the crew. Payloads located in the mid-deck could also be stowed on board shortly before launch and removed quickly after landing.

Space Shuttle Operations

Although each Space Shuttle mission was unique, Space Shuttle missions followed a prescribed sequence of activities that was common to all flights. The following sections describe the typical activities preceding launch, the launch and ascent activities, on-orbit events, and events surrounding descent and landing. Figure 3–6 shows the typical sequence of mission events.

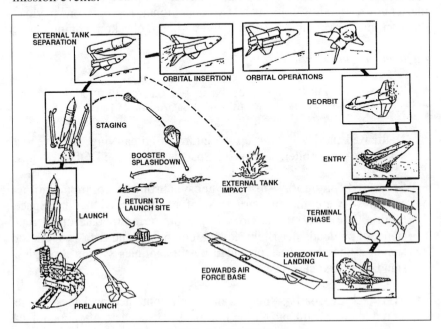

Figure 3–6. Typical STS Flight Profile

Prelaunch Activities

Space Shuttle components were gathered from various locations throughout the country and brought to Launch Complex 39 facilities at the Kennedy Space Center. There, technicians assembled the components—the orbiter, solid rocket booster, and external tank—into an integrated Space Shuttle vehicle, tested the vehicle, rolled it out to the launch pad, and ultimately launched it into space.

Each of the components that comprised the Shuttle system underwent processing prior to launch. NASA used similar processing procedures for new and reused Shuttle flight hardware. In general, new orbiters underwent more checkouts before being installed. In addition, the main engines underwent test firing on the launch pad. Called the Flight Readiness Firing, the test verified that the main propulsion system worked properly. For orbiters that had already flown, turnaround processing procedures included various postflight deservicing and maintenance functions, which were carried out in parallel with payload removal and the installation of equipment needed for the next mission.

If changes are made in external tank design, the tank usually required a tanking test in which it was loaded with liquid oxygen and hydrogen just as it was before launch. This confidence check verified the tank's ability to withstand the high pressures and super cold temperatures of the cryogenics.

The processing of each major flight component consisted of independent hardware checks and servicing in an operation called standalone processing. Actual Shuttle vehicle integration started with the stacking of the solid rocket boosters on a Mobile Launcher Platform in one of the high bays of the Vehicle Assembly Building. Next, the external tank was moved from its Vehicle Assembly Building location to the Mobile Launcher Platform and was mated with the solid rocket boosters. The orbiter, having completed its prelaunch processing and after horizontally integrated payloads had been installed, was towed from the Orbiter Processing Facility to the Vehicle Assembly Building and hoisted into position alongside the solid rocket boosters and the external tank. It was then mated to the external tank/solid rocket booster assembly. After mating was completed, the erection slings and load beams that had been holding the orbiter in place were removed, and the platforms and stands were positioned for orbiter/external tank/solid rocket booster access.

After the orbiter had been mated to the external tank/solid rocket booster assembly and all umbilicals were connected, technicians performed an electrical and mechanical verification of the mated interfaces to verify all critical vehicle connections. The orbiter underwent a Space Shuttle interface test using the launch processing system to verify Shuttle vehicle interfaces and Shuttle vehicle-to-ground interfaces. After completion of interface testing, ordnance devices were installed, but not electrically connected. Final ordnance connection and flight close-out were completed at the pad.

When the Vehicle Assembly Building prelaunch preparations were completed, the crawler transporter, an enormous tracked vehicle that NASA originally used during the Apollo and Skylab programs, lifted the assembled Space Shuttle and the Mobile Launcher Platform and rolled them slowly down a crawlerway to the launch pad at Launch Complex 39. Loaded, the vehicle moved at a speed of one mile an hour. The move took about six hours. At the pad, vertically integrated payloads were loaded into the payload bay. Then, technicians performed propellant servicing and needed ordnance tasks.

After the Space Shuttle had been rolled out to the launch pad on the Mobile Launcher Platform, all prelaunch activities were controlled from the Launch Control Center using the Launch Processing System. On the launch pad, the Rotating Service Structure was placed around the Shuttle and power for the vehicle was activated. The Mobile Launcher Platform and the Shuttle were then electronically and mechanically mated with support launch pad facilities and ground support equipment. An extensive series of validation checks verified that the numerous interfaces were functioning properly. Meanwhile, in parallel with prelaunch pad activities, cargo operations began in the Rotating Service Structure's Payload Changeout Room.

Vertically integrated payloads were delivered to the launch pad before the Space Shuttle was rolled out and stored in the Payload Changeout Room until the Shuttle was ready for cargo loading. Once the Rotating Service Structure was in place around the orbiter, the payload bay doors were opened and the cargo installed. Final cargo and payload bay close-outs were completed in the Payload Changeout Room, and the payload bay doors were closed for flight.

Propellant Loading. Initial Shuttle propellant loading involved pumping hypergolic propellants into the orbiter's aft and forward OMS and RCS storage tanks, the orbiter's hydraulic Auxiliary Power Units, and the solid rocket booster hydraulic power units. These were hazardous operations, and while they were under way, work on the launch pad was suspended. Because these propellants were hypergolic—they ignite on contact with one another—oxidizer and fuel loading operations were carried out serially, never in parallel.

Dewar tanks on the Fixed Service Structure were filled with liquid oxygen and liquid hydrogen, which would be loaded into the orbiter's Power Reactant and Storage Distribution tanks during the launch countdown. Before the formal Space Shuttle launch countdown began, the vehicle was powered down while pyrotechnic devices were installed or hooked up. The extravehicular mobility units—spacesuits—were stored on board along with other items of flight crew equipment.

Launch Processing System. The Launch Processing System made Space Shuttle processing, checkout, and countdown procedures more automated and streamlined than those of earlier human spaceflight programs. The countdown for the Space Shuttle took only about forty hours, compared with more than eighty hours usually needed for a

Saturn/Apollo countdown. Moreover, the Launch Processing System called for only about ninety people to work in the firing room during launch operations, compared with about 450 needed for earlier human missions. This system automatically controlled and performed much of the Shuttle processing from the arrival of individual components and their integration to launch pad operations and, ultimately, the launch itself. The system consisted of three basic subsystems: the Central Data Subsystem located on the second floor of the Launch Control Center, the Checkout, Control and Monitor Subsystem located in the firing rooms, and the Record and Playback Subsystem.

Complex 39 Launch Pad Facilities. The Kennedy Space Center's Launch Complex 39 had two identical launch pads, which were originally designed and built for the Apollo lunar landing program. The pads, built in the 1960s, were used for all of the Apollo/Saturn V missions and the Skylab space station program. Between 1967 and 1975, twelve Apollo/Saturn V vehicles, one Skylab/Saturn V workshop, three Apollo/Saturn 1B vehicles for Skylab crews, and one Apollo/Saturn 1B for the joint U.S.-Soviet Apollo Soyuz Test Project were launched from these pads.

The pads underwent major modifications to accommodate the Space Shuttle vehicle. Initially, Pad A modifications were completed in mid-1978, while Pad B was finished in 1985 and first used for the ill-fated STS 51-L mission in January 1986. The modifications included the construction of new hypergolic fuel and oxidizer support areas at the southwest and southeast corners of the pads, the construction of new Fixed Service Structures, the addition of a Rotating Service Structure, the addition of 1,135,620-liter water towers and associated plumbing, and the replacement of the original flame deflectors with Shuttle-compatible deflectors.

Following the flight schedule delays resulting from the STS 51-L accident, NASA made an additional 105 pad modifications. These included the installation of a sophisticated laser parking system on the Mobile Launcher Platform to facilitate mounting the Shuttle on the pad and emergency escape system modifications to provide emergency egress for up to twenty-one people. The emergency shelter bunker also was modified to allow easier access from the slidewire baskets.

Systems, facilities, and functions at the complex included:

- Fixed Service Structure
- Orbiter Access Arm
- External Tank Hydrogen Vent Line and Access Arm
- External Tank Gaseous Oxygen Vent Arm
- Emergency Exit System
- Lightning Mast
- Rotating Service Structure
- Payload Changeout Room

- Orbiter Midbody Umbilical Unit
- Hypergolic Umbilical System
- Orbital Maneuvering System Pod Heaters
- Sound Suppression Water System
- Solid Rocket Booster Overpressure Suppression System
- Main Engine Hydrogen Burnoff System
- Pad Surface Flame Detectors
- Pad-Propellant Storage and Distribution

Launch Sites. NASA used the Kennedy Space Center in Florida for launches that placed the orbiter in equatorial orbits (around the equator). The Vandenberg Air Force Base launch site in California was intended for launches that placed the orbiter in polar orbit missions, but it was never used and has been inactive since 1987.

NASA's prime landing site was at Kennedy. Additional landing sites were provided at Edwards Air Force Base in California and White Sands, New Mexico. Contingency landing sites were also provided in the event the orbiter must return to Earth in an emergency.

Kennedy Space Center launches had an allowable path no less than thirty-five degrees northeast and no greater than 120 degrees southeast. These were azimuth degree readings based on due east from Kennedy as ninety degrees. These two azimuths—thirty-five and 120 degrees—represented the launch limits from Kennedy. Any azimuth angles farther north or south would launch a spacecraft over a habitable land mass, adversely affect safety provisions for abort or vehicle separation conditions, or present the undesirable possibility that the solid rocket booster or external tank could land on foreign land or sea space.

Launch and Ascent

At launch, the three SSMEs were ignited first. When the proper engine thrust level was verified, a signal was sent to ignite the solid rocket boosters. At the proper thrust-to-weight ratio, initiators (small explosives) at eight hold-down bolts on the solid rocket boosters were fired to release the Space Shuttle for liftoff. All this took only a few seconds.

Maximum dynamic pressure was reached early in the ascent, approximately sixty seconds after liftoff. Approximately a minute later (two minutes into the ascent phase), the two solid rocket boosters had consumed their propellant and were jettisoned from the external tank at an altitude of 48.27 kilometers. This was triggered by a separation signal from the orbiter.

The boosters briefly continued to ascend to an altitude of 75.6 kilometers, while small motors fired to carry them away from the Space Shuttle. The boosters then turned and descended, and at a predetermined altitude, parachutes were deployed to decelerate them for a safe splashdown in the ocean. Splashdown occurred approximately 261 kilometers from the launch site.

When a free-falling booster descended to an altitude of about 4.8 kilometers, its nose cap was jettisoned and the solid rocket booster pilot parachute popped open. The pilot parachute then pulled out the 16.5-meter diameter, 499-kilogram drogue parachute. The drogue parachute stabilized and slowed the descent to the ocean.

At an altitude of 1,902 meters, the frustum, a truncated cone at the top of the solid rocket booster where it joined the nose cap, separated from the forward skirt, causing the three main parachutes to pop out. These parachutes were thirty-five meters in diameter and had a dry weight of about 680 kilograms each. When wet with sea water, they weighed about 1,361 kilograms.

At six minutes and forty-four seconds after liftoff, the spent solid rocket boosters, weighing about 7,484 kilograms, had slowed their descent speed to about 100 kilometers per hour, and splashdown took place in the predetermined area. There, a crew aboard a specially designed recovery vessel recovered the boosters and parachutes and returned them to the Kennedy Space Center for refurbishment. The parachutes remained attached to the boosters until they were detached by recovery personnel.

Meanwhile, the orbiter and external tank continued to climb, using the thrust of the three SSMEs. Approximately eight minutes after launch and just short of orbital velocity, the three engines were shut down (main engine cutoff, or MECO), and the external tank was jettisoned on command from the orbiter.

The forward and aft RCS engines provided attitude (pitch, yaw, and roll) and the translation of the orbiter away from the external tank at separation and return to attitude hold prior to the OMS thrusting maneuver. The external tank continued on a ballistic trajectory and entered the atmosphere, where it disintegrated. Its projected impact was in the Indian Ocean (except for fifty-seven-degree inclinations) for equatorial orbits.

Aborts. An ascent abort might become necessary if a failure that affects vehicle performance, such as the failure of an SSME or an OMS. Other failures requiring early termination of a flight, such as a cabin leak, might also require an abort.

Space Shuttle missions had two basic types of ascent abort modes: intact aborts and contingency aborts. Intact aborts were designed to provide a safe return of the orbiter to a planned landing site. Contingency aborts were designed to permit flight crew survival following more severe failures when an intact abort was not possible. A contingency abort would generally result in a ditch operation.

Intact Aborts. There were four types of intact aborts: abort-to-orbit, abort-once-around, transatlantic landing, and return-to-launch-site (Figure 3–7):

- The **abort-to-orbit** (ATO) mode was designed to allow the vehicle to achieve a temporary orbit that was lower than the nominal orbit. This mode required less performance and allowed time to evaluate problems and then choose either an early deorbit maneuver or an OMS thrusting maneuver to raise the orbit and continue the mission.

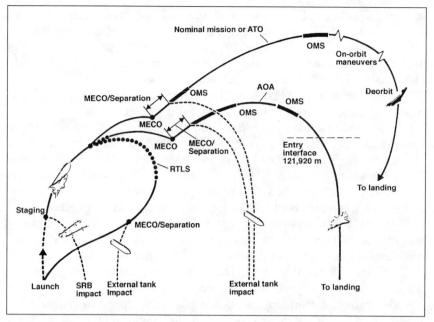

Figure 3–7. Types of Intact Aborts

- The **abort-once-around** (AOA) mode was designed to allow the vehicle to fly once around the Earth and make a normal entry and landing. This mode generally involved two OMS thrusting sequences, with the second sequence being a deorbit maneuver. The entry sequence would be similar to a normal entry. This abort mode was used on STS 51-F and was the only abort that took place.

- The **transatlantic landing** mode was designed to permit an intact landing on the other side of the Atlantic Ocean. This mode resulted in a ballistic trajectory, which did not require an OMS maneuver.

- The **return-to-launch-site** (RTLS) mode involved flying downrange to dissipate propellant and then turning around under power to return directly to a landing at or near the launch site.

A definite order of preference existed for the various abort modes. The type of failure and the time of the failure determined which type of abort is selected. In cases where performance loss was the only factor, the preferred modes would be abort-to-orbit, abort-once-around, transatlantic landing, and return-to-launch-site, in that order. The mode chosen was the highest one that could be completed with the remaining vehicle performance. In the case of some support system failures, such as cabin leaks or vehicle cooling problems, the preferred mode might be the one that would end the mission most quickly. In those cases, transatlantic landing or return-to-launch-site might be preferable to abort-once-around or abort-to-orbit. A contingency abort was never chosen if another abort option existed.

The Mission Control Center in Houston was "prime" for calling these aborts because it had a more precise knowledge of the orbiter's position than the crew could obtain from on-board systems. Before MECO, Mission Control made periodic calls to the crew to tell them which abort mode was (or was not) available. If ground communications were lost, the flight crew had on-board methods, such as cue cards, dedicated displays, and display information, to determine the current abort region.

Contingency Aborts. Contingency aborts would occur when there was a loss of more than one main engine or other systems fail. Loss of one main engine while another was stuck at a low thrust setting might also require a contingency abort. Such an abort would maintain orbiter integrity for in-flight crew escape if a landing could not be achieved at a suitable landing field.

Contingency aborts caused by system failures other than those involving the main engines would normally result in an intact recovery of vehicle and crew. Loss of more than one main engine might, depending on engine failure times, result in a safe runway landing. However, in most three-engine-out cases during ascent, the orbiter would have to be ditched. The in-flight crew escape system would be used before ditching the orbiter.

Orbit Insertion. An orbit could be accomplished in two ways: the conventional OMS insertion method called "standard" (which was last used with STS-35 in December 1990) and the direct insertion method. The standard insertion method involved a brief burn of the OMS engines shortly after MECO, placing the orbiter into an elliptical orbit. A second OMS burn was initiated when the orbiter reached apogee in its elliptical orbit. This brought the orbiter into a near circular orbit. If required during a mission, the orbit could be raised or lowered by additional firings of the OMS thrusters.

The direct insertion technique used the main engines to achieve the desired orbital apogee, or high point, thus saving OMS propellant. Only one OMS burn was required to circularize the orbit, and the remaining OMS fuel could then be used for frequent changes in the operational orbit, as called for in the flight plan. The first direct insertion orbit took place during the STS 41-C mission in April 1984, when *Challenger* was placed in a 463-kilometer-high circular orbit where its flight crew successfully captured, repaired, and redeployed the Solar Maximum Satellite (Solar Max).

The optimal orbital altitude of a Space Shuttle depended on the mission objectives and was determined before launch. The nominal altitude varied between 185 to 402 kilometers. During flight, however, problems, such as main engine and solid rocket booster performance loss and OMS propellant leaks or certain electrical power system failures, might prevent the vehicle from achieving the optimal orbit. In these cases, the OMS burns would be changed to compensate for the failure by selecting a delayed OMS burn, abort-once-around, or abort-to-orbit option.

Tables 3–12 and 3–13 show the events leading up to a typical launch and the events immediately following launch.[5]

On-Orbit Events. Once the orbiter achieved orbit, the major guidance, navigation, and control tasks included achieving the proper position, velocity, and attitude necessary to accomplish the mission objectives. To do this, the guidance, navigation, and control computer maintained an accurate state vector, targeted and initiated maneuvers to specified attitudes and positions, and pointed a specified orbiter body vector at a target. These activities were planned with fuel consumption, vehicle thermal limits, payload requirements, and rendezvous/proximity operations considerations in mind. The Mission Control Center, usually referred to as "Houston," controlled Space Shuttle flights.

Maneuvering in Orbit. Once the Shuttle orbiter went into orbit, it operated in the near gravity-free vacuum of space. However, to maintain proper orbital attitude and to perform a variety of maneuvers, the Shuttle used an array of forty-six large and small rocket thrusters—the OMS and RCS that was used to place the Shuttle in orbit. Each of these thrusters burned a mixture of nitrogen tetroxide and monoethylhydrazine, a combination of fuels that ignited on contact with each other.

Descent and Landing Activities

On-Orbit Checkout. The crew usually performed on-orbit checkout of the orbiter systems that were used during reentry the day before deorbit. System checkout had two parts. The first part used one auxiliary power unit/hydraulic system. It repositioned the left and right main engine nozzles for entry and cycled the aerosurfaces, hydraulic motors, and hydraulic switching valves. After the checkout was completed, the auxiliary power unit was deactivated. The second part checked all the crew-dedicated displays; self-tested the microwave scan beam landing system, tactical air navigation, accelerometer assemblies, radar altimeter, rate gyro assemblies, and air data transducer assemblies; and checked the hand controllers, rudder pedal transducer assemblies, speed brake, panel trim switches, RHC trim switches, speed brake takeover push button, and mode/sequence push button light indicators.

Shuttle Landing Operations. When a mission accomplished its planned in-orbit operations, the crew began preparing the vehicle for its return to Earth. Usually, the crew devoted the last full day in orbit to activities, such as stowing equipment, cleaning up the living areas, and

---

[5]The terms "terminal count," "first stage," and "second stage" are commonly used when describing prelaunch, launch, and ascent events. The terminal phase extends from T minus twenty minutes where "T" refers to liftoff time. First-stage ascent extends from solid rocket booster ignition through solid rocket booster separation. Second-stage ascent begins at solid rocket booster separation and extends through MECO and external tank separation.

making final systems configurations that would facilitate postlanding processing.

The crew schedule was designed so that crew members were awake and into their "work day" six to eight hours before landing. About four hours before deorbit maneuvers were scheduled, the crew and flight controllers finished with the Crew Activity Plan for the mission. They then worked from the mission's *Deorbit Prep Handbook,* which covered the major deorbit events leading to touchdown. Major events included the "go" from Mission Control Center to close the payload bay doors and final permission to perform the deorbit burn, which would return the orbiter to Earth.

Before the deorbit burn took place, the orbiter was turned to a tail-first attitude—that is, the aft end of the orbiter faced the direction of travel. At a predesignated time, the OMS engines were fired to slow the orbiter and to permit deorbit. The RCS thrusters were then used to return the orbiter into a nose-first attitude. These thrusters were used during much of the reentry pitch, roll, and yaw maneuvering until the orbiter's aerodynamic, aircraft-like control surfaces encountered enough atmospheric drag to control the landing. This was called Entry Interface and usually occurred thirty minutes before touchdown at about 122 kilometers altitude. At this time, a communications blackout occurred as the orbiter was enveloped in a sheath of plasma caused by electromagnetic forces generated from the high heat experienced during entry into the atmosphere.

Guidance, navigation, and control software guided and controlled the orbiter from this state (in which aerodynamic forces were not yet felt) through the atmosphere to a precise landing on the designated runway. All of this must be accomplished without exceeding the thermal or structural limits of the orbiter. Flight control during the deorbit phase was similar to that used during orbit insertion.

Orbiter Ground Turnaround. Approximately 160 Space Shuttle Launch Operations team members supported spacecraft recovery operations at the nominal end-of-mission landing site. Beginning as soon as the spacecraft stopped rolling, the ground team took sensor measurements to ensure that the atmosphere in the vicinity of the spacecraft was not explosive. In the event of propellant leaks, a wind machine truck carrying a large fan moved into the area to create a turbulent airflow that broke up gas concentrations and reduced the potential for an explosion.

A ground support equipment air-conditioning purge unit was attached to the righthand orbiter T-0 umbilical so cool air could be directed through the orbiter to dissipate the heat of entry. A second ground support equipment ground cooling unit was connected to the lefthand orbiter T-0 umbilical spacecraft Freon coolant loops to provide cooling for the flight crew and avionics during the postlanding and system checks. The flight crew then left the spacecraft, and a ground crew powered down the spacecraft.

Meanwhile, at the Kennedy Space Center, the orbiter and ground support equipment convoy moved from the runway to the Orbiter Processing

Facility. If the spacecraft landed at Edwards Air Force Base, the same procedures and ground support equipment applied as at Kennedy after the orbiter had stopped on the runway. The orbiter and ground support equipment convoy moved from the runway to the orbiter mate and demate facility. After detailed inspection, the spacecraft was prepared to be ferried atop the Shuttle carrier aircraft from Edwards to Kennedy.

Upon its return to the Orbiter Processing Facility at Kennedy, a ground crew safed the orbiter, removed its payload, and reconfigured the orbiter payload bay for the next mission. The orbiter also underwent any required maintenance and inspections while in the Orbiter Processing Facility. The spacecraft was then towed to the Vehicle Assembly Building and mated to the new external tank, beginning the cycle again.

Mission Control

The Mission Control Center at Johnson Space Center in Houston controlled all Shuttle flights. It has controlled more than sixty NASA human spaceflights since becoming operational in June 1965 for the Gemini IV mission. Two flight control rooms contained the equipment needed to monitor and control the missions.

The Mission Control Center assumed mission control functions when the Space Shuttle cleared the service tower at Kennedy's Launch Complex 39. Shuttle systems data, voice communications, and television traveled almost instantaneously to the Mission Control Center through the NASA Ground and Space Networks, the latter using the orbiting TDRS. The Mission Control Center retained its mission control function until the end of a mission, when the orbiter landed and rolled to a stop. At that point, Kennedy again assumed control.

Normally, sixteen major flight control consoles operated during a Space Shuttle mission. Each console was identified by a title or "call sign," which was used when communicating with other controllers or the astronaut flight crew. Teams of up to thirty flight controllers sat at the consoles directing and monitoring all aspects of each flight twenty-four hours a day, seven days a week. A flight director headed each team, which typically worked an eight-hour shift. Table 3–14 lists the mission command and control positions and responsibilities.

During Spacelab missions, an additional position, the command and data management systems officer, had primary responsibility for the data processing of the Spacelab's two main computers. To support Spacelab missions, the electrical, environmental, and consumables systems engineer and the data processing systems engineer both worked closely with the command and data management systems officer because the missions required monitoring additional displays involving almost 300 items and coordinating their activities with the Marshall Space Flight Center's Payload Operations Control Center (POCC).

The Mission Control Center's display/control system was one of the most unusual support facilities. It consisted of a series of projected screen displays that showed the orbiter's real-time location, live television pictures of crew activities, Earth views, and extravehicular activities. Other displays included mission elapsed time as well as time remaining before a maneuver or other major mission event. Many decisions or recommendations made by the flight controllers were based on information shown on the display/control system displays

Eventually, it was planned that modern state-of-the-art workstations with more capability to monitor and analyze vast amounts of data would replace the Apollo-era consoles. Moreover, instead of driving the consoles with a single main computer, each console would eventually have its own smaller computer, which could monitor a specific system and be linked into a network capable of sharing the data.

The POCCs operated in conjunction with the Flight Control Rooms. They housed principal investigators and commercial users who monitored and controlled payloads being carried aboard the Space Shuttle. One of the most extensive POCCs was at the Marshall Space Flight Center in Huntsville, Alabama, where Spacelab missions were coordinated with the Mission Control Center. It was the command post, communications hub, and data relay station for the principal investigators, mission managers, and support teams. Here, decisions on payload operations were made, coordinated with the Mission Control Center flight director, and sent to the Spacelab or Shuttle.

The POCC at the Goddard Space Flight Center controlled free-flying spacecraft that were deployed, retrieved, or serviced by the Space Shuttle. Planetary mission spacecraft were controlled from the POCC at NASA's Jet Propulsion Laboratory in Pasadena, California. Finally, private sector payload operators and foreign governments maintained their own POCCs at various locations for the control of spacecraft systems under their control.

NASA Centers and Responsibilities

Several NASA centers had responsibility for particular areas of the Space Shuttle program. NASA's Kennedy Space Center in Florida was responsible for all launch, landing, and turnaround operations for STS missions requiring equatorial orbits. Kennedy had primary responsibility for prelaunch checkout, launch, ground turnaround operations, and support operations for the Shuttle and its payloads. Kennedy's Launch Operations had responsibility for all mating, prelaunch testing, and launch control ground activities until the Shuttle vehicle cleared the launch pad tower.

Responsibility was then turned over to NASA's Mission Control Center at the Johnson Space Center in Houston. The Mission Control Center's responsibility included ascent, on-orbit operations, entry, approach, and landing until landing runout completion, at which time the orbiter was handed over to the postlanding operations at the landing site for turnaround and relaunch. At the launch site, the solid rocket boosters and external tank were processed for launch and the solid rocket boosters were recycled for reuse. The Johnson Space Center was responsible for the integration of the complete Shuttle vehicle and was the central control point for Shuttle missions.

NASA's Marshall Space Flight Center in Huntsville, Alabama, was responsible for the SSMEs, external tanks, and solid rocket boosters. NASA's National Space Technology Laboratories at Bay St. Louis, Mississippi, was responsible for testing the SSMEs. NASA's Goddard Space Flight Center in Greenbelt, Maryland, operated a worldwide tracking station network.

Crew Selection, Training, and Related Services

Crew Selection

NASA selected the first group of astronauts—known as the Mercury seven—in 1959. Since then, NASA has selected eleven other groups of astronaut candidates. Through the end of 1987, 172 individuals have graduated from the astronaut program.

NASA selected the first thirty-five astronaut candidates for the Space Shuttle program in January 1978. They began training at the Johnson Space Center the following June. The group consisted of twenty mission specialists and fifteen pilots and included six women and four members of minority groups. They completed their one-year basic training program in August 1979.

NASA accepted applications from qualified individuals—both civilian and military—on a continuing basis. Upon completing the course, successful candidates became regular members of the astronaut corps. Usually, they were eligible for a flight assignment about one year after completing the basic training program.

Pilot Astronauts. Pilot astronauts served as either commanders or pilots on Shuttle flights. During flights, commanders were responsible for the vehicle, the crew, mission success, and safety. The pilots were second in command; their primary responsibility was to assist the Shuttle commander. During flights, commanders and pilots usually assisted in spacecraft deployment and retrieval operations using the Remote Manipulator System (RMS) arm or other payload-unique equipment aboard the Shuttle.

To be selected as a pilot astronaut candidate, an applicant must have a bachelor's degree in engineering, biological science, physical science, or mathematics. A graduate degree was desired, although not essential. The applicant must have had at least 1,000 hours flying time in jet aircraft. Experience as a test pilot was desirable, but not required. All pilots and missions specialists must be citizens of the United States.

Mission Specialist Astronauts. Mission specialist astronauts, working closely with the commander and pilot, were responsible for coordinating on-board operations involving crew activity planning, use, and monitoring of the Shuttle's consumables (fuel, water, food, and so on), as well as conducting experiment and payload activities. They must have a detailed knowledge of Shuttle systems and the operational characteristics, mission requirements and objectives, and supporting systems for each of the experiments to be conducted on the assigned missions. Mission specialists performed on-board experiments, spacewalks, and payload-handling functions involving the RMS arm.

Academically, applicants must have a bachelor's degree in engineering, biological science, physical science, or mathematics, plus at least three years of related and progressively responsible professional experience. An advanced degree could substitute for part or all of the experience requirement—one year for a master's degree and three years for a doctoral degree.

Payload Specialists. This newest category of Shuttle crew member, the payload specialist, was a professional in the physical or life sciences or a technician skilled in operating Shuttle-unique equipment. The payload sponsor or customer selected a payload specialist for a particular mission. For NASA-sponsored spacecraft or experiments requiring a payload specialist, the investigator nominated the specialist who was approved by NASA.

Payload specialists did not have to be U.S. citizens. However, they must meet strict NASA health and physical fitness standards. In addition to intensive training for a specific mission assignment at a company plant, a university, or government agency, the payload specialist also must take a comprehensive flight training course to become familiar with Shuttle systems, payload support equipment, crew operations, housekeeping techniques, and emergency procedures. This training was conducted at the Johnson Space Center and other locations. Payload specialist training might begin as much as two years before a flight.

Astronaut Training

Astronaut training was conducted under the auspices of Johnson's Mission Operations Directorate. Initial training for new candidates consisted of a series of short courses in aircraft safety, including instruction in ejection, parachute, and survival to prepare them in the event their aircraft is disabled and they have to eject or make an emergency landing. Pilot and mission specialist astronauts were trained to fly T-38 high-performance jet aircraft, which were based at Ellington Field near Johnson. Flying these aircraft, pilot astronauts could maintain their flying skills and mission specialists could become familiar with high-performance jets. They also took formal science and technical courses

Candidates obtained basic knowledge of the Shuttle system, including payloads, through lectures, briefings, textbooks, mockups, and flight operations manuals. They also gained one-on-one experience in the single systems trainers, which contained computer databases with software allowing students to interact with controls and displays similar to those of a Shuttle crew station. Candidates learned to function in a weightless or environment using the KC-135 four-engine jet transport and in an enormous neutral buoyancy water tank called the Weightless Environment Training Facility at Johnson.

Because the orbiter landed on a runway much like a high-performance aircraft, pilot astronauts used conventional and modified T-38 trainers and the KC-135 aircraft to simulate actual landings. They also used a modified Grumman Gulfstream II, known as the Shuttle Training Aircraft, which was configured to simulate the handling characteristics of the orbiter for landing practice.

Advanced training included sixteen different course curricula covering all Shuttle-related crew training requirements. The courses ranged from guidance, navigation, and control systems to payload deployment and retrieval systems. This advanced training was related to systems and phases. Systems training provided instruction in orbiter systems and was not related to a specific mission or its cargo. It was designed to familiarize the trainee with a feel for what it was like to work and live in space. Generally, systems training was completed before an astronaut is assigned to a mission. Phase-related training concentrated on the specific skills an astronaut needed to perform successfully in space. This training was conducted in the Shuttle Mission Simulator. Phase-related training continued after a crew was assigned to a specific mission, normally about seven months to one year before the scheduled launch date.

At that time, crew training became more structured and was directed by a training management team that was assigned to a specific Shuttle flight. The training involved carefully developed scripts and scenarios for the mission and was designed to permit the crew to operate as a closely integrated team, performing normal flight operations according to a flight timeline.

About 10 weeks before a scheduled launch, the crew began "flight-specific integrated simulations, designed to provide a dynamic testing ground for mission rules and flight procedures." Simulating a real mission, the crew worked at designated stations interacting with the flight control team members, who staffed their positions in the operationally configured Mission Control Center.

These final prelaunch segments of training were called integrated and joint integrated simulations and normally included the payload users' operations control centers. Everything from extravehicular activity (EVA) operations to interaction with the tracking networks could be simulated during these training sessions.

Shuttle Mission Simulator. The Shuttle Mission Simulator was the primary system for training Space Shuttle crews. It was the only high-fidelity simulator capable of training crews for all phases of a mission beginning at T-minus thirty minutes, including such simulated events as launch, ascent, abort, orbit, rendezvous, docking, payload handling, undocking, deorbit, entry, approach, landing, and rollout.

The unique simulator system could duplicate main engine and solid rocket booster performance, external tank and support equipment, and interface with the Mission Control Center. The Shuttle Mission Simulator's construction was completed in 1977 at a cost of about $100 million.

Crew-Related Services

In support of payload missions, crew members provided unique ancillary services in three specific areas: EVA, intravehicular activity (IVA), and in-flight maintenance. EVAs, also called spacewalks, referred to activities in which crew members put on pressurized spacesuits and life support systems (spacepaks), left the orbiter cabin, and performed various payload-related activities in the vacuum of space, frequently outside the payload bay. (Each mission allowed for at least two crew members to be training for EVA.) EVA was an operational requirement when satellite repair or equipment testing was called for on a mission. However, during any mission, two crew members must be ready to perform a contingency EVA if, for example, the payload bay doors failed to close properly and must be closed manually, or equipment must be jettisoned from the payload bay.

The first Space Shuttle program contingency EVA occurred in April 1985, during STS 51-D, a *Discovery* mission, following deployment of the Syncom IV-3 (Leasat 3) communications satellite. The satellite's sequencer lever failed, and initiation of the antenna deployment and spin-up and perigee kick motor start sequences did not take place. The flight was extended two days to give mission specialists Jeffrey Hoffman and David Griggs an opportunity to try to activate the lever during EVA operations, which involved using the RMS. The effort was not successful, but was accomplished on a later mission. Table 3–15 lists all of the operational and contingency EVAs that have taken place through 1988.

IVA included all activities during which crew members dressed in spacesuits and using life support systems performed hands-on operations inside a customer-supplied crew module. (IVAs performed in the Spacelab did not require crew members to dress in spacesuits with life support systems.)

Finally, in-flight maintenance was any off-normal, on-orbit maintenance or repair action conducted to repair a malfunctioning payload. In-flight maintenance procedures for planned payload maintenance or repair were developed before a flight and often involved EVA.

Space Shuttle Payloads

Space Shuttle payloads were classified as either "attached" or "free-flying." Attached payloads such as Spacelab remained in the cargo bay or elsewhere on the orbiter throughout the mission. Free-flying payloads were released to fly alone. Some free-flyers were meant to be serviced or retrieved by the Shuttle. Others were boosted into orbits beyond the Shuttle's reach.

Attached Payloads

Spacelab

Spacelab was an orbiting laboratory built by the ESA for use with the STS. It provided the scientific community with easy, economical access to space and an opportunity for scientists worldwide to conduct experiments in space concerning astronomy, solar physics, space plasma physics, atmospheric physics, Earth observations, life sciences, and materials sciences.

Spacelab was constructed from self-contained segments or modules. It had two major subsections: cylindrical, pressurized crew modules and U-shaped unpressurized instrument-carrying pallets. The crew modules provided a "shirtsleeve" environment where payload specials worked as they would in a ground-based laboratory. Pallets accommodated experiments for direct exposure to space. They could be combined with another small structure called an igloo.

Crew modules and pallets were completely reusable; they were designed for multi-use applications and could be stacked or fitted together in a variety of configurations to provide for completely enclosed, completely exposed, or a combination of both enclosed-exposed facilities. The Spacelab components got all their electric, cooling, and other service requirements from the orbiter. An instrument pointing system, also part of the Spacelab, provided pointing for the various Spacelab experiment telescopes and cameras.

The crew module maintained an oxygen-nitrogen atmosphere identical to that in the orbiter crew compartment. Depending on mission requirements, crew modules consisted of either one segment (short module) or two segments (long module). The short module was four and two-tenths meters long; the long module measured seven meters. All crew modules were four meters in diameter. Most of the equipment housed in the short module controlled the pallet-mounted experiments. Spacelab missions used the long module when more room was needed for laboratory-type investigations. Equipment inside the crew modules was mounted in fifty-centimeter-wide racks. These racks were easily removed between flights so module-mounted experiments could be changed quickly.

The U-shaped pallet structure accommodated experiment equipment for direct exposure to the space environment when the payload bay doors were opened. It provided hardpoints for mounting heavy experiments and inserts for supporting light payloads. Individual payload segments were three meters long and four meters wide. The orbiter keel attachment fitting provided lateral restraint for the pallet when installed in the orbiter (Figure 3–8).

The igloo was a pressurized cylindrical canister 1,120 millimeters in diameter and 2,384 millimeters in height and with a volume of two and two-tenths cubic meters (Figure 3–9). It consisted of a primary structure, a secondary structure, a removable cover, and an igloo mounting structure and housed the following components:

SPACE TRANSPORTATION/HUMAN SPACEFLIGHT

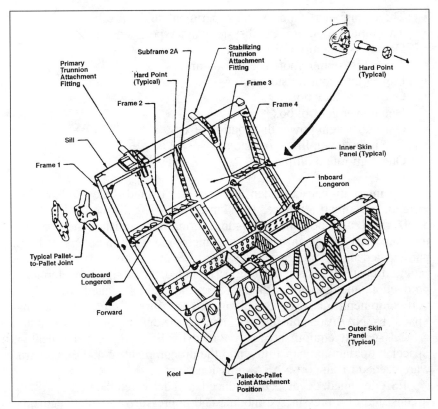

Figure 3–8. Pallet Structure and Panels

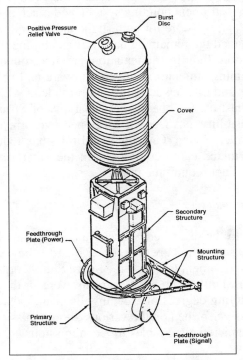

Figure 3–9. Spacelab Igloo Structure

- Three computers (subsystem, experiment, and backup)
- Two input-output units (subsystem and experiment)
- One mass memory unit
- Two subsystem remote acquisition units
- Eleven interconnect stations
- One emergency box
- One power control box
- One subsystem power distribution box
- One remote amplification and advisory box
- One high-rate multiplexer

An international agreement between the United States and Austria, Belgium, Denmark, France, Germany, Italy, The Netherlands, Spain, Switzerland, and the United Kingdom formally established the Spacelab program. Ten European nations, of which nine were members of ESA, participated in the program. NASA and ESA each bore their respective program costs. ESA responsibilities included the design, development, production, and delivery of the first Spacelab and associated ground support equipment to NASA, as well as the capability to produce additional Spacelabs. NASA responsibilities included the development of flight and ground support equipment not provided by ESA, the development of Spacelab operational capability, and the procurement of additional hardware needed to support NASA's missions.

ESA designed, developed, produced, and delivered the first Spacelab. It consisted of a pressurized module and unpressurized pallet segments, command and data management, environmental control, power distribution systems, an instrument pointing system, and much of the ground support equipment and software for both flight and ground operations.

NASA provided the remaining hardware, including the crew transfer tunnel, verification flight instrumentation, certain ground support equipment, and a training simulator. Support software and procedures development, testing, and training activities not provided by ESA, which were needed to demonstrate the operational capability of Spacelab, were also NASA's responsibility. NASA also developed two principal versions of the Spacelab pallet system. One supported missions requiring the igloo and pallet in a mixed cargo configuration; the other version supported missions that did not require the igloo.

Scientific Experiments

In addition to the dedicated Spacelab missions, nearly all STS missions had some scientific experiments on board. They used the unique microgravity environment found on the Space Shuttle or the environment surrounding the Shuttle. These experiments were in diverse disciplines and required varying degrees of crew involvement. Details of the scientific experiments performed on the various Shuttle missions are found in the "mission characteristics" tables for each mission.

Get-Away Specials

The Get-Away Specials were small self-contained payloads. Fifty-three Get-Away Special payloads had flown on Space Shuttle missions through 1988. The idea for the program arose in the mid-1970s when NASA began assigning major payloads to various Shuttle missions. It soon became apparent that most missions would have a small amount of space available after installing the major payloads. NASA's discussion of how best to use this space led to the Small Self-Contained Payloads program, later known as the Get-Away Special program.

This program gave anyone, including domestic and international organizations, an opportunity to perform a small space experiment. NASA hoped that by opening Get-Away Specials to the broadest community possible, it could further the goals of encouraging the use of space by all, enhancing education with hands-on space research opportunities, inexpensively testing ideas that could later grow into major space experiments, and generating new activities unique to space.

In October 1976, NASA's Associate Administrator for Space Flight, John Yardley, announced the beginning of the Get-Away Special program. Immediately, R. Gilbert Moore purchased the first Get-Away Special payload reservation. Over the next few months, NASA defined the program's boundaries. Only payloads of a scientific research and development nature that met NASA's safety regulations were acceptable. Payloads were to be self-contained, supplying their own power, means of data collection, and event sequencing. Keeping safety in mind and the varying technical expertise of Get-Away Special customers, NASA designed a container that could contain potential hazards. Three payload options evolved:

- A 0.07-cubic-meter container for payloads up to twenty-seven kilograms costing $3,000
- A 0.07-cubic-meter container for payloads weighing twenty-eight to forty-five kilograms for $5,000
- A 0.14-cubic-meter container for payloads up to ninety kilograms costing $10,000

Early in 1977, NASA assigned the Get-Away Special program to the Sounding Rocket Division, later renamed the Special Payloads Division, at the Goddard Space Flight Center. Meanwhile, news of the Get-Away Special program had passed informally throughout the aerospace community. With no publicity since Yardley's initial announcement the previous year, NASA had already issued more than 100 payload reservation numbers.

The Get-Away Special team did not anticipate flying a Get-Away Special payload before STS-5. However, the weight of a Get-Away Special container and its adapter beam was needed as ballast for STS-3's aft cargo bay. Thus, the Get-Away Special program and the Flight Verification Payload received an early go-ahead for the STS-3 flight in

March 1982. The first official Get-Away Special, a group of experiments developed by Utah State University students, flew on STS-4. Details of this Get-Away Special and the other Get-Away Special experiments can be found in the detailed STS mission tables that follow.

Shuttle Student Involvement Program

The Shuttle Student Involvement Program (SSIP) was a joint venture of NASA and the National Science Teachers Association (NSTA). It was designed to stimulate the study of science and technology in the nation's secondary schools. To broaden participation in the program, NASA solicited industrial firms and other groups to sponsor the development of the student experiments. Sponsors were asked to assign a company scientist to work with the student; fund the development of the experiment, including the necessary hardware; provide travel funds to take the student to appropriate NASA installations during experiment development; and provide assistance in analyzing postflight data and preparing a final report. Students proposed and designed the payloads associated with the program.

NASA and the NSTA held contests to determine which student experiments would fly on Space Shuttle missions. Following the mission, NASA returned experiment data to the student for analysis. Most Shuttle missions had at least one SSIP experiment; some missions had several experiments on board. Hardware developed to support the student experiments was located in the mid-deck of the orbiter. As a general rule, no more than one hour of crew time was to be devoted to the student experiment.

The first SSIP project took place during the 1981–82 school year as a joint venture of NASA's Academic Affairs Division and the NSTA. The NSTA announced the program, which resulted in the submission of 1,500 proposals and the selection of 191 winners from ten regions. Ten national winners were selected in May 1991. NASA then matched the finalists with industrial or other non-NASA sponsors who would support the development and postflight analysis of their experiments. Winners who were not matched with a sponsor had their experiments supported by NASA. Details of individual SSIP experiments can be found in the detailed STS mission tables that follow.

Free-Flying Payloads

Free-flying payloads are released from the Space Shuttle. Most have been satellites that were boosted into a particular orbit with the help of a inertial upper stage or payload assist module. Most free-flying payloads had lifetimes of several years, with many performing long past their anticipated life span. Some free-flying payloads sent and received communications data. These communications satellites usually belonged to companies that were involved in the communications industry. Other free-flying payloads contained sensors or other instruments to read atmospheric conditions. The data gathered by the sensors was transmitted to Earth either directly to a ground station or by way of a TDRS. Scientists on Earth interpreted the data gathered by the instruments. Examples of this kind of satellite were meteorological satellites and planetary probes. These satellites frequently were owned and operated by NASA or another government agency, although private industry could participate in this type of venture.

Other free-flying payloads were meant to fly for only a short time period. They were then retrieved by a robot arm and returned to the Shuttle's cargo bay. Individual free-flying payload missions are discussed in Chapter 4, "Space Science," in this volume and Chapter 2, "Space Applications," in Volume VI of the *NASA Historical Data Book*.

Payload Integration Process

The payload integration process began with the submission of a Request for Flight Assignment form by the user organization—a private or governmental organization—to NASA Headquarters. If NASA approved the request, a series of actions began that ultimately led to spaceflight. These actions included signing a launch services agreement, developing a payload integration plan, and preparing engineering designs and analyses, safety analysis, and a flight readiness plan. An important consideration was the weight of the payload.

For orbiters *Discovery* (OV-103), *Atlantis* (OV-104), and *Endeavour* (OV-105), the abort landing weight constraints could not exceed 22,906 kilograms of allowable cargo on the so-called simple satellite deployment missions. For longer duration flights with attached payloads, the allowable cargo weight for end-of-mission or abort situations was limited to 11,340 kilograms. For *Columbia* (OV-102), however, these allowable cargo weights were reduced by 3,810.2 kilograms.

In November 1987, NASA announced that the allowable end-of-mission total landing weight for Space Shuttle orbiters had been increased from the earlier limit of 95,709.6 kilograms to 104,328 kilograms. The higher limit was attributed to an ongoing structural analysis and additional review of forces encountered by the orbiter during maneuvers just before touchdown. This new capability increased the performance capability between lift capacity to orbit and the allowable return weight during reentry and landing. Thus, the Shuttle would be able to carry a cumulative weight in excess of 45,360 kilograms of additional cargo through 1993. This additional capability was expected to be an important factor in delivering materials for construction of the space station. Moreover, the new allowable landing weights were expected to aid in relieving the payload backlog that resulted from the STS 51-L *Challenger* accident.

Space Shuttle Missions

The following sections describe each STS mission beginning with the first four test missions. Information on Space Shuttle missions is extremely well documented. The pre- and postflight Mission Operations Reports (MORs) that NASA was required to submit for each mission provided the majority of data. At a minimum, these reports listed the mission objectives, described mission events and the payload in varying degrees of detail, listed program/project management, and profiled the crew. NASA usually issued the preflight MOR a few weeks prior to the scheduled launch date.

The postflight MOR was issued following the flight. It assessed the mission's success in reaching its objectives and discussed anomalies and unexpected events. It was signed by the individuals who had responsibility for meeting the mission objectives.

NASA also issued press kits prior to launch. These documents included information of special interest to the media, the information from the prelaunch MORs, and significant background of the mission. Other sources included NASA Daily Activity Reports, NASA News, NASA Fact Sheets, and other STS mission summaries issued by NASA. Information was also available on-line through NASA Headquarters and various NASA center home pages.

Mission Objectives

Mission objectives may seem to the reader to be rather general and broad. These objectives usually focused on what the vehicle and its components were to accomplish rather than on what the payload was to accomplish. Because one main use of the Space Shuttle was as a launch vehicle, deployment of any satellites on board was usually a primary mission objective. A description of the satellite's objectives (beyond a top level) and a detailed treatment of its configuration would be found in the MOR for that satellite's mission. For instance, the mission objectives for the Earth Radiation Budget Satellite would be found in the MOR for that mission rather than in the MOR for STS 41-G, the launch vehicle for the satellite. In addition, missions with special attached payloads, such as Spacelab or OSTA-1, issued individual MORs. These described the scientific and other objectives of these payloads and on-board experiments or "firsts" to be accomplished in considerable detail.

The Test Missions: STS-1 Through STS-4

Overview

Until the launch of STS-1 in April 1981, NASA had no proof of the Space Shuttle as an integrated Space Transportation System that could reach Earth orbit, perform useful work there, and return safely to the ground. Thus, the purpose of the Orbital Flight Test (OFT) program was to verify the Shuttle's performance under real spaceflight conditions and to establish its readiness for operational duty. The test program would expand the Shuttle's operational range toward the limits of its design in careful increments. During four flights of *Columbia,* conducted from April 1981 to July 1982, NASA tested the Shuttle in its capacities as a launch vehicle, habitat for crew members, freight handler, instrument platform, and aircraft. NASA also evaluated ground operations before, during, and after each launch. Each flight increased the various structural and thermal stresses on the vehicle, both in space and in the atmosphere, by a planned amount. The OFT phase of the STS program demonstrated the flight system's ability to safely perform launch orbital operations, payload/scientific operations, entry, approach, landing, and turnaround operations. Table 3–16 provides a summary of STS-1 through STS-4.

Following the landing of STS-4 on July 4, 1982, NASA declared the OFT program a success, even though further testing and expansion of the Shuttle's capabilities were planned on operational flights. The OFT program consisted of more than 1,100 tests and data collections. NASA tested many components by having them function as planned—if an engine valve or an insulating tile worked normally, then its design was verified. Other components, such as the RMS arm, went through validation runs to check out their different capabilities. Final documentation of Shuttle performance during OFT considered the reports from astronaut crews, ground observations and measurements, and data from orbiter instruments and special developmental flight instrumentation that collected and recorded temperatures and accelerations at various points around the vehicle and motion from points around the Shuttle.

The first OFT flights were designed to maximize crew and vehicle safety by reducing ascent and entry aerodynamic loads on the vehicle as much as possible. The missions used two-person crews, and the orbiter was equipped with two ejection seats until satisfactory performance, reliability, and safety of the Space Shuttle had been demonstrated. Launch operations were controlled from the Kennedy Space Center and flight operations from the Johnson Space Center.

At the end of OFT, *Columbia*'s main engines had been demonstrated successfully up to 100 percent of their rated power level (upgraded engines throttled to 109 percent of this level on later flights) and down to 65 percent. Designed to provide 1.67 million newtons of thrust each at sea level for an estimated fifty-five missions, the engines were on target to meeting these guidelines at the end of the test program. They met all requirements for start and cutoff timing, thrust direction control, and the flow of propellants.

Launch Phase

NASA tested the Space Shuttle in its launch phase by planning increasingly more demanding ascent conditions for each test flight, and then by comparing predicted flight characteristics with data returned from Aerodynamic Coefficient Identification Package and developmental flight instrumentation instruments and ground tracking. *Columbia* lifted slightly heavier payloads into space on each mission. The altitudes and speeds at which the solid rocket boosters and external tank separated were varied, as was the steepness of the vehicle's climb and main engine throttling times. All of these changes corresponded to a gradual increasing during the test program in the maximum dynamic pressure, or peak aerodynamic stress, inflicted on the vehicle. At no time did *Columbia* experience any significant problems with the aerodynamic or heat stresses of ascent.

A major milestone in the test program was the shift (after STS-2) from using wind tunnel data for computing *Columbia*'s ascent path to using aerodynamic data derived from the first two flights. On STS-1 and STS-2, the Shuttle showed a slight lofting—about 3,000 meters at main engine cutoff—above its planned trajectory. This was caused by the inability of wind tunnel models to simulate the afterburning of hot exhaust gasses in the real atmosphere. Beginning with the third flight, the thrust of the booster rockets was reoriented slightly to reduce this lofting.

On STS-3 and STS-4, however, the trajectory was considered too shallow, in part because of a slower than predicted burn rate for the solid rocket boosters that had also been observed on the first two flights. Engineers continued to use OFT data after STS-4 to refine their predictions of this solid propellant burn rate so that ascent trajectories could be planned as accurately as possible on future missions. In all cases, the combined propulsion of main engines, solid boosters, and OMS engines delivered the Shuttle to its desired orbit.

STS-4 was the first mission to orbit at a twenty-eight-and-a-half-degree inclination to the equator. The first flights flew more steeply inclined orbits (thirty-eight to forty degrees) that took them over more ground tracking stations. The more equatorial STS-4 inclination was favored because it gave the vehicle a greater boost from the rotating Earth at launch. The first two flights also verified that the vehicle had enough energy for an emergency landing in Spain or Senegal, as abort options, should two main engines fail during ascent. After STS-5, the crew ejection seats were removed from *Columbia,* eliminating the option to eject and ending the need for astronauts to wear pressure suits during launch.

Solid Rocket Boosters. On each test flight, the twin solid rocket boosters provided evenly matched thrust, shut off at the same times, and separated as planned from the external tank, then parachuted down to their designated recovery area in the Atlantic Ocean for towing back to the mainland and reloading with solid propellant. Each booster had three main parachutes that inflated fully about twenty seconds before water impact. Prior to the test flights, these parachutes were designed to separate automatically from the boosters by means of explosive bolts when the rockets hit the water, because it was thought that recovery would be easier if the chutes were not still attached.

On the first and third flights, however, some parachutes sank before recovery. Then, on STS-4, the separation bolts fired prematurely because of strong vibrations, the parachutes detached from the rockets before water impact, and the rockets hit the water at too great a speed and sank. They were not recovered. As a result of these problems, NASA changed the recovery hardware and procedures beginning with STS-5. Instead of separating automatically with explosives, the parachutes remained attached to the boosters through water impact, and were detached by the recovery team. Sections of the boosters were also strengthened as a result of water impact damage seen on the test flights.

External Tank. The Space Shuttle's external fuel tank met all performance standards for OFT. Heat sensors showed ascent temperatures to be moderate enough to allow for planned reductions in the thickness and weight of the tank's insulation. Beginning with STS-3, white paint on the outside of the tank was left off to save another 243 kilograms of weight, leaving the tank the brown color of its spray-on foam insulation.

Onboard cameras showed flawless separation of the tank from the orbiter after the main engines cut off on each flight, and Shuttle crews reported that this separation was so smooth that they could not feel it happening. To assist its breakup in the atmosphere, the tank had a pyrotechnic device that set it tumbling after separation rather than skipping along the atmosphere like a stone. This tumble device failed on STS-1, but it worked perfectly on all subsequent missions. On all the test flights, radar tracking of the tank debris showed that the pieces fell well within the planned impact area in the Indian Ocean.

Orbital Maneuvering System. Shortly after it separated from the fuel tank, the orbiter fired its two aft-mounted OMS engines for additional boosts to higher and more circularized orbits. At the end of orbital operations, these engines decelerated the vehicle, beginning the orbiter's fall to Earth. The engines performed these basic functions during OFT with normal levels of fuel consumption and engine wear. Further testing included startups after long periods of idleness in vacuum and low gravity (STS-1 and STS-2), exposure to cold (STS-3), and exposure to the Sun (STS-4). Different methods of distributing the system's propellants were also demonstrated. Fuel from the left tank was fed to the right tank, and vice versa, and from the OMS tanks to the smaller RCS thrusters. On STS-2, the engine cross-feed was performed in the middle of an engine burn to simulate engine failure.

Orbital Operations

Once in space, opening the two large payload bay doors with their attached heat radiators was an early priority. If the doors did not open in orbit, the Shuttle could not deploy payloads or shed its waste heat. If they failed to close at mission's end, reentry through the atmosphere would be impossible.

The STS-1 crew tested the payload bay doors during *Columbia*'s first few hours in space. The crew members first unlatched the doors from the bulkheads and from each other. One at a time, they were opened in the manual drive mode. The movement of the doors was slightly more jerky and hesitant in space than in Earth-gravity simulations, but this was expected and did not affect their successful opening and closing. The crew members closed and reopened the doors again one day into the STS-1 mission as a further test, then closed them for good before reentry. The crew verified normal alignment and latching of the doors, as did the STS-2 crew during their door cycling tests, including one series in the automatic mode.

The crew also tested door cycling after prolonged exposure to heat and cold. The doors were made of a graphite-epoxy composite material, while the orbiter itself was made of aluminum. It was therefore important to understand how they would fit together after the aluminum expanded or contracted in the temperature extremes of space. At the beginning of STS-3 orbital operations, the doors opened as usual. The payload bay was then exposed to cold shadow for a period of twenty-three hours. When the crew closed the port-side door at the end of this "coldsoak," the door failed to latch properly, as it did after a similar cold exposure on the STS-4 mission. Apparently, the orbiter warped very slightly with nose and tail bent upward toward each other, accounting in part for the doors' inability to clear the aft bulkhead.

The crew solved the problem by holding the orbiter in a top-to-Sun position for fifteen minutes to warm the cargo bay, then undergoing a short "barbecue roll" to even out vehicle temperatures, allowing the doors to close and latch normally. In addition, hardware changes to the doors and to the aft bulkhead improved their clearance.

Thermal Tests. Thermal tests accounted for hundreds of hours of OFT mission time. The temperatures of spacecraft structures changed dramatically in space, depending on their exposure to the Sun. Temperatures on the surface of payload bay insulation on STS-3, for example, went from a low of –96° C to a peak of 127° C. The Space Shuttle kept its components within their designed temperature limits through its active thermal control system, which included two coolant loops that transported waste heat from the orbiter and payload electronics to the door-mounted radiator panels for dumping into space, and through the use of insulation and heaters. Figure 3–10 shows the insulating materials used on the orbiter.

The OFT program tested the orbiter's ability to keep cool and keep warm under conditions much more extreme than that of the average mission. STS-3 and STS-4 featured extended thermal "soaks," where parts of the orbiter were deliberately heated up or cooled down by holding certain attitudes relative to the Sun for extended time periods. These long thermal soaks were separated by shorter periods of "barbecue roll" for even heating. On STS-4, the thermal soak tests continued with long tail-to-Sun and bottom-to-Sun exposures.

Overall, these hot and cold soak tests showed that the Shuttle had a better than predicted thermal stability. STS-3 readings showed that the orbiter's skin kept considerably warmer during coldsoaks than had been expected and that many critical systems, such as the orbital maneuvering engines, were also warmer. Most vehicle structures also tended to heat up or cool down more slowly than expected. The active thermal control sys-

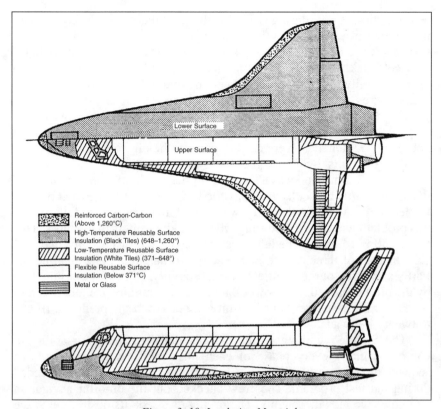

Figure 3–10. Insulating Materials

tem, with its coolant loops and space radiators, proved capable of handling Shuttle heat loads in orbit, even under extreme conditions.

The crew tested the space radiators with all eight panels deployed, and they proved capable of shedding most heat loads with only four panels deployed. During ascent, another part of the thermal control system, the Shuttle's flash evaporators, transferred heat from circulating coolant to water, beginning about two minutes into the ascent when the vehicle first required active cooling. These flash evaporators normally worked until the space radiators were opened in orbit. Then, during reentry, the flash evaporators were reactivated and used down to an altitude of approximately 36,000 meters. From that altitude down to the ground, the Shuttle shed heat by boiling ammonia rather than water. During OFT, the crew members successfully tested these methods of cooling as backups to each other.

Subsystems. All crews for the flight test program tested and retested the Space Shuttle's main subsystems under varying conditions. On the four OFT flights, virtually every system—hydraulic, electrical, navigation and guidance, communications, and environmental control—performed up to design standards or better.

The hydraulic subsystem that controlled the movement of the Shuttle's engine nozzles, its airplane-like control flaps, and its landing gear functioned well during OFT launches and reentries. The crew tested the hydraulic system successfully on STS-2 by cycling the eleven control surfaces while in orbit. On STS-4, the hydraulics were evaluated after a long coldsoak, and the crew found that the circulation pumps needed to operate at only minimal levels to keep the hydraulic fluids above critical temperatures, thus saving on electric power usage.

Although an oil filter clog in the hydraulic system's auxiliary power units delayed the launch of STS-2 by more than a week, the problem did not recur. Tighter seals were used to prevent the oil from being contaminated by the units' hydrazine fuel.

The STS-2 mission was also cut short because of the failure of one of the three Shuttle fuel cells that converted cryogenic hydrogen and oxygen to electricity. A clog in the cell's water flow lines caused the failure, and this problem was remedied during OFT by adding filters to the pipes. This failure allowed an unscheduled test of the vehicle using only two fuel cells instead of three, which were enough to handle all electrical needs. Partly as a result of the Shuttle's thermal stability, electricity consumption by the orbiter proved to be lower than expected, ranging from fourteen to seventeen kilowatts per hour in orbit as opposed to the predicted fifteen to twenty kilowatts.

The Shuttle's computers successfully demonstrated their ability to control virtually every phase of each mission, from final countdown sequencing to reentry, with only minor programming changes needed during the test program. The crew checked out the on-orbit navigation and guidance aids thoroughly. The orbiter "sensed" its position in space by means of three inertial measurement units, whose accuracy was checked and periodically updated by a star tracker located on the same navigation base in the flight deck. The crew tested this star tracker/inertial measurement unit alignment on the first Shuttle mission, including once when the vehicle was rolling. The star tracker could find its guide stars in both darkness and daylight. Its accuracy was better than expected, and the entire navigation instrument base showed stability under extreme thermal conditions.

Radio and television communication was successful on all four flights, with only minimal hardware and signal acquisition problems at ground stations. Specific tests checked different transmission modes, radio voice through the Shuttle's rocket exhaust during ascent, and UHF transmission as a backup to the primary radio link during launch and operations in space. All were successful. Tests on STS-4 also evaluated how different orbiter attitudes affected radio reception in space.

The closed-circuit television system inside the orbiter and out in the cargo bay gave high-quality video images of operations in orbit. In sunlight and in artificial floodlighting of the payload bay, they showed the necessary sensitivity, range of vision, remote control, and video-recording capabilities.

Attitude Control. When in orbit, the Shuttle used its RCS to control its attitude and to make small-scale movements in space. The thrusting power and propellant usage of both types of RCS jets were as expected, with the smaller verniers more fuel-efficient than expected. Two of the four vernier jets in *Columbia*'s tail area had a problem with the downward direction of their thrust. The exhaust hit the aft body flap and eroded some of its protective tiles, which also reduced the power of the jets. One possible solution considered was to reorient these jets slightly on future orbiters.

The orbiter demonstrated its ability to come to rest after a maneuver. At faster rates, it proved nearly impossible to stop the vehicle's motion without overshooting, then coming back to the required "stop" position, particularly with the large primary engines. Both types of thrusters were used to keep the orbiter steady in "attitude hold" postures. The small thrusters were particularly successful and fuel-efficient, holding the vehicle steady down to one-third of a degree of drift at normal rates of fuel use, which was three times their required sensitivity.

Further tests of the RCS assessed how well *Columbia* could hold steady without firing its jets when differential forces of gravity tended to tug the vehicle out of position. The results of these tests looked promising for the use of "passive gravity gradient" attitudes for future missions where steadiness for short periods of time was required without jet firings.

Remote Manipulator System. Ground simulators could not practice three-dimensional maneuvers because the remote manipulator system (RMS) arm was too fragile to support its own weight in Earth gravity. Therefore, one of the most important as well as most time-consuming of all OFT test series involved the fifteen-meter mechanical arm. This Canadian-built device, jointed as a human arm at the shoulder, elbow, and wrist, attached to the orbiter at various cradle points running the length of the inside of the cargo bay. In place of a hand, the arm had a cylindrical end effector that grappled a payload and held it rigid with wire snares. A crew member controlled the arm from inside the orbiter. The arm could be moved freely around the vehicle in a number of modes, with or without help from the Shuttle's computers.

The crew tested all manual and automatic drive modes during OFT. They also tested the arm's ability to grab a payload firmly, remove it from a stowed position, then reberth it precisely and securely. Lighting and television cameras also were verified—the crew relied on sensitive elbow and wrist cameras as well as cameras mounted in the payload bay to monitor operations. For the test program, special data acquisition cameras in the cargo bay documented arm motion.

STS-2 was the first mission to carry the arm. Although the crew did not pick up a payload with the arm, the astronauts performed manual approaches to a grapple fixture in the cargo bay, and they found the arm to control smoothly. The crew also began tests to see how the arm's movement interacted with orbiter motions. The crew reported that firings of the small vernier thrusters did not influence arm position, nor did arm motions necessitate attitude adjustment firings by the orbiter.

STS-3 tests evaluated the arm with a payload. The end effector grappled the 186-kilogram Plasma Diagnostics Package (PDP), removed it manually from its berth in the cargo bay, and maneuvered it automatically around the orbiter in support of OFT space environment studies. Pilot Gordon Fullerton deployed and reberthed the package. Before one such deployment, the arm automatically found its way to within 3.8 centimeters of the grapple point in accordance with preflight predictions. The crew also verified the computer's ability to automatically stop an arm joint from rotating past the limit of its mobility. The third crew completed forty-eight hours of arm tests, including one unplanned demonstration of the elbow camera's ability to photograph *Columbia*'s nose area during an on-orbit search for missing tiles.

Television cameras provided excellent views of arm operations in both sunshine and darkness, and the STS-4 crew reported that nighttime operations, although marginal, were still possible after three of the six payload bay cameras failed. The third and fourth crews continued evaluating vehicle interactions with arm motion by performing roll maneuvers as the arm held payloads straight up from the cargo bay. This was done with the PDP on STS-3 and with the Induced Environment Contamination Monitor on STS-4, which weighed twice as much. In both cases, the crew noted a slight swaying of the arm when the vehicle stopped, which was expected.

The RMS was designed to move a payload of 29,250 kilograms, but it was tested only with masses under 450 kilograms during OFT. Future arm tests would graduate to heavier payloads, some with grapple points fixed to simulate the inertias of even more massive objects.

The Shuttle Environment. In addition to these hardware checkouts, the test program also assessed the Space Shuttle environment. This was important for planning future missions that would carry instruments sensitive to noise, vibration, radiation, or contamination. During OFT, *Columbia* carried two sensor packages for examining the cargo bay environment. The Dynamic, Acoustic and Thermal Environment experiment—a group of accelerometers, microphones, and heat and strain gauges—established that noise and stress levels inside the bay were generally lower than predicted. The Induced Environment Contamination Monitor, normally secured in the cargo bay, was also moved around by the manipulator arm to perform an environmental survey outside the orbiter on STS-4.

The Contamination Monitor and the Shuttle-Spacelab Induced Atmosphere Experiment and postlanding inspections of the cargo bay backed up the Induced Environment Contamination Monitor's survey of polluting particles and gasses. These inspections revealed minor deposits and some discoloration of films and painted surfaces in the bay, which were still being studied after OFT. A new payload bay lining was added after STS-4.

The PDP measured energy fields around the orbiter on STS-3. The PDP, used in conjunction with the Vehicle Charging and Potential Experiment, mapped the distribution of charged particles around the spacecraft. These readings showed a vehicle that was relatively "quiet" electrically—it moved through the Earth's energy fields with interference levels much lower than the acceptable limits. The crew also discovered a soft glow around some of the Shuttle's surfaces that appeared in several nighttime photographs. An experiment added to STS-4 to identify the glow's spectrum supported a tentative explanation that the phenomenon resulted from the interaction with atomic oxygen in the thin upper atmosphere.

Inside the Shuttle, the cabin and mid-deck areas proved to be livable and practical working environments for the crew members. The test flight crews monitored cabin air quality, pressure, temperature, radiation, and noise levels and filmed their chores and activities in space to document the Shuttle's "habitability." The crews reported that their mobility inside *Columbia* was excellent, and they found that anchoring themselves in low gravity was easier than expected. There was almost no need for special foot restraints, and the crew members could improvise with ordinary duct tape attached to their shoes to hold themselves in place.

Descent and Landing

At the end of its time in orbit, the Space Shuttle's payload bay doors were closed, and the vehicle assumed a tail-first, upside-down posture and retrofired its OMS engines to drop out of orbit. It then flipped to a nose-up attitude and began its descent through the atmosphere back to Earth. Figure 3–11 shows the STS-1 entry flight profile.

The Shuttle's insulation needed to survive intact the burning friction of reentry to fly on the next mission. *Columbia*'s aluminum surface was covered with several different types of insulation during the test program, with their distribution based on predicted heating patterns. These included

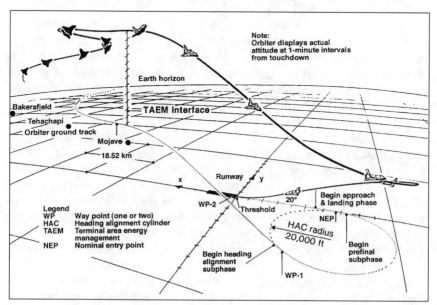

Figure 3-11. STS-1 Entry Flight Profile

more than 30,000 rigid silica tiles of two types (black for high temperatures, white for lower) that accounted for over 70 percent of the orbiter's surface area.

Television cameras viewing the outside of the Shuttle clearly revealed that several tiles had shaken loose during the vehicle's ascent and were missing from the aft engine pods. These tiles had not been densified—a process that strengthened the bond between tile and orbiter—as had all the tiles in critical areas and every tile installed after October 1979. No densified tiles were lost during the test flights.

On each flight, there was some damage to tile surfaces during launch and reentry. Vehicle inspection revealed hundreds of pits and gouges after STS-1 and STS-2. While the damage was not critical, many tiles needed to be replaced. Crew reports, launch pad cameras, and cockpit films recorded chunks of ice and/or insulation falling from the external tank; during ascent and launch, pad debris flew up and hit the orbiter, and these impacts were blamed for most of the tile damage. During the test program, NASA instituted a general cleanup of the pad before launch, and the removal of a particular insulation that had come loose from the booster rockets reduced debris significantly. On the external tank, certain pieces of ice-forming hardware were removed. As a result, impact damage to the tiles was greatly reduced. While some 300 tiles needed to be replaced after STS-1, fewer than forty were replaced after STS-4.

Weather also damaged some tiles during the test program. Factory waterproofing of new tiles did not survive the heat of reentry, and *Columbia* had to be sprayed with a commercial waterproofing agent after each mission so as not to absorb rainwater on the pad. The waterproofing agent was found to loosen tile bonds where it formed puddles, though, and STS-3 lost some tiles as a result.

Then, while STS-4 sat on the pad awaiting launch, a heavy hail and rainstorm allowed an estimated 540 kilograms of rainwater to be absorbed into the porous tiles through pits made by hailstones. This water added unwanted weight during ascent and later caused motion disturbances to the vehicle when the water evaporated into space. Shuttle engineers planned to use an injection procedure to waterproof the interior of the tiles for future missions.

As a whole, the thermal protection system kept the orbiter's skin within required limits during the OFT flights, even during the hottest periods of reentry. For the test program's last three flights, the crews performed short-duration maneuver changes in the vehicle's pitch angle that tested the effects of different attitudes on heating. Heating on the control surfaces was increased over the four flights, and on STS-3 and STS-4, the angle of entry into the atmosphere was flown more steeply to collect data under even more demanding conditions. Sensors on the orbiter reported temperatures consistent with preflight predictions. Notable exceptions were the aft engine pods, where some low-temperature flexible insulation was replaced with high-temperature black tiles after STS-1 showed high temperatures and scorching.

Aerodynamic Tests

The major objective of aerodynamic testing was to verify controlled flight over a wide range of altitudes (beginning at 120,000 meters where the air is very thin) and velocities, from hypersonic to subsonic. In both manual and automatic control modes, the vehicle flew very reliably and agreed with wind tunnel predictions.

Each flight crew also conducted a number of maneuvers either as programmed inputs by the guidance computer or as control stick commands by the crew in which the vehicles flaps and rudder were positioned to bring about more demanding flight conditions or to fill data gaps where wind tunnel testing was not adequate. These corrections were executed perfectly. In the thin upper atmosphere, the Space Shuttle used its reaction control thrusters to help maintain its attitude. Over the four test flights, these thrusters showed a greater-than-expected influence on the vehicle's motion. The orbiter's navigation and guidance equipment also performed well during reentry. Probes that monitored air speeds were successfully deployed at speeds below approximately Mach 3, and navigational aids by which the orbiter checks its position relative to the ground worked well with only minor adjustments.

Unlike returning Apollo capsules, the Space Shuttle had some cross-range capability—it could deviate from a purely ballistic path by gliding right or left of its aim point and so, even though it had no powered thrust during final approach, it did have a degree of control over where it landed. The largest cross-range demonstrated during the test program was 930 kilometers on STS-4.

The Space Shuttle could return to Earth under full computer control from atmospheric entry to the runway. During the test program, however, *Columbia*'s approach and landing were partly manual. The STS-1 approach and landing was fully manual. On STS-2, the auto-land control was engaged at 1,500 meters altitude, and the crew took over at ninety meters. Similarly, STS-3 flew on auto-land from 3,000 meters down to thirty-nine meters before the commander took stick control. It was decided after an error in nose attitude during the STS-3 landing that the crew should not take control of the vehicle so short a time before touchdown. The STS-4 crew therefore took control from the auto-land as *Columbia* moved into its final shallow glide slope at 600 meters. Full auto-land capability remained to be demonstrated after STS-5, as did a landing with a runway cross-wind.

Stress gauges on the landing gear and crew reports indicated that a Shuttle landing was smoother than most commercial airplanes. Rollout on the runway after touchdown fell well within the 4,500-meter design limit on each landing, but the actual touchdown points were all considerably beyond the planned touchdown points. This was because the Shuttle had a higher ratio of lift to drag near the ground than was expected, and it "floated" farther down the runway.

Ground Work

The OFT program verified thousands of ground procedures, from mating the vehicle before launch to refurbishing the solid rocket boosters and ferrying the orbiter from landing site to launch pad. As the test program progressed, many ground operations were changed or streamlined. Certain tasks that had been necessary for an untried vehicle before STS-1 could be eliminated altogether. As a result of this learning, the "turnaround" time between missions was shortened dramatically—from 188 days for STS-2 to seventy-five days between STS-4 and STS-5. Major time-saving steps included:

- Leaving cryogenic fuels in their on-board storage tanks between flights rather than removing them after landing
- Alternating the use of primary and backup systems on each flight rather than checking out both sets of redundant hardware on the ground before each launch
- Reducing the number of tests of critical systems as they proved flightworthy from mission to mission

The OFT program verified the soundness of the STS and its readiness for future scientific, commercial, and defense applications.

Orbiter Experiments Program

Many of the experiments that flew on the first four Shuttle missions were sponsored by the Office of Aeronautics and Space Technology (Code R) through its Orbiter Experiments Program. NASA used the data gathered from these experiments to verify the accuracy of wind tunnel and other ground-based simulations made prior to flight, ground-to-flight extrapolation methods, and theoretical computational methods.

The prime objective of these experiments was to increase the technology reservoir for the development of future (twenty-first century) space transportation systems, such as single-stage-to-orbit, heavy-lift launch vehicles and orbital transfer vehicles that could deploy and service large, automated, person-tended, multifunctional satellite platforms and a staffed, permanent facility in Earth orbit. The Orbiter Experiments Program experiments included:

- Aerodynamic Coefficient Identification Package
- Shuttle Entry Air Data System
- Shuttle Upper Atmospheric Mass Spectrometer
- Data Flight Instrumentation Package
- Dynamic, Acoustic and Thermal Environment Experiment
- Infrared Imagery of Shuttle
- Shuttle Infrared Leeside Temperature Sensing
- Tile Gap Heating Effects Experiment
- Catalytic Surface Effects

Each of these experiments, plus the others listed in Table 3–16, is discussed as part of the individual "mission characteristics" tables (Tables 3–17 through 3–20).

Mission Characteristics of the Test Missions (STS-1 Through STS-4)

STS-1

Objective. The mission objective was to demonstrate a safe ascent and return of the orbiter and crew.

Overview. *Columbia* reported on spacecraft performance and the stresses encountered during launch, flight, and landing. The flight successfully demonstrated two systems: the payload bay doors with their attached heat radiators and the RCS thrusters used for attitude control in orbit. John W. Young and Robert L. Crippen tested all systems and conducted many engineering tests, including opening and closing the cargo bay doors. Opening these doors is critical to deploy the radiators that release the heat that builds up in the crew compartment. Closing them is necessary for the return to Earth.

Young and Crippen also documented their flight in still and motion pictures. One view of the cargo bay that they telecast to Earth indicated that all or part of sixteen heat shielding tiles were lost. The loss was not considered critical as these pods were not subjected to intense heat, which could reach $1,650°$ C while entering the atmosphere. More than 30,000 tiles did adhere. A detailed inspection of the tiles, carried out later, however, revealed minor damage to approximately 400 tiles. About 200 would require replacement, 100 as a result of flight damage and 100 identified prior to STS-1 as suitable for only one flight.

Observations revealed that the water deluge system designed to suppress the powerful acoustic pressures of liftoff needed to be revised, after the shock from the booster rockets was seen to be much larger than anticipated. In the seconds before and after liftoff, a "rainbird" deluge system had poured tens of thousands of gallons of water onto the launch platform and into flame trenches beneath the rockets to absorb sound energy that might otherwise damage the orbiter or its cargo. Strain gauges and microphones measured the acoustic shock, and they showed up to four times the predicted values in parts of the vehicle closest to the launch pad.

Although *Columbia* suffered no critical damage, the sound suppression system was modified before the launch of STS-2. Rather than dumping into the bottom of the flame trenches, water was injected directly into the exhaust plumes of the booster rockets at a point just below the exhaust nozzles at the time of ignition. In addition, energy-absorbing water troughs were placed over the exhaust openings. The changes were enough to reduce acoustic pressures to 20 to 30 percent of STS-1 levels for the second launch.

STS-2

Objectives. NASA's mission objectives for STS-2 were to:

- Demonstrate the reusability of the orbiter vehicle
- Demonstrate launch, on-orbit, and entry performance under conditions more demanding than STS-1
- Demonstrate orbiter capability to support scientific and applications research with an attached payload
- Conduct RMS tests

Overview. Originally scheduled for five days, the mission was cut short because one of *Columbia*'s three fuel cells that converted supercold (cryogenic) hydrogen and oxygen to electricity failed shortly after the vehicle reached orbit. Milestones were the first tests of the RMS's fifteen-meter arm and the successful operation of Earth-viewing instruments in the cargo bay. The mission also proved the Space Shuttle's reusability.

In spite of the shortened mission, approximately 90 percent of the major test objectives were successfully accomplished, and 60 percent of the tests requiring on-orbit crew involvement were completed. The performance of lower priority tests were consistent with the shortened mission, and 36 percent of these tasks were achieved.

The mission's medical objectives were to provide routine and contingency medical support and to assure the health and well-being of flight personnel during all phases of the STS missions. This objective was achieved through the careful planning, development, training, and implementation of biomedical tests and procedures compatible with STS operations and the application of principles of general preventive medicine. It was also discovered that shortened sleep periods, heavy work loads, inadequate time allocation for food preparation and consumption, and estimated lower water intake were just sufficient for a fifty-four-hour mission. A plan was therefore developed to restructure in-flight timelines and institute corrective health maintenance procedures for longer periods of flight.

OSTA-1 was the major on-board mission payload. Sponsored by the Office of Space and Terrestrial Applications, it is addressed in Chapter 2, "Space Applications," in Volume VI of the *NASA Historical Data Book*.

STS-3

Objectives. The NASA mission objectives for STS-3 were to:

- Demonstrate ascent, on-orbit, and entry performance under conditions more demanding than STS-2 conditions
- Extend orbital flight duration
- Conduct long-duration thermal soak tests
- Conduct scientific and applications research with an attached payload

Overview. NASA designated OSS-1 as the attached payload on STS-3. The Office of Space Science sponsored the mission. This mission is discussed in Chapter 4, "Space Science."

The crew performed tests of the robot arm and extensive thermal testing of *Columbia* itself during this flight. Thermal testing involved exposing the tail, nose, and tip to the Sun for varying periods of time, rolling it ("barbecue roll") in between tests to stabilize temperatures over the entire body. The robot arm tested satisfactorily, moving the PDP experiment around the orbiter.

STS-4

Objectives. The NASA mission objectives for STS-4 were to:

- Demonstrate ascent, on-orbit, and entry performance under conditions more demanding than STS-3 conditions
- Conduct long-duration thermal soak tests
- Conduct scientific and applications research with attached payloads

Overview. This was the first Space Shuttle launch that took place on time and with no schedule delays. The mission tested the flying, handling, and operating characteristics of the orbiter, performed more exercises with the robot arm, conducted several scientific experiments in orbit, and landed at Edwards Air Force Base for the first time on a concrete runway of the same length as the Shuttle Landing Facility at the Kennedy Space Center. *Columbia* also planned to conduct more thermal tests by exposing itself to the Sun in selected altitudes, but these plans were changed because of damage caused by hail, which fell while *Columbia* was on the pad. The hail cut through the protective coating on the tiles and let rainwater inside. In space, the affected area on the underside of the orbiter was turned to the Sun. The heat of the Sun vaporized the water and prevented further possible tile damage from freezing.

The only major problem on this mission was the loss of the two solid rocket booster casings. The main parachutes failed to function properly, and the two casings hit the water at too high a velocity and sank. They were later found and examined by remote camera, but not recovered.

During the mission, the crew members repeated an STS-2 experiment that required the robot arm to move an instrument called the Induced Environmental Contamination Monitor around the orbiter to gather data on any gases or particles being released by the orbiter. They also conducted the Continuous Flow Electrophoresis System experiment, which marked the first use of the Shuttle by a commercial concern, McDonnell Douglas (Figure 3–12). In addition to a classified Air Force payload in the cargo bay, STS-4 carried the first Get-Away Special—a series of nine experiments prepared by students from Utah State University.

The payload bay was exposed to cold shadow for several hours after opening of the doors. When the port-side door was closed at the end of

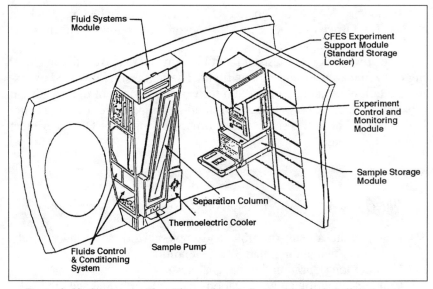

Figure 3–12. Continuous Flow Electrophoresis System Mid-deck Gallery Location

the "coldsoak," it failed to latch properly, as it did during the STS-3 mission. The solution on both flights was the same and was adopted as the standard procedure for closing the doors following a long cold exposure: the orbiter would hold a top-to-Sun position for fifteen minutes to warm the cargo bay, then undergo a short "barbecue roll" to even out vehicle temperatures, allowing the doors to close normally.

Mission Characteristics of the Operational Missions (STS-5 Through STS-27)

The Space Transportation System became operational in 1982, after completing the last of four orbital flight tests. These flights had demonstrated that the Space Shuttle could provide flexible, efficient transportation into space and back for crew members, equipment, scientific experiments, and payloads. From this point, payload requirements would take precedence over spacecraft testing. Table 3–21 summarizes Shuttle mission characteristics. The narrative and tables that follow (Tables 3–22 through 3–44) provide more detailed information on each Shuttle mission.

STS-5

STS-5 was the first operational Space Shuttle mission. The crew adopted the theme "We Deliver" as it deployed two commercial communications satellites: Telesat-E (Anik C-3) for Telesat Canada and SBS-C for Satellite Business Systems. Each was equipped with the Payload Assist Module-D (PAM-D) solid rocket motor, which fired about forty-five minutes after deployment, placing each satellite into highly elliptical orbits.

The mission carried the first crew of four, double the number on the previous four missions. It also carried the first mission specialists—individuals qualified in satellite deployment payload support, EVAs, and the operation of the RMS. This mission featured the first Shuttle landing on the 15,000-foot-long concrete runway at Edwards Air Force Base in California. NASA canceled the first scheduled EVA, or spacewalk, in the Shuttle program because of a malfunction in the spacesuits.

Experiments on this mission were part of the Orbiter Experiments Program, managed by NASA's Office of Aeronautics and Space Technology (OAST). The primary objective of this program was to increase the technology reservoir for the development of future space transportation systems to be used by the Office of Space Flight for further certification of the Shuttle and to expand its operational capabilities. Figure 3–13 shows the STS-5 payload configuration, and Table 3–22 lists the mission's characteristics.

STS-6

STS-6, carrying a crew of four, was the first flight of *Challenger*, NASA's second operational orbiter. The primary objective of this mission was the deployment of the first Tracking and Data Relay Satellite (TDRS-1) to provide improved tracking and data acquisition services to spacecraft in low-Earth orbit. It was to be injected into a geosynchronous transfer orbit by a two-stage inertial upper stage. The first stage fired as planned, but the second stage cut off after only seventy seconds of a

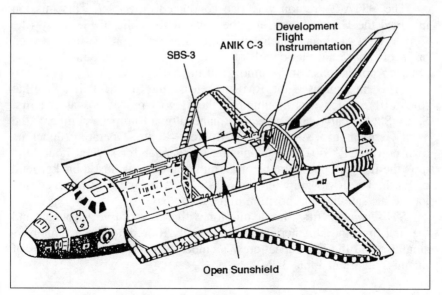

Figure 3–13. STS-5 Payload Configuration
(The payload was covered by a sunshield to protect against thermal extremes when the orbiter bay doors were open. The sunshield, resembling a two-piece baby buggy canopy, was constructed of tubular aluminum and mylar sheeting.)

planned 103-second burn. TDRS entered an unsatisfactory elliptical orbit. Excess propellant was used over the next several months to gradually circularize the orbit, using the spacecraft's own attitude control thrusters. The maneuver was successful, and TDRS-1 reached geosynchronous orbit and entered normal service.

This mission featured the first successful spacewalk of the Space Shuttle program, which was performed by astronauts Donald H. Peterson and F. Story Musgrave. It lasted about four hours, seventeen minutes. The astronauts worked in the cargo bay during three orbits, testing new tools and equipment-handling techniques.

This mission used the first lightweight external tank and lightweight solid rocket booster casings. The lightweight external tank was almost 4,536 kilograms lighter than the external tank on STS-1, with each weighing approximately 30,391 kilograms. The lightweight solid rocket booster casings increased the Shuttle's weight-carrying capability by about 363 kilograms. Each booster's motor case used on STS-6 and future flights weighed about 44,453 kilograms, approximately 1,814 kilograms less than those flown on previous missions. Table 3–23 identifies the characteristics of STS-6.

STS-7

STS-7 deployed two communications satellites, Telesat-F (Anik C-2) and Palapa-B1 into geosynchronous orbit. Also, the Ku-band antenna used with the TDRS was successfully tested.

The OSTA-2 mission was also conducted on STS-7. This mission involved the United States and the Federal Republic of Germany (the former West Germany) in a cooperative materials processing research project in space. Further details of the OSTA-2 mission are in Chapter 2, "Space Applications," in Volume VI of the *NASA Historical Data Book*.

This mission used the RMS to release the Shuttle Pallet Satellite (SPAS-01), which was mounted in the cargo bay. SPAS was the first Space Shuttle cargo commercially financed by a European company, the West German firm Messerschmitt-Bolkow-Blohm. Operating under its own power, SPAS-01 flew alongside *Challenger* for several hours and took the first full photographs of a Shuttle in orbit against a background of Earth. The RMS grappled the SPAS-01 twice and then returned and locked the satellite into position in the cargo bay.

STS-7 was the first Shuttle mission with a crew of five astronauts and the first flight of an American woman, Sally Ride, into space. This mission also had the first repeat crew member—Robert Crippen. Details of the mission are in Table 3–24.

STS-8

STS-8's primary mission objectives were to deploy Insat 1B, complete RMS loaded arm testing using the payload flight test article (PFTA),

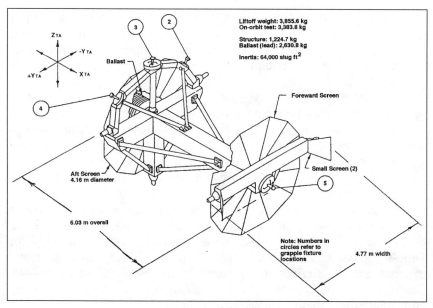

Figure 3–14. Payload Flight Test Article

accomplish TDRS/Ku-band communications testing, and achieve assigned experiments and test objectives. The RMS carried its heaviest loads to date, and the PFTA had several grapple points to simulate the inertias of even heavier cargoes. Figure 3–14 illustrates the PFTA configuration.

STS-8 was the first Space Shuttle mission launched at night. The tracking requirements for the Indian Insat 1B satellite, the primary payload, dictated the time of launch. STS-8 also had the first night landing.

The crew performed the first tests of Shuttle-to-ground communications using TDRS. Launched into geosynchronous orbit on STS-6, TDRS was designed to improve communications between the spacecraft and the ground by relaying signals between the spacecraft and the ground, thus preventing the loss of signal that occurred when using only ground stations.

This mission carried the first African-American astronaut, Guion S. Bluford, to fly in space. Details of STS-8 are listed in Table 3–25.

STS-9

STS-9 carried the first Spacelab mission (Spacelab 1), which was developed by ESA, and the first astronaut to represent ESA, Ulf Merbold of Germany. It successfully implemented the largest combined NASA and ESA partnership to date, with more than 100 investigators from eleven European nations, Canada, Japan, and the United States. It was the longest Space Shuttle mission up to that time in the program and was the first time six crew members were carried into space on a single vehicle. The crew included payload specialists selected by the science community.

The primary mission objectives were to verify the Spacelab system and subsystem performance capability, to determine the Spacelab/orbiter interface capability, and to measure the induced environment. Secondary mission objectives were to obtain valuable scientific, applications, and technology data from a U.S.-European multidisciplinary payload and to demonstrate to the user community the broad capability of Spacelab for scientific research.

ESA and NASA jointly sponsored Spacelab 1 and conducted investigations on a twenty-four-hour basis, demonstrating the capability for advanced research in space. Spacelab was an orbital laboratory with an observations platform composed of cylindrical pressurized modules and U-shaped unpressurized pallets, which remained in the orbiter's cargo bay during flight. It was the first use of a large-scale space airlock for scientific experiments.

Altogether, seventy-three separate investigations were carried out in astronomy and physics, atmospheric physics, Earth observations, life sciences, materials sciences, space plasma physics, and technology—the largest number of disciplines represented on a single mission. These experiments are described in Chapter 4, "Space Science," in Table 4–45. Spacelab 1 had unprecedented large-scale direct interaction of the flight crew with ground-based science investigators.

All of the mission objectives for verifying Spacelab's modules were met, and Earth-based scientists communicated directly with the orbiting space crew who performed their experiments, collected data immediately, and offered directions for the experiments. Table 3–26 list the characteristics of this mission.

STS 41-B

The primary goal of STS 41-B was to deploy into orbit two commercial communications satellites—Western Union's Westar VI and the Indonesian Palapa-B2. (Failure of the PAM-D rocket motors left both satellites in radical low-Earth orbits.) The crew devoted the remainder of STS 41-B to a series of rendezvous maneuvers using an inflatable balloon as the target, the test flights of two Manned Maneuvering Units (Figure 3–15), and the checkout of equipment and procedures in preparation for *Challenger*'s flight (41-C) in April, which would be the Solar Maximum satellite repair mission.

Commander Vance D. Brand led the five-person crew for this mission. He had previously commanded the first operational flight of the Space Shuttle, STS-5. The other crew members, pilot Robert L. "Hoot" Gibson and three mission specialists (Bruce McCandless II, Ronald E. McNair, and Robert L. Stewart), flew in space for the first time.

This mission featured the first untethered spacewalks. Gas-powered backpacks were used to demonstrate spacewalk techniques important for the successful retrieval and repair of the disabled Solar Maximum spacecraft. The crew members also tested several pieces of specialized equipment during the two five-hour EVAs. The Manipulator Foot Restraint, a portable workstation, was attached to the end of and maneuvered by the

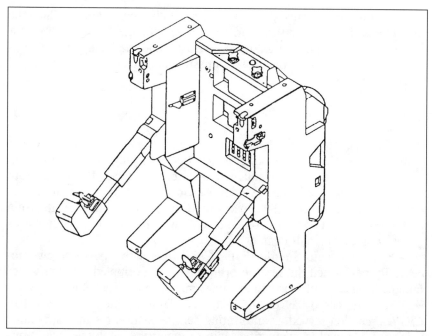

Figure 3–15. Manned Maneuvering Unit

RMS arm. Attached to the foot restraint, an astronaut could use the robot arm as a space-age "cherry picker" to reach and work on various areas of a satellite.

The RMS, just over fifteen meters long and built for the Space Shuttle by the National Research Council of Canada, was to be used to deploy the SPAS as a target for Manned Maneuvering Unit-equipped astronauts to perform docking maneuvers. However, the SPAS remained in the payload bay because of an electrical problem with the RMS. SPAS was to be used as a simulated Solar Maximum satellite. The astronauts were to replace electrical connectors attached to the SPAS during one of the spacewalks to verify procedures that astronauts would perform on the actual repair mission. The Manned Maneuvering Unit-equipped astronauts were also to attempt to dock with the pallet satellite, thereby simulating maneuvers needed to rendezvous, dock, and stabilize the Solar Maximum satellite.

The crew members conducted two days of rendezvous activities using a target balloon (Integrated Rendezvous Target) to evaluate the navigational ability of *Challenger*'s on-board systems, as well as the interaction among the spacecraft, flight crew, and ground control. The activities obtained data from *Challenger*'s various sensors (the rendezvous radar, star tracker, and crew optical alignment sight) required for rendezvous and exercised the navigation and maneuvering capabilities of the on-board software. The rendezvous occurred by maneuvering the orbiter to within 244 meters of its target from a starting distance of approximately 193.1 kilometers. In the process, sensors gathered additional performance data.

This mission initiated the new Shuttle numbering system in which the first numeral stood for the year, the second for the launch site (1 for Kennedy, 2 for Vandenberg Air Force Base), and the letter for the original order of the assignment. The mission characteristics are listed in Table 3–27.

STS 41-C

STS 41-C launched *Challenger* into its highest orbit yet so it could rendezvous with the wobbling, solar flare-studying Solar Maximum satellite, which had been launched in February 1980. Its liftoff from Launch Complex 39's Pad A was the first to use a "direct insertion" ascent technique that put the Space Shuttle into an elliptical orbit with a high point of about 461.8 kilometers and an inclination to the equator of twenty-eight and a half degrees.

On the eleventh Shuttle flight, *Challenger*'s five-person crew successfully performed the first on-orbit repair of a crippled satellite. After failed rescue attempts early in the mission, the robot arm hauled the Solar Max into the cargo bay on the fifth day of the mission (Figure 3–16). *Challenger* then served as an orbiting service station for the astronauts, using the Manned Maneuvering Unit, to repair the satellite's fine-pointing system and to replace the attitude control system and coronagraph/polarimeter electronics box during two six-hour spacewalks.

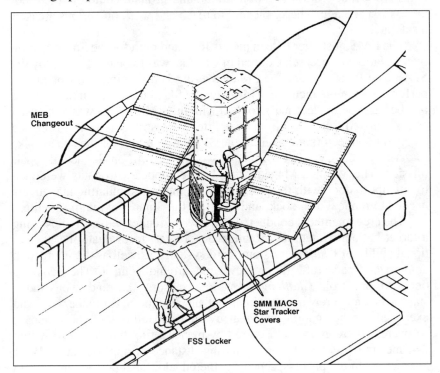

Figure 3–16. Solar Max On-Orbit Berthed Configuration

The robot arm then returned the Solar Max to orbit to continue its study of the violent nature of the Sun's solar activity and its effects on Earth. The successful in-orbit repair demonstrated the STS capability of "in-space" payload processing, which would be exploited on future missions.

Challenger's RMS released the Long Duration Exposure Facility into orbit on this mission (Figure 3–17). Carrying fifty-seven diverse, passive experiments on this mission, it was to be left in space for approximately one year but was left in space for almost six years before being retrieved by STS-32 in January 1990.

Cinema 360 made its second flight, mounted in the cargo bay. The 35mm movie camera recorded the Solar Max rescue mission. A second film camera, IMAX, flew on the Shuttle to record the event on 70mm film designed for projection on very large screens. Table 3–28 contains the details of this mission.

STS 41-D

Discovery made its inaugural flight on this mission, the twelfth flight in the Space Shuttle program. The mission included a combination cargo from some of the payloads originally manifested to fly on STS 41-D and STS 41-F. The decision to remanifest followed the aborted launch of *Discovery* on June 26 and provided for a minimum disruption to the launch schedule.

Failures of the PAM on earlier missions prompted an exhaustive examination of production practices by the NASA-industry team. This team established new test criteria for qualifying the rocket motors. The new testing procedures proved satisfactory when the Shuttle successfully deployed two communications satellites equipped with PAMs, SBS-4 and Telstar 3-C, into precise geosynchronous transfer orbits. A third satellite, Syncom IV-2 (also called Leasat-2), was equipped with a unique upper

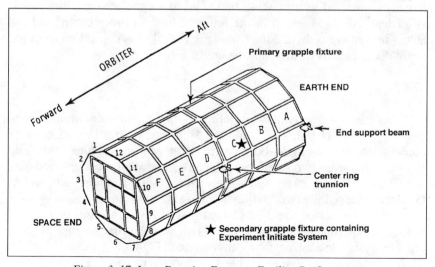

Figure 3–17. Long Duration Exposure Facility Configuration

stage. This satellite was the first built especially for launch from the Shuttle.

NASA's Office of Aeronautics and Space Technology (OAST) sponsored this mission, designated OAST-1. Details of this mission are located in Chapter 3, "Aeronautics and Space Research and Technology," in Volume VI of the *NASA Historical Data Book*. Payload specialist Charles Walker, a McDonnell Douglas employee, was the first commercial payload specialist assigned by NASA to a Shuttle crew. At 21,319.2 kilograms, this mission had the heaviest payload to date. Details of STS 41-D are in Table 3–29.

STS 41-G

This mission was the first with seven crew members and featured the first flight of a Canadian payload specialist, the first to include two women, the first spacewalk by an American woman (Sally Ride), the first crew member to fly a fourth Space Shuttle mission, the first demonstration of a satellite refueling technique in space, and the first flight with a reentry profile crossing the eastern United States. OSTA-3 was the primary payload and was the second in a series of Shuttle payloads that carried experiments to take measurements of Earth. Details of the payload can be found in Chapter 2, "Space Applications," in Volume VI of the *NASA Historical Data Book*.

This mission deployed the Earth Radiation Budget Satellite less than nine hours into flight. This satellite was the first of three planned sets of orbiting instruments in the Earth Radiation Budget Experiment. Overall, the program aimed to measure the amount of energy received from the Sun and reradiated into space and the seasonal movement of energy from the tropics to the poles.

The Orbital Refueling System experiment demonstrated the possibility of refueling satellites in orbit. This experiment required spacesuited astronauts working in the cargo bay to attach a hydrazine servicing tool, already connected to a portable fuel tank, to a simulated satellite panel. After leak checks, the astronauts returned to the orbiter cabin, and the actual movement of hydrazine from tank to tank was controlled from the flight deck. Details of this mission are in Table 3–30.

STS 51-A

This mission deployed two satellites—the Canadian communications satellite Telesat H (Anik-D2) and the Hughes Syncom IV-1 (Leasat-1) communications satellite—both destined for geosynchronous orbit. The crew also retrieved two satellites, Palapa B-2 and Westar 6, deployed during STS 41-B in February 1984. Astronauts Joseph P. Allen and Dale A. Gardner retrieved the two malfunctioning satellites during a spacewalk.

Discovery carried the 3-M Company's Diffusive Mixing of Organic Solutions experiment in the mid-deck. This was the first attempt to grow organic crystals in a microgravity environment. Figure 3–18 shows the STS 51-A cargo configuration, and Table 3–31 lists the mission's characteristics.

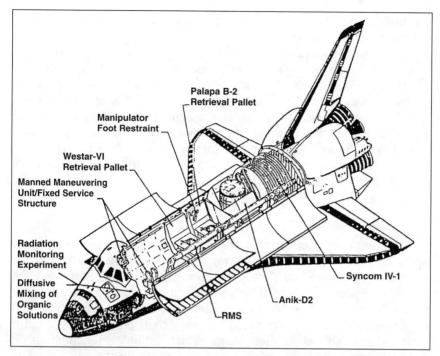

Figure 3–18. STS 51-A Cargo Configuration

STS 51-C

STS 51-C was the first mission dedicated to DOD. A U.S. Air Force inertial upper stage booster was deployed and met the mission objectives.

The Aggregation of Red Cells mid-deck payload tested the capability of NASA's Ames Research Center apparatus to study some characteristics of blood and their disease dependencies under microgravity conditions. NASA's Microgravity Science and Applications Division of the Office of Space Science and Applications sponsored this experiment, which was a cooperative effort between NASA and the Department of Science and Technology of the government of Australia. For details of this mission, see Table 3–32.

STS 51-D

STS 51-D deployed the Telesat-1 (Anik C-1) communications satellite attached to PAM-D motor. The Syncom IV-3 (also called Leasat-3) was also deployed, but the spacecraft sequencer failed to initiate antenna deployment, spinup, and ignition of the perigee kick motor. The mission was extended two days to ensure that the sequencer start levers were in the proper position. Astronauts S. David Griggs and Jeffrey A. Hoffman performed a spacewalk to attach "flyswatter" devices to the RMS. Astronaut M. Rhea Seddon engaged the Leasat lever using the RMS, but the postdeployment sequence failed to begin, and the satellite continued to drift in a low-Earth orbit.

This mission also involved the first public official, Senator Jake Garn from Utah, flying on a Space Shuttle mission; Garn carried out a number of medical experiments. The crew members conducted three mid-deck experiments as part of NASA's microgravity science and applications and space science programs: American Flight Echocardiograph, Phase Partitioning Experiment, and Protein Crystal Growth. Another payload was "Toys in Space," an examination of simple toys in a weightless environment, with the results to be made available to students. The mission's characteristics are in Table 3–33.

STS 51-B

The first operational flight of the Spacelab took place on STS 51-B. Spacelab 3 provided a high-quality microgravity environment for delicate materials processing and fluid experiments. (Table 4–46 describes the individual Spacelab 3 experiments.) The primary mission objective was to conduct science, application, and technology investigations (and acquire intrinsic data) that required the low-gravity environment of Earth orbit and an extended-duration, stable vehicle attitude with emphasis on materials processing. The secondary mission objective was to obtain data on research in materials sciences, life sciences, fluid mechanics, atmospheric science, and astronomy. This mission was the first in which a principal investigator flew with his experiment in space.

The NUSAT Get-Away Special satellite was successfully deployed. The Global Low Orbiting Message Relay satellite failed to deploy from its Get-Away Special canister and was returned to Earth. Details of this mission are in Table 3–34.

STS 51-G

During this mission, NASA flew the first French and Arabian payload specialists. The mission's cargo included domestic communications satellites from the United States, Mexico, and Saudi Arabia—all successfully deployed.

STS 51-G also deployed and retrieved the Spartan-1, using the RMS. The Spartan, a free-flyer carrier developed by NASA's Goddard Space Flight Center, could accommodate scientific instruments originally developed for the sounding rocket program. The Spartan "family" of short-duration satellites were designed to minimize operational interfaces with the orbiter and crew. All pointing sequences and satellite control commands were stored aboard the Spartan in a microcomputer controller. All data were recorded on a tape recorder. No command or telemetry link was provided. Once the Spartan satellite completed its observing sequence, it "safed" all systems and placed itself in a stable attitude to permit its retrieval and return to Earth. NASA's Astrophysics Division within the

Office of Space Science and Applications sponsored the Spartan with a scientific instrument on this mission provided by the Naval Research Laboratory. The mission mapped the x-ray emissions from the Perseus Cluster, the nuclear region of the Milky Way galaxy, and the Scorpius X-2.

In addition, the mission conducted a Strategic Defense Initiative experiment called the High Precision Tracking Experiment. STS 51-G included two French biomedical experiments and housed a materials processing furnace named the Automated Directional Solidification Furnace. Further details are in Table 3–35.

STS 51-F

STS 51-F was the third Space Shuttle flight devoted to Spacelab. Spacelab 2 was the second of two design verification test flights required by the Spacelab Verification Flight Test program. (Spacelab 1 flew on STS-9 in 1983.) Its primary mission objectives were to verify the Spacelab system and subsystem performance capabilities and to determine the Spacelab-orbiter and Spacelab-payload interface capabilities. Secondary mission objectives were to obtain scientific and applications data from a multidisciplinary payload and to demonstrate to the user community the broad capability of Spacelab for scientific research. The monitoring of mission activities and a quick-look analysis of data confirmed that the majority of Verification Flight Test functional objectives were properly performed in accordance with the timeline and flight procedures.

NASA developed the Spacelab 2 payload. Its configuration included an igloo attached to a lead pallet, with the instrument point subsystem mounted on it, a two-pallet train, and an experiment special support structure. The instrument point subsystem—a gimbaled platform attached to a pallet that provides precision pointing for experiments requiring greater pointing accuracy and stability than is provided by the orbiter—flew for the first time on Spacelab 2. The Spacelab system supported and accomplished the experiment phase of the mission. The Spacelab 2 experiments are listed in Table 4–47, and the overall mission characteristics are in Table 3–36.

STS 51-I

STS 51-I deployed three communications satellites, ASC-1, Aussat-1, and Syncom IV-4 (Leasat-4). It also retrieved, repaired, and redeployed Syncom IV-3 (Leasat-3) so that it could be activated from the ground. Astronauts William F. Fisher and James D.A. van Hoften performed two EVAs totaling eleven hours, fifty-one minutes. Part of the time was spent retrieving, repairing, and redeploying the Syncom IV-3, which was originally deployed on STS 51-D.

Physical Vapor Transport of Organic Solids was the second microgravity-based scientific experiment to fly aboard the Space Shuttle. (The first was the Diffusive Mixing of Organic Solutions, which flew on

STS 51-A in November 1984.) Physical Vapor Transport of Organic Solids consisted of nine independent experimental cells housed in an experimental apparatus container mounted on the aft bulkhead in the middeck area. The crew interface was through a handheld keyboard and display terminal. Using this terminal, the crew selected and activated the experiment cells, monitored cell temperatures and power levels, and performed diagnostic tests. Table 3–37 includes the details of STS 51-I.

STS 51-J

STS 51-J was the second Space Shuttle mission dedicated to DOD. *Atlantis* flew for the first time on this mission. Details are in Table 3–38.

STS 61-A

The "Deutschland Spacelab Mission D-1" was the first of a series of dedicated West German missions on the Space Shuttle. The Federal German Aerospace Research Establishment (DFVLR) managed Spacelab D-1 for the German Federal Ministry of Research and Technology. DFVLR provided the payload and was responsible for payload analytical and physical integration and verification, as well as payload operation on orbit. The Spacelab payload was assembled by MBB/ERNO over a five-year period at a cost of about $175 million. The D-1 was used by German and other European universities, research institutes, and industrial enterprises, and it was dedicated to experimental scientific and technological research.

This mission included 75 experiments, most performed more than once (see Chapter 4). These included basic and applied microgravity research in the fields of materials science, life sciences and technology, and communications and navigation. Weightlessness was the common denominator of the experiments carried out aboard Spacelab D-1. Scientific operations were controlled from the German Space Operations Center at Oberpfaffenhofen near Munich.

The mission was conducted in the long module configuration, which featured the Vestibular Sled designed to provide scientists with data on the functional organization of human vestibular and orientation systems. The Global Low Orbiting Message Relay satellite was also deployed from a Get-Away Special canister. Figure 3–19 shows the STS 61-A cargo configuration, and Table 3–39 lists the mission's characteristics.

STS 61-B

Three communications satellites were deployed on this mission: Morelos-B, AUSSAT-2 and Satcom KU-2. The crew members conducted two experiments to test the assembling of erectable structures in space: Experimental Assembly of Structures in Extravehicular Activity and Assembly Concept for Construction of Erectable Space Structure (EASE/ACCESS), shown in Figure 3–20. These experiments required two

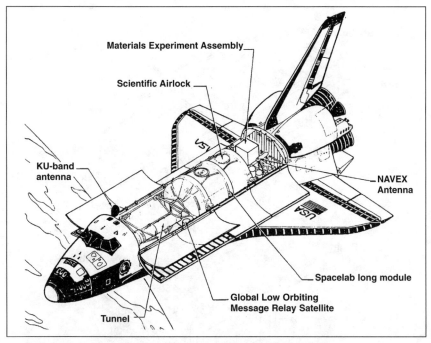

Figure 3–19. STS 61-A Cargo Configuration

spacewalks by Astronauts Sherwood C. Spring and Jerry L. Ross lasting five hours, thirty-two minutes and six hours, thirty-eight minutes, respectively.

This flight featured the first Mexican payload specialist, the first flight of the PAM-D2, the heaviest PAM payload yet (on the Satcom), and the first assembly of a structure in space. Table 3–40 contains STS 61-B's characteristics.

STS 61-C

This mission used the Hitchhiker, a new payload carrier system in the Space Shuttle's payload bay, for the first time. This Hitchhiker flight carrier contained three experiments in the Small Payload Accommodation program: particle analysis cameras to study particle distribution within the Shuttle bay environment, coated mirrors to test the effect of the Shuttle's environment, and a capillary pumped loop heat acquisition and transport system.

Columbia successfully deployed the Satcom KU-1 satellite/PAM-D. However, the Comet Halley Active Monitoring Program experiment, a 35mm camera that was to photograph Comet Halley, did not function properly because of battery problems. This mission also carried Materials Science Laboratory-2 (MSL-2), whose configuration is shown in Figure 3–21.

Franklin R. Chang-Diaz was the first Hispanic American to journey into space. He produced a videotape in Spanish for live distribution to

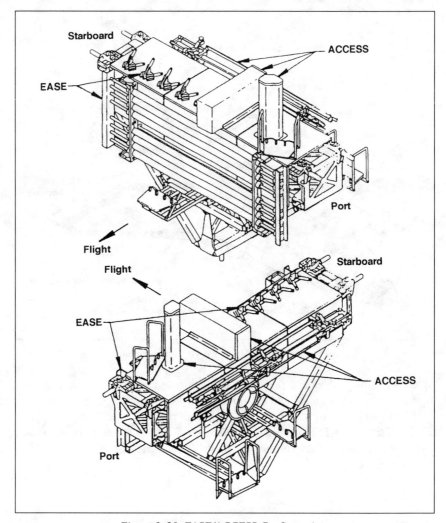

Figure 3–20. EASE/ACCESS Configuration

audiences in the United States and Latin America via the NASA Select television circuit. Details of this mission are in Table 3–41.

STS 51-L

The planned objectives of STS 51-L were the deployment of TDRS-2 and the flying of Shuttle-Pointed Tool for Astronomy (SPARTAN-203)/Halley's Comet Experiment Deployable, a free-flying module designed to observe the tail and coma of Halley's comet with two ultraviolet spectrometers and two cameras. Other cargo included the Fluid Dynamics Experiment, the Comet Halley Active Monitoring Program experiment, the Phase Partitioning Experiment, three SSIP experiments, and a set of lessons for the Teacher in Space Project.

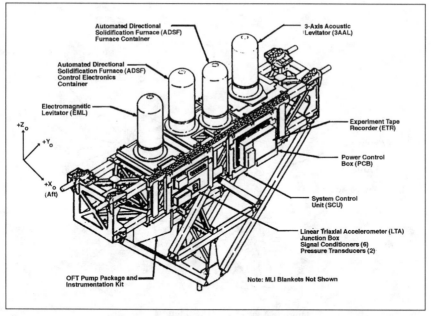

Figure 3–21. Integrated MSL-2 Payload

See the following major section on the *Challenger* accident for detailed information about this mission. STS 51-L's characteristics are listed in Table 3–42.

STS-26

This mission marked the resumption of Space Shuttle flights after the 1986 STS 51-L accident. The primary objective was to deliver TDRS-3 to orbit (Figure 3–22). Meeting this objective, the satellite was boosted to geosynchronous orbit by its inertial upper stage. TDRS-3 was the third TDRS advanced communications spacecraft to be launched from the Shuttle. (TDRS-1 was launched during *Challenger*'s first flight in April 1983. The second, TDRS-2, was lost during the 1986 *Challenger* accident.)

Secondary payloads on *Discovery* included the Physical Vapor Transport of Organic Solids, the Protein Crystal Growth Experiment, the Infrared Communications Flight Experiment, the Aggregation of Red Blood Cells Experiment, the Isoelectric Focusing Experiment, the Mesoscale Lightning Experiment, the Phase Partitioning Experiment, the Earth-Limb Radiance Experiment, the Automated Directional Solidification Furnace, and two SSIP experiments. Special instrumentation was also mounted in the payload bay to record the environment experienced by *Discovery* during the mission. The Orbiter Experiments Autonomous Supporting Instrumentation System-1 (OASIS-1) collected and recorded a variety of environmental measurements during the orbiter's in-flight phases. The data were used to study the effects on the

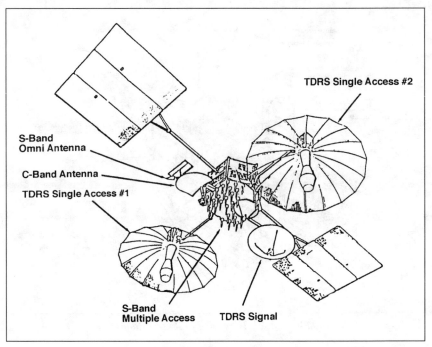

Figure 3–22. Tracking and Data Relay Satellite On-Orbit Configuration

orbiter of temperature, pressure, vibration, sound, acceleration, stress, and strain.

See the section below on the *Challenger* accident and subsequent return to space for information on changes to the Space Shuttle implemented for this mission. STS-26's characteristics are listed in Table 3–43.

STS-27

This was the third STS mission dedicated to DOD. Details of STS-27 are listed in Table 3–44.

The Challenger Accident and Return to Flight

Until the explosion that ended the STS 51-L mission on January 28, 1986, few had been aware of the flaws in the various systems and operations connected with the Space Shuttle. The investigations that followed the accident, which interrupted the program for more than two years, disclosed that long-standing conditions and practices had caused the accident. The following section focuses on the activities of the commission that investigated the explosion, the findings of the various investigations that revealed problems with the Shuttle system in general and with *Challenger* in particular, and the changes that took place in the Shuttle program as a result of the investigations.

The following documents have provided most of the data for this section, and they provide a fascinating look at the events surrounding the accident. The reader might consult them for additional insight about this part of NASA history.

- *STS 51-L Data and Design Analysis Task Force—Historical Summary,* June 1986
- *Report of the Presidential Commission on the Space Shuttle Challenger Accident, Vol. I–IV,* June 6, 1986
- *Report to the President—Actions to Implement the Recommendations of the Presidential Commission on the Space Shuttle Challenger Accident,* July 14, 1986
- "Statement by Dr. James Fletcher," NASA administrator, regarding revised Shuttle manifest, October 3, 1986
- *Report to the President—Implementation of the Recommendations of the Presidential Commission on the Space Shuttle Challenger Accident,* June 30, 1987

Immediately after the *Challenger* explosion, a series of events began that occupied NASA for the next two years, culminating with the launch of *Discovery* on the STS-26 mission. Table 3–45 summarizes the activities that took place from January 28, 1986, through September 29, 1988, the Shuttle's return to flight.

Presidential Commission

Formation and Activities of the Rogers Commission

On February 3, 1986, President Ronald Reagan appointed an independent commission chaired by William P. Rogers, former secretary of state and attorney general, and composed of persons not connected with the mission to investigate the accident. The commission's mandate was to:

1. Review the circumstances surrounding the accident to establish the probable cause or causes of the accident
2. Develop recommendations for corrective or other action based upon the commission's findings and determinations[6]

Immediately after its establishment, the commission began its investigation and, with the full support of the White House, held public hearings on the facts leading up to the accident.

---

[6]*Report at a Glance, Report to the President by the Presidential Commission on the Space Shuttle Challenger Accident* (Washington, DC: U.S. Government Printing Office, 1986), Preface (no page number).

The commission construed its mandate to include recommendations on safety matters that were not necessarily involved in this accident but required attention to make future flights safer. Careful attention was given to concerns expressed by astronauts. However, the commission felt that its mandate did not require a detailed investigation of all aspects of the Space Shuttle program nor a review of budgetary matters. Nor did the commission wish to interfere with or displace Congress in the performance of its duties. Rather, the commission focused its attention on the safety aspects of future flights based on the lessons learned from the investigation, with the objective of returning to safe flight. Congress recognized the desirability of having a single investigation of this accident and agreed to await the commission's findings before deciding what further action it might find necessary.

For the first several days after the accident—possibly because of the trauma resulting from the accident—NASA seemed to be withholding information about the accident from the public. After the commission began its work, and at its suggestion, NASA began releasing much information that helped to reassure the public that all aspects of the accident were being investigated and that the full story was being told in an orderly and thorough manner.

Following the suggestion of the commission, NASA also established several teams of persons not involved with the 51-L launch process to supported the commission and its panels. These NASA teams cooperated with the commission and contributed to what was a comprehensive and complete investigation.

Following their swearing in on February 6, 1986, commission members immediately began a series of hearings during which NASA officials outlined agency procedures covering the Space Shuttle program and the status of NASA's investigation of the accident. On February 10, Dr. Alton G. Keel, Jr., associate director of the Office of Management and Budget, was appointed executive director. Dr. Keel gathered a staff of fifteen experienced investigators from various government agencies and the military services, as well as administrative personnel to support commission activities.

Testimony began on February 10 in a closed session, when the commission began to learn of the troubled history of the solid rocket motor joint and seals. Commission members discovered the first indication that the contractor, Morton Thiokol, initially recommended against the launch on January 27, 1986, the night before the launch of STS 51-L, because of concerns regarding the effects of low temperature on the joint and seal. Additional evidence supplied to the commission on February 13 and 14 provided the first evidence that the solid rocket motor joint and seal might have malfunctioned, initiating the accident. The session on February 14 included NASA and contractor participants who had been involved in the discussion on January 27 about whether to launch *Challenger*. Following that session, Chairman Rogers issued a statement noting that "the process [leading to the launch of *Challenger*] may have been flawed" and that NASA's Acting

Administrator Dr. William Graham had been asked "not to include on the investigating teams at NASA, persons involved in that process."[7]

The commission itself thus assumed the role of investigators and divided itself into four investigative panels:

1. Development and Production, responsible for investigating the acquisition and test and evaluation processes for Space Shuttle elements
2. Pre-Launch Activities, responsible for assessing the Shuttle system processing, launch readiness process, and prelaunch security
3. Mission Planning and Operations, responsible for investigating mission planning and operations, schedule pressures, and crew safety areas
4. Accident Analysis, charged with analyzing the accident data and developing both an anomaly tree and accident scenarios

After the panels were finalized and the new approach described before Congress, the working groups went to the Marshall Space Flight Center, the Kennedy Space Center, and Morton Thiokol to begin analyzing data relating to the accident.

A series of public hearings on February 25, 26, and 27 presented additional information about the launch decision obtained from testimony by Thiokol, Rockwell, and NASA officials. At that time, details about the history of problems with the then-suspect solid rocket motor joints and seals also began emerging and focused the commission's attention on the need to document fully the extent of knowledge and awareness about the problems within both Thiokol and NASA.

Following these hearings, separate panels conducted much of the commission's investigative efforts in parallel with full commission hearings. Panel members made numerous trips to Kennedy, Marshall, the Johnson Space Center, and Thiokol facilities in Utah to hold interviews and gather and analyze data relating to their panels' respective responsibilities.

At the same time, a general investigative staff held a series of individual interviews to document fully the teleconference between NASA and Thiokol officials the night before the launch; the history of joint design and O-ring problems; NASA safety, reliability, and quality assurance functions; and the assembly of the right solid rocket booster for STS 51-L. Subsequent investigations by this group were directed toward the effectiveness of NASA's organizational structure, particularly the Shuttle program structure, and allegations that there had been external pressure on NASA to launch on January 28.

Members of the commission and its staff interviewed more than 160 individuals and held more than thirty-five formal panel investigations, which generated almost 12,000 pages of transcript. Almost 6,300 documents, totaling more than 122,000 pages, and hundreds of

---

[7]*Ibid.*, Appendix A, Commission Activities, p. 206.

photographs were examined and became part of the commission's permanent data base and archives. These sessions and all the data gathered added to the 2,800 pages of hearing transcripts generated by the commission in both closed and open sessions.

In addition to the work of the commission and its staff, more than 1,300 employees from all NASA facilities were involved in the investigation and were supported by more than 1,600 people from other government agencies and more than 3,100 from NASA's contractor organizations. Particularly significant were the activities of the military, the Coast Guard, and the National Transportation Safety Board in the salvage and analysis of the Shuttle wreckage.

Description of the Accident

The flight of *Challenger* on STS 51-L began at 11:38 a.m., Eastern Standard Time, on January 28, 1986. It ended 73 seconds later with the explosion and breakup of the vehicle. All seven members of the crew were killed. They were Francis R. Scobee, commander; Michael J. Smith, pilot; mission specialists Judith A. Resnik, Ellison Onizuka, and Ronald E. McNair; and payload specialists Gregory Jarvis of Hughes Aircraft and S. Christa McAuliffe, a New Hampshire teacher—the first Space Shuttle passenger/observer participating in the NASA Teacher in Space Program. She had planned to teach lessons during live television transmissions.

The primary cargo was the second TDRS. Also on board was a SPARTAN free-flying module that was to observe Halley's comet.

The commission determined the sequence of flight events during the 73 seconds before the explosion and 37 seconds following the explosion based on visual examination and image enhancement of film from NASA-operated cameras and telemetry data transmitted by the Shuttle to ground stations. Table 3–46 lists this sequence of events.

The launch had been the first from Pad B at Kennedy's Launch Complex 39. The flight had been scheduled six times earlier but had been delayed because of technical problems and bad weather.

Investigation and Findings of the Cause of the Accident

Throughout the investigation, the commission focused on three critical questions:

1. What circumstances surrounding mission 51-L contributed to the catastrophic termination of that flight in contrast to twenty-four successful flights preceding it?
2. What evidence pointed to the right solid rocket booster as the source of the accident as opposed to other elements of the Space Shuttle?
3. Finally, what was the mechanism of failure?

Using mission data, subsequently completed tests and analyses, and recovered wreckage, the commission identified all possible faults that could originate in the respective flight elements of the Space Shuttle that might have led to loss of *Challenger*. The commission examined the launch pad, the external tank, the Space Shuttle main engines, the orbiter and related equipment, payload/orbiter interfaces, the payload, the solid rocket boosters, and the solid rocket motors. They also examined the possibility of and ruled out sabotage.

The commission eliminated all elements except the right solid rocket motor as a cause of the accident. Four areas related to the functioning of that motor received detailed analysis to determine their part in the accident:

1. Structural loads were evaluated, and the commission determined that these loads were well below the design limit loads and were not considered the cause of the accident.
2. Failure of the case wall (case membrane) was considered, with the conclusion that the assessments did not support a failure that started in the membrane and progressed slowly to the joint or one that started in the membrane and grew rapidly the length of the solid rocket motor segment.
3. Propellant anomalies were considered, with the conclusion that it was improbable that propellant anomalies contributed to the STS 51-L accident.
4. The remaining area relating to the functioning of the right solid rocket motor, the loss of the pressure seal at the case joint, was determined to be the cause of the accident.

The commission released its report and findings on the cause of the accident on June 9, 1986. The consensus of the commission and participating investigative agencies was that the loss of *Challenger* was caused by a failure in the joint between the two lower segments of the right solid rocket motor. The specific failure was the destruction of the seals that were intended to prevent hot gases from leaking through the joint during the propellant burn of the rocket motor. The evidence assembled by the commission indicated that no other element of the Space Shuttle system contributed to this failure.

In arriving at this conclusion, the commission reviewed in detail all available data, reports, and records, directed and supervised numerous tests, analyses, and experiments by NASA, civilian contractors, and various government agencies, and then developed specific scenarios and the range of most probable causative factors. The commission released the following sixteen findings:

1. *A combustion gas leak through the right solid rocket motor aft field joint initiated at or shortly after ignition eventually weakened and/or penetrated the external tank initiating vehicle structural breakup and loss of the Space Shuttle* Challenger *during STS mission 51-L.*

2. The evidence shows that no other STS 51-L Shuttle element or the payload contributed to the causes of the right solid rocket motor aft field joint combustion gas leak. Sabotage was not a factor.
3. Evidence examined in the review of Space Shuttle material, manufacturing, assembly, quality control, and processing on nonconformance reports found no flight hardware shipped to the launch site that fell outside the limits of Shuttle design specifications.
4. Launch site activities, including assembly and preparation, from receipt of the flight hardware to launch were generally in accord with established procedures and were not considered a factor in the accident.
5. Launch site records show that the right solid rocket motor segments were assembled using approved procedures. However, significant out-of-round conditions existed between the two segments joined at the right solid rocket motor aft field joint (the joint that failed).
    a. While the assembly conditions had the potential of generating debris or damage that could cause O-ring seal failure, these were not considered factors in this accident.
    b. The diameters of the two solid rocket motor segments had grown as a result of prior use.
    c. The growth resulted in a condition at time of launch wherein the maximum gap between the tang and clevis in the region of the joint's O-rings was no more than 0.008 inch (0.2032 millimeter) and the average gap would have been 0.004 inch (0.1016 millimeter).
    d. With a tang-to-clevis gap of 0.004 inch (0.1016 millimeter), the O-ring in the joint would be compressed to the extent that it pressed against all three walls of the O-ring retaining channel.
    e. The lack of roundness of the segments was such that the smallest tang-to-clevis clearance occurred at the initiation of the assembly operation at positions of 120 degrees and 300 degrees around the circumference of the aft field joint. It is uncertain if this tight condition and the resultant greater compression of the O-rings at these points persisted to the time of launch.
6. The ambient temperature at time of launch was 36 degrees F, or 15 degrees lower than the next coldest previous launch.
    a. The temperature at the 300-degree position on the right aft field joint circumference was estimated to be 28 degrees plus or minus 5 degrees F. This was the coldest point on the joint.
    b. Temperature on the opposite side of the right solid rocket booster facing the sun was estimated to be about 50 degrees F.
7. Other joints on the left and right solid rocket boosters experienced similar combinations of tang-to-clevis gap clearance and temperature. It is not known whether these joints experienced distress during the flight of 51-L.

8. *Experimental evidence indicates that due to several effects associated with the solid rocket booster's ignition and combustion pressures and associated vehicle motions, the gap between the tang and the clevis will open as much as 0.017 and 0.029 inches (0.4318 and 0.7366 millimeters) at the secondary and primary O-rings, respectively.*
   a. *This opening begins upon ignition, reaches its maximum rate of opening at about 200–300 milliseconds, and is essentially complete at 600 milliseconds when the solid rocket booster reaches its operating pressure.*
   b. *The external tank and right solid rocket booster are connected by several struts, including one at 310 degrees near the aft field joint that failed. This strut's effect on the joint dynamics is to enhance the opening of the gap between the tang and clevis by about 10–20 percent in the region of 300–320 degrees.*
9. *O-ring resiliency is directly related to its temperature.*
   a. *A warm O-ring that has been compressed will return to its original shape much quicker than will a cold O-ring when compression is relieved. Thus, a warm O-ring will follow the opening of the tang-to-clevis gap. A cold O-ring may not.*
   b. *A compressed O-ring at 75 degrees F is five times more responsive in returning to its uncompressed shape than a cold O-ring at 30 degrees F.*
   c. *As a result it is probable that the O-rings in the right solid booster aft field joint were not following the opening of the gap between the tang and clevis at time of ignition.*
10. *Experiments indicate that the primary mechanism that actuates O-ring sealing is the application of gas pressure to the upstream (high-pressure) side of the O-ring as it sits in its groove or channel.*
    a. *For this pressure actuation to work most effectively, a space between the O-ring and its upstream channel wall should exist during pressurization.*
    b. *A tang-to-clevis gap of 0.004 inch (0.1016 millimeter), as probably existed in the failed joint, would have initially compressed the O-ring to the degree that no clearance existed between the O-ring and its upstream channel wall and the other two surfaces of the channel.*
    c. *At the cold launch temperature experienced, the O-ring would be very slow in returning to its normal rounded shape. It would not follow the opening of the tang-to-clevis gap. It would remain in its compressed position in the O-ring channel and not provide a space between itself and the upstream channel wall. Thus, it is probable the O-ring would not be pressure actuated to seal the gap in time to preclude joint failure due to blow-by and erosion from hot combustion gases.*

11. The sealing characteristics of the solid rocket booster O-rings are enhanced by timely application of motor pressure.
    a. Ideally, motor pressure should be applied to actuate the O-ring and seal the joint prior to significant opening of the tang-to-clevis gap (100 to 200 milliseconds after motor ignition).
    b. Experimental evidence indicates that temperature, humidity and other variables in the putty compound used to seal the joint can delay pressure application to the joint by 500 milliseconds or more.
    c. This delay in pressure could be a factor in initial joint failure.
12. Of 21 launches with ambient temperatures of 61 degrees F or greater, only four showed signs of O-ring thermal distress; i.e., erosion or blow-by and soot. Each of the launches below 61 degrees F resulted in one or more O-rings showing signs of thermal distress.
    a. Of these improper joint sealing actions, one-half occurred in the aft field joints, 20 percent in the center field joints, and 30 percent in the upper field joints. The division between left and right solid rocket boosters was roughly equal.
    b. Each instance of thermal O-ring distress was accompanied by a leak path in the insulating putty. The leak path connects the rocket's combustion chamber with the O-ring region of the tang and clevis. Joints that actuated without incident may also have had these leak paths.
13. There is a possibility that there was water in the clevis of the STS 51-L joints since water was found in the STS-9 joints during a destack operation after exposure to less rainfall than STS 51-L. At time of launch, it was cold enough that water present in the joint would freeze. Tests show that ice in the joint can inhibit proper secondary seal performance.
14. A series of puffs of smoke were observed emanating from the 51-L aft field joint area of the right solid rocket booster between 0.678 and 2.500 seconds after ignition of the Shuttle solid rocket motors.
    a. The puffs appeared at a frequency of about three puffs per second. This roughly matches the natural structural frequency of the solids at lift off and is reflected in slight cyclic changes of the tang-to-clevis gap opening.
    b. The puffs were seen to be moving upward along the surface of the booster above the aft field joint.
    c. The smoke was estimated to originate at a circumferential position of between 270 degrees and 315 degrees on the booster aft field joint, emerging from the top of the joint.
15. This smoke from the aft field joint at Shuttle lift off was the first sign of the failure of the solid rocket booster O-ring seals on STS 51-L.
16. The leak was again clearly evident as a flame at approximately 58 seconds into the flight. It is possible that the leak was continuous but unobservable or non-existent in portions of the intervening period. It is possible in either case that thrust vectoring and normal vehicle response to wind shear as well as planned maneuvers reinitiated

or magnified the leakage from a degraded seal in the period preceding the observed flames. The estimated position of the flame, centered at a point 307 degrees around the circumference of the aft field joint, was confirmed by the recovery of two fragments of the right solid rocket booster.
   a. *A small leak could have been present that may have grown to breach the joint in flame at a time on the order of 58 to 60 seconds after lift off.*
   b. *Alternatively, the O-ring gap could have been resealed by deposition of a fragile buildup of aluminum oxide and other combustion debris. This resealed section of the joint could have been disturbed by thrust vectoring, Space Shuttle motion and flight loads inducted by changing winds aloft.*
   c. *The winds aloft caused control actions in the time interval of 32 seconds to 62 seconds into the flight that were typical of the largest values experienced on previous missions.*

Conclusion. *In view of the findings, the commission concluded that the cause of the* Challenger *accident was the failure of the pressure seal in the aft field joint of the right solid rocket booster. The failure was due to a faulty design unacceptably sensitive to a number of factors. These factors were the effects of temperature, physical dimensions, the character of materials, the effects of reusability, processing and the reaction of the joint to dynamic loading.*[8]

Contributing Causes of the Accident

In addition to the failure of the pressure seal as the primary cause of the accident, the commission identified a contributing cause of the accident having to do with the decision to launch. The commission concluded that the decision-making process was flawed in several ways. The testimony revealed failures in communication, which resulted in a decision to launch based on incomplete and sometimes misleading information, a conflict between engineering data and management judgments, and a NASA management structure that permitted internal flight safety problems to bypass key Shuttle managers.

The decision to launch concerned two problem areas. One was the low temperature and its effect on the O-ring. The second was the ice that formed on the launch pad. The commission concluded that concerns regarding these issues had either not been communicated adequately to senior management or had not been given sufficient weight by those who made the decision to launch.

O-Ring Concerns. Formal preparations for launch, consisting of the Level I Flight Readiness Review and Certification of Flight Readiness to

---

[8]*Ibid.*, Findings, pp. 70–72.

the Level II program manager at the Johnson Space Center, were followed in a procedural sense for STS 51-L. However, the commission concluded that relevant concerns of Level III NASA personnel and element contractors had not been, in critical areas, adequately communicated to the NASA Levels I and II management responsible for the launch. In particular, objections to the launch voiced by Morton Thiokol engineers about the detrimental effect of cold temperatures on the performance of the solid rocket motor joint seal and the degree of concern of Thiokol and the Marshall Space Flight Center about the erosion of the joint seals in prior Shuttle flights, notably STS 51-C and 51-B, were not communicated sufficiently.

Since December 1982, the O-rings had been designated a "Criticality 1" feature of the solid rocket booster design, meaning that component failure without backup could cause a loss of life or vehicle. In July 1985, after a nozzle joint on STS 51-B showed secondary O-ring erosion, indicating that the primary seal failed, a launch constraint was placed on flight STS 51-F and subsequent launches. These constraints had been imposed and regularly waived by the solid rocket booster project manager at Marshall, Lawrence B. Mulloy. Neither the launch constraint, the reason for it, nor the six consecutive waivers prior to STS 51-L were known to Associate Administrator for Space Flight Jesse W. Moore (Level I), Aldrich Arnold, the manager of space transportation programs at the Johnson Space Center (Level II), or James Thomas, the deputy director of launch and landing operations at the Kennedy Space Center at the time of the Flight Readiness Review process for STS 51-L.

In addition, no mention of the O-ring problems appeared in the Certification of Flight Readiness for the solid rocket booster set designated BI026 signed for Thiokol on January 9, 1986, by Joseph Kilminster. Similarly, no mention appeared in the certification endorsement, signed on January 15, 1986, by Kilminster and Mulloy. No mention appeared in the entire chain of readiness reviews for STS 51-L, contrary to testimony by Mulloy, who claimed that concern about the O-ring was "in the Flight Readiness Review record that went all the way to the L-I review."[9]

On January 27 and through the night to January 28, NASA and contractor personnel debated the wisdom of launching on January 28, in light of the O-ring performance under low temperatures. Table 3–47 presents the chronology of discussions relating to temperature and the decision to launch. Information is based on testimony and documents provided to the commission through February 24, 1986. Except for the time of launch, all times are approximate.

According to the commission, the decision to launch *Challenger* was flawed. Those who made that decision were unaware of the recent history of problems concerning the O-rings and the joints and were unaware

---

[9]*Ibid.*, p. 85, from Commission Hearing Transcript, May 2, 1986, pp. 2610–11.

of the initial written recommendation of the contractor advising against the launch at temperatures below 53 degrees F and the continuing opposition of the engineers at Thiokol after management reversed its position. If the decision makers had known all of the facts, it is highly unlikely that they would have decided to launch STS 51-L on January 28, 1986. The commission revealed the following four findings:

1. The commission concluded that there was a serious flaw in the decision-making process leading up to the launch of flight 51-L. A well-structured and managed system emphasizing safety would have flagged the rising doubts about the solid rocket booster joint seal. Had these matters been clearly stated and emphasized in the flight readiness process in terms reflecting the views of most of the Thiokol engineers and at least some of the Marshall engineers, it seems likely that the launch of 51-L might not have occurred when it did.
2. The waiving of launch constraints seems to have been at the expense of flight safety. There was no system that mandated that launch constraints and waivers of launch constraints be considered by all levels of management.
3. The commission noted what seemed to be a propensity of management at Marshall to contain potentially serious problems and to attempt to resolve them internally rather than communicate them forward. This tendency, the commission stated, was contrary to the need for Marshall to function as part of a system working toward successful flight missions, interfacing and communicating with the other parts of the system that worked to the same end.
4. The commission concluded that Thiokol management reversed its position and recommended the launch of 51-L at the urging of Marshall and contrary to the views of its engineers in order to accommodate a major customer.

Ice on the Launch Pad. The commission also found that decision makers did not clearly understand Rockwell's concern that launching was unsafe because of ice on the launch pad and whether Rockwell had indeed recommended the launch. They expressed concern about three aspects of this issue:

1. An analysis of all of the testimony and interviews established that Rockwell's recommendation on launch was ambiguous. The commission found it difficult, as did Aldrich, to conclude that there was a no-launch recommendation. Moreover, all parties were asked specifically to contact Aldrich or other NASA officials after the 9:00 a.m. Mission Management Team meeting and subsequent to the resumption of the countdown.
2. The commission was also concerned about NASA's response to Rockwell's position at the 9:00 a.m. meeting. The commission was not convinced Levels I and II appropriately considered Rockwell's

concern about the ice. However ambiguous as Rockwell's position was, it was clear that Rockwell did tell NASA that the ice was an unknown condition. Given the extent of the ice on the pad, the admitted unknown effect of the solid rocket motor and Space Shuttle main engines' ignition on the ice, as well as the fact that debris striking the orbiter was a potential flight safety hazard, the commission found the decision to launch questionable. In this situation, NASA seemed to be requiring a contractor to prove that it was *not* safe to launch, rather than proving it *was* safe. Nevertheless, the commission determined that the ice was not a cause of the 51-L accident and did not conclude that NASA's decision to launch specifically overrode a no-launch recommendation by an element contractor.

3. The commission concluded that the freeze protection plan for Launch Pad 39-B was inadequate. The commission believed that the severe cold and presence of so much ice on the fixed service structure made it inadvisable to launch and that margins of safety were whittled down too far. Additionally, access to the crew emergency slide wire baskets was hazardous due to icy conditions. Had the crew been required to evacuate the orbiter on the launch pad, they would have been running on an icy surface. The commission believed that the crew should have been told of the condition and that greater consideration should have been given to delaying the launch.

Precursor to the Accident

Earlier events helped set the stage for the conditions that caused the STS 51-L accident. The commission stated that the Space Shuttle's solid rocket booster problem began with the faulty design of its joint and increased as both NASA and contractor management first failed to recognize the problem, then failed to fix it, and finally treated it as an acceptable flight risk.

Morton Thiokol did not accept the implication of tests early in the program that the design had a serious and unanticipated flaw. NASA did not accept the judgment of its engineers that the design was unacceptable, and as the joint problems grew in number and severity, NASA minimized them in management briefings and reports. Thiokol's stated position was that "the condition is not desirable but is acceptable."[10]

Neither Thiokol nor NASA expected the rubber O-rings sealing the joints to be touched by hot gases of motor ignition, much less to be partially burned. However, as tests and then flights confirmed damage to the sealing rings, the reaction by both NASA and Thiokol was to increase the amount of damage considered "acceptable." At no time, the commission found, did management either recommend a redesign of the joint or call for the Shuttle's grounding until the problem was solved.

---

[10]*Ibid.,* p. 120, from Report, "STS-3 Through STS-25 Flight Readiness Reviews to Level III Center Board," NASA.

The commission stated that the genesis of the *Challenger* accident—the failure of the joint of the right solid rocket motor—began with decisions made in the design of the joint and in the failure by both Thiokol and NASA's solid rocket booster project office to understand and respond to facts obtained during testing. The commission concluded that neither Thiokol nor NASA responded adequately to internal warnings about the faulty seal design. Furthermore, Thiokol and NASA did not make a timely attempt to develop and verify a new seal after the initial design was shown to be deficient. Neither organization developed a solution to the unexpected occurrences of O-ring erosion and blow-by, even though this problem was experienced frequently during the Shuttle's flight history. Instead, Thiokol and NASA management came to accept erosion and blow-by as unavoidable and an acceptable flight risk. Specifically, the commission found that:

1. The joint test and certification program was inadequate. There was no requirement to configure the qualifications test motor as it would be in flight, and the motors were static-tested in a horizontal position, not in the vertical flight position.
2. Prior to the accident, neither NASA nor Thiokol fully understood the mechanism by which the joint sealing action took place.
3. NASA and Thiokol accepted escalating risk apparently because they "got away with it last time." As Commissioner Richard Feynman observed, the decision making was "a kind of Russian roulette. . . . [The Shuttle] flies [with O-ring erosion] and nothing happens. Then it is suggested, therefore, that the risk is no longer so high for the next flights. We can lower our standards a little bit because we got away with it last time. . . . You got away with it, but it shouldn't be done over and over again like that."[11]
4. NASA's system for tracking anomalies for Flight Readiness Reviews failed in that, despite a history of persistent O-ring erosion and blow-by, flight was still permitted. It failed again in the sequence of six consecutive launch constraint waivers prior to 51-L, permitting it to fly without any record of a waiver, or even of an explicit constraint. Tracking and continuing only anomalies that are "outside the data base" of prior flight allowed major problems to be removed from and lost by the reporting system.
5. The O-ring erosion history presented to Level I at NASA Headquarters in August 1985 was sufficiently detailed to require corrective action prior to the next flight.
6. A careful analysis of the flight history of O-ring performance would have revealed the correlation of O-ring damage and low temperature. Neither NASA nor Thiokol carried out such an analysis; consequently, they were unprepared to properly evaluate the risks of launching

---

[11]*Ibid.*, p. 148, from Commission Hearing Testimony, April 3, 1986, p. 2469.

the 51-L mission in conditions more extreme than they had encountered before.

NASA's Safety Program

The commission found surprising and disturbing the lack of reference to NASA's safety staff. Individuals who testified before the commission did not mention the quality assurance staff, and no reliability and quality assurance engineer had been asked to participate in the discussions that took place prior to launch.

The commission concluded that "the extensive and redundant safety assurance functions" that had existed "during and after the lunar program to discover any safety problems" had become ineffective between that period and 1986. This loss of effectiveness seriously degraded the checks and balances essential for maintaining flight safety.[12] Although NASA had a safety program in place, communications failures relating to safety procedures did not operate properly during STS 51-L.

On April 3, 1986, Arnold Aldrich, the Space Shuttle program manager, appeared before the commission at a public hearing in Washington, D.C. He described five different communications or organizational failures that affected the launch decision on January 28, 1986. Four of those failures related directly to faults within the safety program: lack of problem reporting requirements, inadequate trend analysis, misrepresentation of criticality, and lack of involvement in critical discussions. A properly staffed, supported, and robust safety organization, he stated, might well have avoided these faults and thus eliminated the communications failures. The commission found that:

1. Reductions in the safety, reliability and quality assurance work force at the Marshall and NASA Headquarters seriously limited capability in those vital functions.
2. Organizational structures at Kennedy and Marshall placed safety, reliability, and quality assurance offices under the supervision of the very organizations and activities whose efforts they are to check.
3. Problem reporting requirements were not concise and failed to get critical information to the proper levels of management.
4. Little or no trend analysis was performed on O-ring erosion and blow-by problems.
5. As the flight rate increased, the Marshall safety, reliability, and quality assurance work force was decreasing, which adversely affected mission safety.
6. Five weeks after the 51-L accident, the criticality of the solid rocket motor field joint had still not been properly documented in the problem reporting system at Marshall.

---

[12]*Ibid.*, p. 152.

Pressures on the System

From the Space Shuttle's inception, NASA had advertised that the Shuttle would make space operations "routine and economical." The implication was that the greater annual number of flights, the more routine Shuttle flights would become. Thus, NASA placed heavy emphasis on the schedule. However, one effect of the agency's determination to meet an accelerated flight rate was the dilution of resources available for any one mission. In addition, NASA had difficulty evolving from its single-flight focus to a system that could support an ongoing schedule of flights. Managers forgot in their insistence on proving it operational, the commission stated, that the Shuttle system was still in its early phase. There might not have been enough preparation for what "operational" entailed. For instance, routine and regular postflight maintenance and inspections, spare parts production or acquisition, and software tools and training facilities developed during a test program were not suitable for the high volume of work required in an operational environment. The challenge was to streamline the processes to provide the needed support without compromising quality.

Mission planning requires establishing the manifest, defining the objectives, constraints, and capabilities of the mission, and translating those into hardware, software, and flight procedures. Within each of these major goals is a series of milestones in which managers decide whether to proceed to the next step. Once a decision has been made to go ahead and the activity begun, if a substantial change occurs, it may be necessary to go back and repeat the preceding process. In addition, if one group fails to meet its due date, the delay cascades throughout the system.

The ambitious flight rate meant that less and less time was available for completing each of the steps in the mission planning and preparation process. In addition, a lack of efficient production processing and manifest changes disrupted the production system. In particular, the commission found that manifest changes, which forced repeating certain steps in the production cycle, sometimes severely affected the entire cycle and placed impossible demands on the system.

The commission found that pressures on the STS to launch at an overambitious rate contributed to severe strains on the system. The flight rate did not seem to be based on an assessment of available resources and capabilities and was not modified to accommodate the capacity of the work force. The commission stated that NASA had not provided adequate resources to support its launch schedule and that the system had been strained by the modest nine missions that had launched in 1985.

After the accident, rumors appeared that persons who made the decision to launch might have been subjected to outside pressures to launch. The commission examined these rumors and concluded that the decision to launch was made solely by the appropriate NASA officials without any outside intervention or pressure.[13] The commission listed the following findings:

---

[13]*Ibid.*, p. 176.

1. The capabilities of the system were stretched to the limit to support the flight rate in the winter of 1985–86. Projections into the spring and summer of 1986 showed that the system, as it existed, would have been unable to deliver crew training software for scheduled flights by the designated dates. The result would have been an unacceptable compression of the time available for the crews to accomplish their required training.
2. Spare parts were in critically short supply. The Space Shuttle program made a conscious decision to postpone spare parts procurements in favor of budget items of perceived higher priority. The lack of spare parts would likely have limited flight operations in 1986.
3. The stated manifesting policies were not enforced. Numerous late manifest changes (after the cargo integration review) were made to both major payloads and minor payloads throughout the Shuttle program. These changes required additional resources and used existing resources more rapidly. They also adversely affected crew training and the development of procedures for subsequent missions.
4. The scheduled flight rate did not accurately reflect the capabilities and resources.
    - The flight rate was not reduced to accommodate periods of adjustment in the capacity of the work force. No margin existed in the system to accommodate unforeseen hardware problems.
    - Resources were primarily directed toward supporting the flights and thus were inadequate to improve and expand facilities needed to support a higher flight rate.
5. Training simulators may be the limiting factor on the flight rate; the two current simulators cannot train crews for more than twelve to fifteen flights per year.
6. When flights come in rapid succession, current requirements do not ensure that critical anomalies occurring during one flight are identified and addressed appropriately before the next flight.

Other Safety Considerations

During its investigation, the commission examined other safety-related issues that had played no part in the STS 51-L accident but nonetheless might lead to safety problems in the future. These safety-related areas were ascent (including abort capabilities and crew escape options), landing (including weather considerations, orbiter tires and brakes, and choice of a landing site), Shuttle elements other than the solid rocket booster, processing and assembly (including record keeping and inspections), capabilities of Launch Pad 39-B, and involvement of the development contractors.

Ascent. The events of flight 51-L illustrated the dangers of the first stage of a Space Shuttle ascent. The accident also focused attention on orbiter abort capabilities and crew escape. The current abort capabilities, options to improve those capabilities, options for crew escape, and the performance of the range safety system were of particular concern to the commission.

The Shuttle's design capabilities allowed for successful intact mission abort (a survivable landing) on a runway after a single main engine failure. The Shuttle's design specifications did not require that the orbiter be able to manage an intact abort if a second main engine should fail. If two or three main engines failed, the Shuttle would land in water in a contingency abort or ditching. This maneuver was not believed to be survivable because of damage incurred at water impact. In addition, the Shuttle system was not designed to survive a failure of the solid rocket boosters. Furthermore, although technically the orbiter had the capability to separate from the external tank during the first stage, analysis had shown that if it were attempted while the solid rocket boosters were still thrusting, the orbiter would "hang up" on its aft attach points and pitch violently, with probable loss of the orbiter and crew. This "fast separation" would provide a useful means of escape during first stage only if solid rocket booster thrust could be terminated first.[14]

Studies identified no viable means of crew escape during first-stage ascent. The commission supported the further study of escape options. However, it concluded that no corrective actions could have been taken that would have saved the *Challenger*'s flight crew.

Landing. The Space Shuttle's entry and landing formed another risky and complicated part of a mission. Because the crew could not divert to an alternate landing site after entry, the landing decision must be both timely and accurate. In addition, the landing gear, including the wheels, tires, and brakes, must function properly.

Although the orbiter tires were designed to support a landing up to 108,864 kilograms at 416.7 kilometers per hour with thirty-seven kilometers per hour of crosswind and have successfully passed testing programs, they had shown excessive wear during landings at Kennedy, especially when crosswinds were involved. The tires were rated as Criticality 1 because the loss of a single tire could cause a loss of control and a subsequent loss of the vehicle and crew. Because actual wear on a runway did not correspond to test results, NASA directed testing to examine actual tire, wheel, and strut failure to better understand this failure case.

The commission found that the brakes used on the orbiter were known to have little or no margin, because they were designed based on the orbiter's design weight. As the actual orbiter's weight grew, the brakes were not redesigned; rather, the runway length was extended. Actual flight experience had shown brake damage on most flights, which required that special crew procedures be developed to ensure successful braking.

The original Shuttle plan called for routine landings at Kennedy to minimize turnaround time and cost per flight and to provide efficient operations for both the Shuttle system and the cargo elements. While those considerations remained important, concerns such as the performance of the orbiter tires and brakes and the difficulty of accurate weather prediction in Florida had called the plan into question.

---

[14]*Ibid.*, p. 180.

When the Shuttle landed at Edwards Air Force Base, approximately six days are added to the turnaround time. The commission stated that although there were valid programmatic reasons for landing the Shuttle routinely at Kennedy, the demanding nature of landing and the impact of weather conditions might dictate the prudence of using Edwards on a regular basis for landing. The cost associated with regular scheduled landing and turnaround operations at Edwards was thus a necessary program cost. Decisions governing Shuttle operations, the commission stated, must coincide with the philosophy that unnecessary risks have to be eliminated.

Shuttle Elements. The Space Shuttle main engine teams at Marshall and Rocketdyne had developed engines that achieved their performance goals and performed extremely well. Nevertheless, according to the commission, the main engines continued to be highly complex and critical components of the Shuttle, with an element of risk principally because important components of the engines degraded more rapidly with flight use than anticipated. Both NASA and Rocketdyne took steps to contain that risk. An important aspect of the main engine program was the extensive "hot fire" ground tests. Unfortunately, the vitality of the test program, the commission found, was reduced because of budgetary constraints.

The number of engine test firings per month had decreased over the two years prior to STS 51-L. Yet this test program had not demonstrated the limits of engine operation parameters or included tests over the full operating envelope to show full engine capability. In addition, tests had not yet been deliberately conducted to the point of failure to determine actual engine operating margins.

The commission also identified one serious potential failure mode related to the disconnect valves between the orbiter and the external tank.

Processing and Assembly. During the processing and assembly of the elements of flight 51-L, the commission found various problems that could bear on the safety of future flights. These involved structural inspections in which waivers were granted on sixty of the 146 required orbiter structural inspections, errors in the recordkeeping for the Space Shuttle main engine/main propulsion system and the orbiter, areas in which items called for by the Operational Maintenance Requirements and Specifications Document were not met and were not formally waived or excepted, the Shuttle processing contractor's policy of using "designated verifiers" to supplement quality assurance personnel, and the lack of accidental damage reporting because technicians were concerned about losing their jobs.

Launch Pad 39-B. The damage to the launch pad from the explosion was considered to be normal or minor, with three exceptions: the loss of the springs and plungers of the booster hold-down posts, the failure of the gaseous hydrogen vent arm to latch, and the loss of bricks from the flame trench.

Involvement of Development Contractors. The commission determined that, although NASA considered the Shuttle program to be operational, it was "clearly a developmental program and must be treated as such by NASA."¹⁵ Using procedures accepted by the transportation industry was only partly valid because each mission expanded system and performance requirements. The Shuttle's developmental status demanded that both NASA and all its contractors maintain a high level of in-house experience and technical ability. The demands of the developmental aspects of the program required:

1. Maintaining a significant engineering design and development capability among the Shuttle contractors and an ongoing engineering capability within NASA
2. Maintaining an active analytical capability so that the evolving capabilities of the Shuttle can be matched to the demands on the Shuttle

Recommendations of the Presidential Commission

The commission unanimously adopted nine recommendations, which they submitted to President Reagan. They also urged NASA's administrator to submit a report to the president on the progress NASA made in implementing the recommendations. These recommendations are restated below.

I

Design. *The faulty solid rocket motor joint and seal must be changed. This could be a new design eliminating the joint or a redesign of the current joint and seal. No design options should be prematurely precluded because of schedule, cost or reliance on existing hardware. All solid rocket motor joints should satisfy the following requirements:*

- *The joints should be fully understood, tested and verified.*
- *The integrity of the structure and of the seals of all joints should be not less than that of the case walls throughout the design envelope.*
- *The integrity of the joints should be insensitive to:*
  - *Dimensional tolerances.*
  - *Transportation and handling.*
  - *Assembly procedures.*
  - *Inspection and test procedures.*
  - *Environmental effects.*
  - *Internal case operating pressure.*
  - *Recovery and reuse effects.*
  - *Flight and water impact loads.*
- *The certification of the new design should include:*
  - *Tests which duplicate the actual launch configuration as closely as possible.*

---

¹⁵*Ibid.*, p. 194.

- Tests over the full range of operating conditions, including temperature.
- Full consideration should be given to conducting static firings of the exact flight configuration in a vertical attitude.

Independent Oversight. The administrator of NASA should request the National Research Council to form an independent solid rocket motor design oversight committee to implement the commission's design recommendations and oversee the design effort. This committee should:

- Review and evaluate certification requirements.
- Provide technical oversight of the design, test program and certification.
- Report to the administrator of NASA on the adequacy of the design and make appropriate recommendations.

II

Shuttle Management Structure. The Shuttle Program Structure should be reviewed. The project managers for the various elements of the Shuttle program felt more accountable to their center management than to the Shuttle program organization. Shuttle element funding, work package definition, and vital program information frequently bypass the National STS (Shuttle) Program Manager.

A redefinition of the Program Manager's responsibility is essential. This redefinition should give the Program Manager the requisite authority for all ongoing STS operations. Program funding and all Shuttle Program work at the centers should be placed clearly under the Program Manager's authority.

Astronauts in Management. The commission observes that there appears to be a departure from the philosophy of the 1960s and 1970s relating to the use of astronauts in management positions. These individuals brought to their positions flight experience and a keen appreciation of operations and flight safety.

- NASA should encourage the transition of qualified astronauts into agency management positions.
- The function of the Flight Crew Operations director should be elevated in the NASA organization structure.

Shuttle Safety Panel. NASA should establish an STS Safety Advisory Panel reporting to the STS Program Manager. The Charter of this panel should include Shuttle operational issues, launch commit criteria, flight rules, flight readiness and risk management. The panel should include representation from the safety organization, mission operations, and the astronaut office.

III

Criticality Review and Hazard Analysis. *NASA and the primary Shuttle contractors should review all Criticality 1, 1R, 2, and 2R items and hazard analyses. This review should identify those items that must be improved prior to flight to ensure mission safety. An Audit Panel, appointed by the National Research Council, should verify the adequacy of the effort and report directly to the administrator of NASA.*

IV

Safety Organization. *NASA should establish an Office of Safety, Reliability and Quality Assurance to be headed by an associate administrator, reporting directly to the NASA administrator. It would have direct authority for safety, reliability, and quality assurance throughout the agency. The office should be assigned the work force to ensure adequate oversight of its functions and should be independent of other NASA functional and program responsibilities.*

The responsibilities of this office should include:

- *The safety, reliability and quality assurance functions as they relate to all NASA activities and programs.*
- *Direction of reporting and documentation of problems, problem resolution and trends associated with flight safety.*

V

Improved Communications. *The commission found that Marshall Space Flight Center project managers, because of a tendency at Marshall to management isolation, failed to provide full and timely information bearing on the safety of flight 51-L to other vital elements of Shuttle program management.*

- *NASA should take energetic steps to eliminate this tendency at Marshall Space Flight Center, whether by changes of personnel, organization, indoctrination or all three.*
- *A policy should be developed which governs the imposition and removal of Shuttle launch constraints.*
- *Flight Readiness Reviews and Mission Management Team meetings should be recorded.*
- *The flight crew commander, or a designated representative, should attend the Flight Readiness Review, participate in acceptance of the vehicle for flight, and certify that the crew is properly prepared for flight.*

VI

Landing Safety. *NASA must take actions to improve landing safety:*

- *The tire, brake and nose wheel steering systems must be improved. These systems do not have sufficient safety margin, particularly at abort landing sites.*
- *The specific conditions under which planned landings at Kennedy would be acceptable should be determined. Criteria must be established for tires, brakes and nose wheel steering. Until the systems meet those criteria in high fidelity testing that is verified at Edwards, landing at Kennedy should not be planned.*
- *Committing to a specific landing site requires that landing area weather be forecast more than an hour in advance. During unpredictable weather periods at Kennedy, program officials should plan on Edwards landings. Increased landings at Edwards may necessitate a dual ferry capability.*

VII

Launch Abort and Crew Escape. *The Shuttle program management considered first-stage abort options and crew escape options several times during the history of the program, but because of limited utility, technical unfeasibility, or program cost and schedule, no systems were implemented. The commission recommends that NASA:*

- *Make all efforts to provide a crew escape system for use during controlled gliding flight.*
- *Make every effort to increase the range of flight conditions under which an emergency runway landing can be successfully conducted in the event that two or three main engines fail early in ascent.*

VIII

Flight Rate. *The nation's reliance on the Shuttle as its principal space launch capability created a relentless pressure on NASA to increase the flight rate. Such reliance on a single launch capability should be avoided in the future.*

NASA must establish a flight rate that is consistent with its resources. A firm payload assignment policy should be established. The policy should include rigorous controls on cargo manifest changes to limit the pressures such changes exert on schedules and crew training.

IX

Maintenance Safeguards. *Installation, test, and maintenance procedures must be especially rigorous for Space Shuttle items designated*

Criticality 1. NASA should establish a system of analyzing and reporting performance trends of such items.

Maintenance procedures for such items should be specified in the Critical Items List, especially for those such as the liquid-fueled main engines, which require unstinting maintenance and overhaul.

With regard to the orbiters, NASA should:

- *Develop and execute a comprehensive maintenance inspection plan.*
- *Perform periodic structural inspections when scheduled and not permit them to be waived.*
- *Restore and support the maintenance and spare parts programs, and stop the practice of removing parts from one orbiter to supply another.*[16]

Concluding Thought

The commission urged that NASA continue to receive the support of the administration and the nation. The agency constitutes a national resource that plays a critical role in space exploration and development. It also provides a symbol of national pride and technological leadership.

The commission applauded NASA's spectacular achievements of the past and anticipated impressive achievements in the future. The findings and recommendations presented in this report were intended to contribute to future NASA successes that the nation both expects and requires as the 21st century approaches.

STS 51-L Investigations and Actions by NASA

Safely Returning the Shuttle to Flight Status

While the Presidential Commission investigated the accident, NASA also conducted an investigation to determine strategies and major actions for safely returning to flight status. In a March 24, 1986, memorandum, Associate Administrator for Space Flight Richard H. Truly defined NASA's comprehensive strategy and major actions that would allow for resuming the Space Shuttle's schedule. He stated that NASA Headquarters (particularly the Office of Space Flight), the Office of Space Flight centers, the NSTS program organization, and its various contractors would use the guidance supplied in the memo to proceed with "the realistic, practical actions necessary to return to the NSTS flight schedule with emphasis on flight safety."[17] In his memo, Truly focused on three areas: actions required prior to the next flight, first flight/first year operations, and development of sustainable safe flight rate.

---

[16]*Ibid.*, p. 196.

[17]Richard H. Truly, NASA Memorandum, "Strategy for Safely Returning the Space Shuttle to Flight Status," March 24, 1986.

Actions Required Prior to the Next Flight. Truly directed NASA to take the following steps before the return to flight:

- Reassess the entire program management structure and operation
- Redesign the solid rocket motor joint (A dedicated solid rocket motor joint design group would be established at Marshall to recommend a program plan to quantify the solid rocket motor joints problem and to accomplish the solid rocket motor joints redesign.)
- Reverify design requirements
- Complete Critical Item List (CIL)/Operations and Maintenance Instructions reviews (NASA would review all Category 1 and 1R critical items and implement a complete reapproval process. Any items not revalidated by this review would be redesigned, certified, and qualified for flight.)
- Complete Operations and Maintenance Requirements and Specifications Document review
- Reassess launch and abort rules and philosophy

First Flight/First Year Operations. The first flight mission design would incorporate:

- Daylight Kennedy launch
- Conservative flight design to minimize transatlantic-abort-launch exposure
- Repeat payload (not a new payload class)
- No waiver on landing weight
- Conservative launch/launch abort/landing weather
- NASA-only flight crew
- Engine thrust within the experience base
- No active ascent/entry Developmental Test Objectives
- Conservative mission rules
- Early, stable flight plan with supporting flight software and training load
- Daylight Edwards Air Force Base landing

The planning for the flight schedule for the first year of operation would reflect a conservative launch rate. The first year of operation would be maintained within the current flight experience base, and any expansion of the base, including new classes of payloads, would be approved only after a very thorough safety review.

Development of Sustainable Safe Flight Rate. This flight rate would be developed using a "bottoms-up" approach in which all required work was identified and that work was optimized, keeping in mind the available work force. Factors with the potential for disrupting schedules as well as the availability of resources would be considered when developing the flight rate.

SPACE TRANSPORTATION/HUMAN SPACEFLIGHT

Design and Development Task Force

Also while the Presidential Commission was meeting, NASA formed the 51-L Data and Design Analysis Task Force. This group supported the Presidential Commission and was responsible for:

1. Determining, reviewing, and analyzing the facts and circumstances surrounding the STS 51-L launch
2. Reviewing all factors relating to the accident determined to be relevant, including studies, findings, recommendations, and other actions that were or might be undertaken by the program offices, field centers, and contractors involved
3. Examining all other factors that could relate to the accident, including design issues, procedures, organization, and management factors
4. Using the full required technical and scientific expertise and resources available within NASA and those available to NASA
5. Documenting task force findings and determinations and conclusions derived from the findings.
6. Providing information and documentation to the commission regarding task force activities.

The task force, which was chaired by Truly, established teams to examine development and production; prelaunch activities; accident analysis; mission planning and operations; and search, recovery, and reconstruction; and a photo and TV support team. Figure 3–23 shows the task force organization.

Each task force team submitted multivolume reports to the Presidential Commission, which included descriptions of the accident as

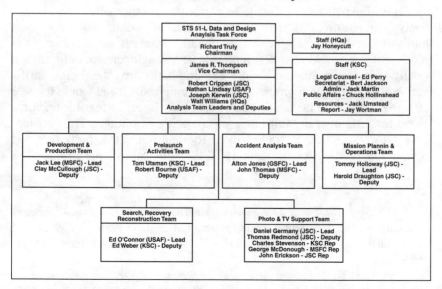

Figure 3–23. *STS 51-L Data and Design Analysis Task Force*

well as numerous corrective measures needed to be taken. Called "Lessons Learned and Collateral Findings," this report contained eight lessons learned and twenty-nine collateral findings, all addressing virtually every aspect of Shuttle planning, processing, launch, and recovery.[18] The task force also briefed members of Congress on its findings.

Actions to Implement Recommendations

After the report of the Presidential Commission was published (on June 9, 1986), President Reagan directed NASA Administrator James Fletcher on June 13 to report to him within 30 days on how and when the commission's recommendations would be implemented. The president said that "this report should include milestones by which progress in the implementation process can be measured."[19] NASA's *Report to the President: Actions to Implement the Recommendations of the Presidential Commission on the Space Shuttle Challenger Accident,* submitted to the president on July 14, 1986, responded to each of the commission's recommendations and included a key milestone schedule that illustrated the planned implementation (Figure 3–24).

The proposed actions and the steps that NASA had already taken when the report was issued follow in the narrative below. Table 3–48 presents an implementation timetable.[20]

Recommendation I
Solid Rocket Motor Design. At NASA's direction, the Marshall Space Flight Center formed a solid rocket motor joint redesign team to include participants from Marshall and other NASA centers and individuals from outside NASA.

The Marshall team evaluated several design alternatives and began analysis and testing to determine the preferred approaches that minimized hardware redesign. To ensure adequate program contingency, the redesign team would also develop, at least through concept definition, a totally new design that did not use existing hardware. The design verification and certification program would be emphasized and would include tests that duplicated the actual launch loads as closely as feasible and provided for tests over the full range of operating conditions. The verification effort included a trade study to determine the preferred test orientation (vertical or horizontal) of the full-scale motor firings. The

---

[18]*STS 51-L Data and Design Analysis Task Force, Historical Summary* (Washington, DC: U.S. Government Printing Office, June 1986), p. 3–90.

[19]Ronald Reagan, Letter to James C. Fletcher, NASA Administrator, June 13, 1986.

[20]*Report to the President: Actions to Implement the Recommendations of the Presidential Commission on the Space Shuttle Challenger Accident* (Washington, DC: U.S. Government Printing Office, July 14, 1986), Executive Summary.

SPACE TRANSPORTATION/HUMAN SPACEFLIGHT 213

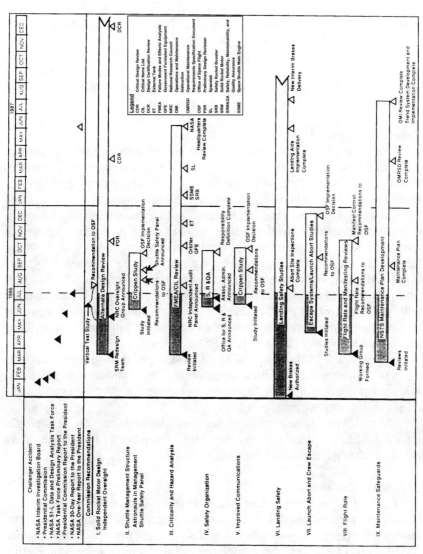

Figure 3–24. Space Shuttle Return to Flight

solid rocket motor redesign and certification schedule was under review to fully understand and plan for the implementation of the design solutions. The schedule would be reassessed after the solid rocket motor Preliminary Design Review in September 1986.

Independent Oversight. In accordance with the commission's recommendation, the National Research Council (NRC) established an Independent Oversight Group chaired by Dr. H. Guyford Stever and reporting to the NASA administrator. The NRC Independent Oversight Group was briefed on Shuttle system requirements, implementation, and control; solid rocket motor background; and candidate modifications. The group established a near-term plan, which included briefings and visits to review inflight loads, assembly processing, redesign status, and other solid rocket motor designs, including participation in the solid rocket motor Preliminary Design Review in September 1986.

Recommendation II

Shuttle Management Structure. The NASA administrator appointed General Samuel C. Phillips to study how NASA managed its programs, including relationships between various field centers and NASA Headquarters and emphasizing the Space Shuttle management structure.

On June 25, 1986, the administrator directed Astronaut Robert L. Crippen to form a fact-finding group to assess the Space Shuttle management structure. The group would report recommendations to the associate administrator for spaceflight by August 15, 1986. Specifically, this group will address the roles and responsibilities of the Space Shuttle program manager to assure that the position had the authority commensurate with its responsibilities. General Phillips and the administrator would review the results of this study with a decision on implementation of the recommendations by October 1, 1986.

Astronauts in Management. The Crippen group would also address ways to stimulate the transition of astronauts into management positions. It would also determine the appropriate position for the flight crew operations directorate within the NASA.

Shuttle Safety Panel. The associate administrator for spaceflight would establish a Shuttle Safety Panel by September 1, 1986, with direct access to the Space Shuttle program manager.

Recommendation III

Critical Item Review and Hazard Analysis. On March 13, 1986, NASA initiated a complete review of all Space Shuttle program failure modes and effects analyses and associated Critical Item Lists. Each Space Shuttle project element and associated prime contractor was conducting separate comprehensive reviews which would culminate in a program-wide review with the Space Shuttle program manager at Johnson Space Center later in 1986. Technical specialists outside the Space Shuttle program were assigned as formal members of each of these review teams. All Criticality 1 and 1R critical item waivers were canceled. The teams reassessed and resubmitted waivers in categories recommended for continued program applicability. Items which could not be revalidated would be redesigned, qualified, and certified for flight. All Criticality 2 and 3 Critical Item Lists were being reviewed for reacceptance and proper categorization. This activity would culminate in a comprehensive final review with NASA Headquarters beginning in March 1987.

As recommended by the commission, the National Research Council agreed to form an Independent Audit Panel, reporting to the NASA administrator, to verify the adequacy of this effort.

Recommendation IV

Safety Organization. The NASA administrator announced the appointment of George A. Rodney to the position of associate administrator for safety, reliability, maintainability, and quality assurance (SRM&QA) on July 8, 1986. This office would oversee the safety, reliability, and quality assurance functions related to all NASA activities and programs and the implementation system for anomaly documentation and resolution, including a trend analysis program. One of Rodney's first actions would be to assess the available resources, including the work force required to ensure adequate execution of the safety organization functions. In addition, he would assure appropriate interfaces between the functions of the new safety organization and the Shuttle Safety Panel, which would be established in response to the commission Recommendation II.

Recommendation V

Improved Communications. Astronaut Robert Crippen's team (formed as part of Recommendation II) developed plans and recommended policies for the following:

- Implementation of effective management communications at all levels
- Standardization of the imposition and removal of STS launch constraints and other operational constraints
- Conduct of Flight Readiness Review and Mission Management Team meetings, including requirements for documentation and flight crew participation

This review of effective communications would consider the activities and information flow at NASA Headquarters and the field centers that supported the Shuttle program. The study team would present findings and recommendations to the associate administrator for spaceflight by August 15, 1986.

Recommendation VI

Landing Safety. A Landing Safety Team was established to review and implement the commission's findings and recommendations on landing safety. All Shuttle hardware and systems were undergoing design reviews to ensure compliance with the specifications and safety concerns. The tires, brakes, and nose wheel steering system were included in this activity, and funding for a new carbon brakes system was approved. Ongoing runway surface tests and landing aid requirement reviews were continuing. Landing aid implementation would be complete by July 1987. The interim brake system would be delivered by August 1987.

Improved methods of local weather forecasting and weather-related support were being developed. Until the Shuttle program demonstrated satisfactory safety margins through high fidelity testing and during actual landings at Edwards Air Force Base, the Kennedy Space Center landing site would not be used for nominal end-of-mission landings.

Recommendation VII

Launch Abort and Crew Escape. On April 7, 1986, NASA initiated a Shuttle Crew Egress and Escape review. The analysis focused on egress and escape capabilities from launch through landing and would analyze concepts, feasibility assessments, cost, and schedules for pad abort, bailout, ejection systems, water landings, and powered flight separation. This review would specifically assess options for crew escape during controlled gliding flight and options for extending the intact abort flight envelope to include failure of two or three main engines during the early ascent phase.

In conjunction with this activity, NASA established a Launch Abort Reassessment Team to review all launch and launch abort rules to ensure that launch commit criteria, flight rules, range safety systems and procedures, landing aids, runway configurations and lengths, performance versus abort exposure, abort and end-of-mission landing weights, runway surfaces, and other landing-related capabilities provided the proper margin of safety to the vehicle and crew. Crew escape and launch abort studies would be complete on October 1, 1986, with an implementation decision in December 1986.

Recommendation VIII

Flight Rate. In March 1986, NASA established a Flight Rate Capability Working Group that studied:

1. The capabilities and constraints that governed the Shuttle processing flows at the Kennedy Space Center
2. The impact of flight specific crew training and software delivery/certification on flight rates

The working group would present flight rate recommendations to the Office of Space Flight by August 15, 1986. Other collateral studies in progress addressed commission recommendations related to spares provisioning, maintenance, and structural inspection. This effort would also consider the NRC independent review of flight rate, which a congressional subcommittee had requested.

The report emphasized NASA's strong support for a mixed fleet to satisfy launch requirements and actions to revitalize the United States expendable launch vehicle capabilities. Additionally, NASA Headquarters was formulating a new cargo manifest policy, which would establish manifest ground rules and impose constraints to late changes. Manifest control policy recommendations would be completed in November 1986.

Recommendation IX

Maintenance Safeguards. A Maintenance Safeguards Team was established to develop a comprehensive plan for defining and implementing actions to comply with the commission recommendations concerning maintenance activities. The team was preparing a Maintenance Plan to ensure that uniform maintenance requirements were imposed on all elements of the Space Shuttle program. The plan would also define organizational responsibilities, reporting, and control requirements for Space Shuttle maintenance activities. The Maintenance Plan would be completed by September 30, 1986.

In addition to the actions described above, a Space Shuttle Design Requirements Review Team headed by the Space Shuttle Systems Integration Office at the Johnson Space Center was reviewing all Shuttle design requirements and associated technical verification. The team focused on each Shuttle project element and on total Space Shuttle system design requirements. This activity was to culminate in a Space Shuttle Incremental Design Certification Review approximately three months before the next Space Shuttle launch.

Because of the number, complexity, and interrelationships among the many activities leading to the next flight, the Space Shuttle program manager at the Johnson Space Center initiated a series of formal Program Management Reviews for the Space Shuttle program. These reviews were to be regular face-to-face discussions among the managers of all major Space Shuttle program activities. Each meeting would focus on progress, schedules, and actions associated with each of the major program review activities and would be tailored directly to current program activity for the time period involved. The first of these meetings was held at the Marshall Space Flight Center on May 5–6, 1986, with the second at the Kennedy Space Center on June 25, 1986. Follow-on reviews will occur approximately every six weeks. Results of these reviews will be reported to the associate administrator for spaceflight and to the NASA administrator.

On June 19, 1986, the NASA administrator announced the termination of the development of the Centaur upper stage for use aboard the Space Shuttle. NASA had planned to use the Centaur upper stage for NASA planetary spacecraft launches as well as for certain national security satellite launches. Major safety reviews of the Centaur system were under way at the time of the *Challenger* accident, and these reviews were intensified to determine whether the program should be continued. NASA decided to terminate because, even with certain modifications identified by the ongoing reviews, the resultant stage would not meet safety criteria being applied to other cargo or elements of the Space Shuttle system.

Revised Manifest

On October 3, 1986, NASA Administrator James C. Fletcher announced NASA's plan to resume Space Shuttle flights on February 18, 1988. He also announced a revised manifest for the thirty-nine months following the resumption of Shuttle flights (Table 3–49). (The manifest was revised several times prior to the resumption of Shuttle flights. Most flights did not launch on the dates listed here.)

Fletcher stated that the manifest was based on a reduced flight rate goal that was "acceptable and prudent" and that complied with presidential policy that limited use of the Shuttle for commercial and foreign payloads to those that were Shuttle-unique or those with national security or foreign policy implications. Prior to the *Challenger* accident, roughly one-third of the Shuttle manifest was devoted to DOD missions, another third to scientific missions, and the remainder to commercial satellites and foreign government missions. Fletcher said that for the seven-year period following resumption of Shuttle flights (through 1994), NASA would use 40 percent of the Shuttle's capability for DOD needs, 47 percent for NASA needs, and 12 percent to accommodate commercial, foreign government, and U.S. government civil space requirements. This reflected the priorities for payload assignments with national security at the top, STS operational capability (TDRS) and dedicated science payloads next, and other science and foreign and commercial needs last. He stated that at the beginning of this seven-year period, DOD would use considerable Shuttle capability to reduce its payload backlog, but for the remaining years, DOD's use would even out at approximately one-third of Shuttle capability.

Fletcher stated that the revised manifest placed a high priority on major NASA science payloads. The Hubble Space Telescope, Ulysses, and Galileo, which had been scheduled for a 1986 launch, would be launched "as expeditiously as possible."[21]

Implementing the Commission's Recommendations

Approximately one year after NASA addressed how it would implement the recommendations of the Presidential Commission, NASA issued a report to the president that described the actions taken by NASA in response to the commission's recommendations on how to return to safe, reliable spaceflight.[22] This report and the accompanying milestone

---

[21]Statement by Dr. James C. Fletcher, Press Briefing, NASA Headquarters, October 3, 1986.

[22]*Report to the President: Implementation of the Recommendations of the Presidential Commission on the Space Shuttle Challenger Accident* (Washington, DC: U.S. Government Printing Office, June 1987).

chart (Figure 3-25) showed the significant progress NASA made in meeting its implementation milestones. The recovery activity, as described in the report, focused on three key aspects: the technical engineering changes being selected and implemented; the new procedures, safeguards, and internal communication processes that had been or were being put in place; and the changes in personnel, organizations, and attitudes that occurred.

Responding to the commission's findings as to the cause of the accident, NASA changed the design of the solid rocket motor. The new design eliminated the weakness that had led to the accident and incorporated of a number of improvements. The new rocket motors were to be tested in a series of full-scale firings before the next Shuttle flight. In addition, NASA reviewed every element of the Shuttle system and added improved hardware and software to enhance safety. Improved or modified items or systems included the landing system, the main liquid-fueled engines, and the flight and ground systems.

NASA implemented new procedures to provide independent SRM&QA functions. A completely new organization, the SRM&QA office, which reported directly to the NASA administrator, now provided independent oversight of all critical flight safety matters. The new office worked directly with the responsible program organization to solve technical problems while still retaining its separate identity as final arbiter of safety and related matters.

NASA completed personnel and organizational changes that had begun immediately after the accident. A new, streamlined management team was put in place at NASA Headquarters, with new people well down within the field centers. Special attention was given to the critical issues of management isolation and the tendency toward technical complacency, which, combined with schedule pressure, led to an erosion in flight safety. This awareness of the risk of spaceflight operations, along with NASA's responsibility to control and contain that risk without claiming its elimination, became the controlling philosophy the Space Shuttle program.

The report addressed the nine recommendations made by the Presidential Commission and other related concerns.

Recommendation I

The commission recommended that the design of the solid rocket motor be changed, that the testing of the new design reflect the operational environment, and that the National Research Council (NRC) form a committee to provide technical oversight of the redesign effort.

NASA thoroughly evaluated the solid rocket motor design. As well as the solid rocket motor field joint, this evaluation resulted in design changes to many components of the motor. The field joint was redesigned to provide high confidence in its ability to seal under all operating conditions (Figure 3-26). In addition, the redesign included a new tang capture latch that controlled movement between the tang and clevis in the joint, a third O-ring seal, insulation design improvements, and an external heater

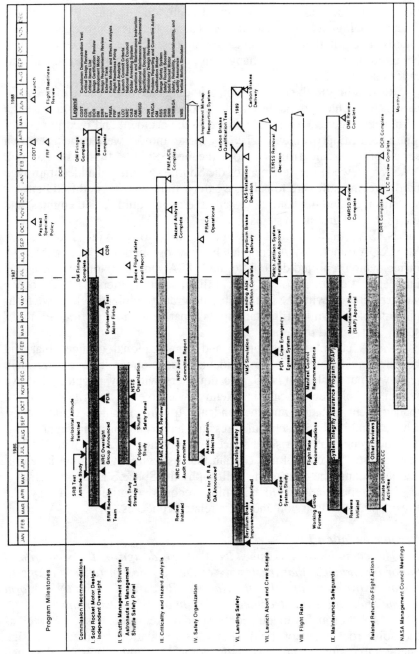

Figure 3–25. Space Shuttle Return to Flight Milestones

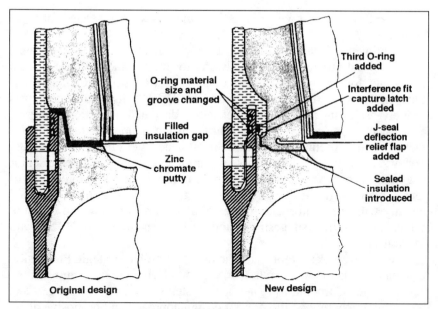

Figure 3-26. Field Joint Redesign

with integral weather seals. The nozzle-to-case joint, the case parts, insulation, and seals were redesigned to preclude seal leakage observed in prior flights. The nozzle metal parts, ablative components, and seals were redesigned to improve redundancy and to provide pressure verification of seals. Other nozzle modifications included improvements to the inlet, cowl/boot, and aft exit assemblies.

Modifications were incorporated into the igniter case chamber and into the factory joints to improve their margins of safety. The igniter case chamber wall thickness was being increased. Additional internal insulation and an external weather seal were added to the factory joint. Ground support equipment was redesigned to minimize case distortion during storage and handling, to improve case measurement and rounding techniques for assembly, and to improve leak testing capabilities.

Component laboratory tests, combined with subscale simulation tests and full-scale tests, were being conducted to meet verification requirements. Several small-scale and full-scale joint tests were successfully completed, confirming insulation designs and joint deflection analyses. One engineering test, two developmental tests, and three qualification full-scale motor test firings were to be completed before STS-26. The engineering test motor was fired on May 27, 1987, and early analysis of the data indicated that the test met its objectives.

NASA selected the horizontal attitude as the optimum position for static firing, and a second test stand, which could introduce dynamic loads at the external tank/solid rocket motor aft attach struts, was constructed. Improved nondestructive evaluation techniques were being developed, in conjunction with the Air Force, to perform ultrasonic inspection and mechanical testing of propellant and insulation bonding surfaces. Complete x-ray testing of all segments were reinstated for near-term flights.

Contingency planning included development of alternate designs, which did not utilize existing hardware, for the field and nozzle-to-case joints and for the rocket motor nozzle. An NRC Solid Rocket Motor Independent Oversight Panel, chaired by Dr. H. Guyford Stever, was actively reviewing the solid rocket motor design, verification analyses, and test planning and was participating in the major program reviews, including the preliminary requirements and the preliminary design reviews. A separate technical advisory group, consisting of twelve senior engineers from NASA and the aerospace industry and a separate group of representatives from four major solid motor manufacturers, worked directly with the solid rocket motor design team to review the redesign status and provide suggestions and recommendations to NASA and Morton Thiokol.

The solid rocket motor manufacturers—Aerojet Strategic Propulsion Company, Atlantic Research Corporation, Hercules Inc., and United Technologies Corporation (Chemical Systems Division)—were reviewing and commenting on the present design approach and proposing alternate approaches that they felt would enhance the design. As a result of these and other studies, NASA initiated a definition study for a new advanced solid rocket motor. Additional details of the redesigned solid rocket motor can be found in Chapter 2 as part of the discussion of the Shuttle's propulsion system.

Recommendations II and V

The commission recommended [II] that the Space Shuttle Program management structure be reviewed, that astronauts be encouraged to make the transition into management positions, and that a flight safety panel be established. The commission recommended [V] that the tendency for management isolation be eliminated, that a policy on launch constraints be developed, and that critical launch readiness reviews be recorded.

In March 1986, Associate Administrator for Space Flight Rear Admiral Richard Truly initiated a review of the Shuttle program management structure and communications. After the commission report was issued, he assigned Captain Robert L. Crippen responsibility for developing the response to commission recommendations II and V. This effort resulted in the establishment of a director, NSTS, reporting directly to the associate administrator for spaceflight, and other changes necessary to strengthen the Shuttle program management structure and improve lines of authority and communication (see Figure 3–1) at the beginning of this chapter. The NSTS funding process was revised, and the director, NSTS, now was given control over program funding at the centers.

Additionally, the flight readiness review and mission management team processes were strengthened. The director of flight crew operations would participate in both of these activities, and the flight crew commander, or a representative, would attend the flight readiness review. These meetings would be recorded and formal minutes published.

Since the accident, several current and former astronauts were assigned to top management positions. These included: the associate administrator for spaceflight; the associate administrator for external affairs; the acting assistant administrator, office of exploration; chief, Headquarters operational safety branch; the deputy director, NSTS operations; the Johnson Space Center deputy center director; the chairman of the Space Flight Safety Panel; and the former chief of the astronaut office as special assistant to the Johnson director for engineering, operations, and safety.

A Space Flight Safety Panel, chaired by astronaut Bryan O'Connor, was established. The panel reported to the associate administrator for SRM&QA. The panel's charter was to promote flight safety for all NASA spaceflight programs involving flight crews, including the Space Shuttle and Space Station programs.

Recommendation III

The Commission recommended that the critical items and hazard analyses be reviewed to identify items requiring improvement prior to flight to ensure safety and that the NRC verify the adequacy of this effort.

The NSTS uses failure modes and effects analyses, critical item lists, and hazard analyses as techniques to identify the potential for failure of critical flight hardware, to determine the effect of the failure on the crew, vehicle, or mission, and to ensure that the criticality of the item is reflected in the program documentation. Several reviews were initiated by program management in March 1986 to reevaluate failure analyses of critical hardware items and hazards. These reviews provided improved analyses and identified hardware designs requiring improvement prior to flight to ensure mission success and enhance flight safety.

A review of critical items, failure modes and effects analyses, and hazard analyses for all Space Shuttle systems was under way. NASA developed detailed instructions for the preparation of these items to ensure that common ground rules were applied to each project element analysis. Each NASA element project office and its prime contractor, as well as the astronaut office and mission operations directorate, were reviewing their systems to identify any areas in which the design did not meet program requirements, to verify the assigned criticality of items, to identify new items, and to update the documentation. An independent contractor was conducting a parallel review for each element. Upon completion of this effort, each element would submit those items with failure modes that could not meet full design objectives to the Program Requirements Control Board, chaired by the director, NSTS. The board would review the documentation, concur with the proposed rationale for safely accepting the item, and issue a waiver to the design requirement, if appropriate.

The NRC Committee on Shuttle Criticality Review and Hazard Analysis Audit, chaired by retired U.S. Air Force General Alton Slay, was responsible for verifying the adequacy of the proposed actions for returning the Space Shuttle to flight status. In its interim report of January 13, 1987, the committee expressed concern that critical items were not adequately prioritized to highlight items that may be most significant. NASA was implementing a critical items prioritization system for the Shuttle program to alleviate the committee's concerns.

Recommendation IV

The commission recommended that NASA establish an Office of Safety, Reliability, and Quality Assurance, reporting to the NASA administrator, with responsibility for related functions in all NASA activities and programs.

The NASA administrator established a new NASA Headquarters organization, the Office of Safety, Reliability, Maintainability, and Quality Assurance (SRM&QA), and appointed George Rodney as associate administrator. The Operational Safety Branch of that office was headed by astronaut Frederick Gregory. The new organization centralized agency policy in its areas of responsibility, provided for NASA-wide standards and procedures, and established an independent reporting line to top management for critical problem identification and analysis. The new office exercised functional management responsibility and authority over the related organizations at all NASA field centers and major contractors.

The new organization was participating in specific NSTS activities, such as the hardware redesign, failure modes and effects analysis, critical item identification, hazard analysis, risk assessment, and spaceflight system assurance. This approach allowed the NSTS program line management at Headquarters and in the field to benefit from the professional safety contributions of an independent office without interrupting the two different reporting lines to top management. Additional safeguards were added by both the line project management and the SRM&QA organization to ensure free, open, rapid communication upward and downward within all agency activities responsible for flight safety. Such robust multiple communications pathways were expected to eliminate the possibility of serious issues not rising to the attention of senior management.

Recommendation VI

The commission recommended that NASA take action to improve landing system safety margin and to determine the criteria under which planned landings at Kennedy would be acceptable.

Several orbiter landing system modifications to improve landing system safety margins would be incorporated for the first flight. These included a tire pressure monitoring system, a thick-stator beryllium brake to increase brake energy margin, a change to the flow rates in the brake hydraulic system, a stiffer main gear axle, and a balanced brake pressure application feature that would decrease brake wear upon landing and provide additional safety margin.

Several other changes were being evaluated to support longer term upgrading of the landing system. A new structural carbon brake, with increased energy capacity, was approved and would be available in 1989. A fail-operational/fail-safe nose wheel steering design, including redundant nose wheel hydraulics capability, was being reviewed by the orbiter project office for later implementation.

The initial Shuttle flights were scheduled to land at the Edwards Air Force Base complex. A total understanding of landing performance data, the successful resolution of significant landing system anomalies, and increased confidence in weather prediction capabilities were preconditions to resuming planned end-of-mission landings at the Kennedy Space Center.

Recommendation VII

The commission recommended that NASA make every effort to increase the capability for an emergency runway landing following the loss of two or three engines during early ascent and to provide a crew escape system for use during controlled gliding flight.

Launch and launch abort mode definition, flight and ground procedures, range safety, weather, flight and ground software, flight rules, and launch commit criteria were reviewed. Changes resulting from this review were being incorporated into the appropriate documentation, including ground operating procedures, and the on-board flight data file. NASA reviewed abort trajectories, vehicle performance, weather requirements, abort site locations, support software, ground and on-board procedures, and abort decision criteria to ensure that the requirements provided for maximum crew safety in the event an abort was required. The review resulted in three actions: the landing field at Ben Guerir, Morocco, was selected as an additional transatlantic abort landing site; ground rules for managing nominal and abort performance were established and the ascent data base was validated and documented; and a permanent Launch Abort Panel was established to coordinate all operational and engineering aspects of ascent-phase contingencies.

Representatives from NASA and the Air Force were reviewing the external tank range safety system. This review readdressed the issue of whether the range safety system is required to ensure propellant dispersal capability in the event of an abort during the critical first minutes of flight. The results of this analysis would be available in early 1988.

Flight rules (which define the response to specific vehicle anomalies that might occur during flight) were being reviewed and updated. The Flight Rules Document was being reformatted to include both the technical and operational rationales for each rule. Launch commit criteria (which define responses to specific vehicle and ground support system anomalies that might occur during launch countdown) were being reviewed and updated. These criteria were being modified to include the technical and operational rationale and to document any procedural workarounds that would allow the countdown to proceed in the event one of the criteria was violated.

Although a final decision to implement a Space Shuttle crew escape capability was not made, the requirements for a system to provide crew egress during controlled gliding flight were established. The requirements for safe egress of up to eight crew members were determined through a review of escape routes, time lines, escape scenarios, and proposed orbiter modifications. The options for crew egress involved manual and powered extraction techniques. Design activities and wind tunnel assessments for each were initiated. The manual egress design would ensure that the crew member did not contact the vehicle immediately after exiting the crew module. Several approaches being assessed for reducing potential contact included a deployable side hatch tunnel that provided sufficient initial velocity to prevent crew/vehicle contact and an extendable rod and/or rope that placed the crew release point in a region of safe exit (Figure 3–27). Both approaches provided for crew egress through the orbiter side hatch.

The director, NSTS, authorized the development of a rocket-powered extraction capability for use in a crew egress/escape system. Crew escape would be initiated during controlled gliding flight at an altitude of 6,096 meters and a velocity of 321.8 kilometers per hour. The system consisted of a jettisonable crew hatch (which has been approved for installation and also applied to the manual bail-out mode) and individual rockets to extract the crew from the vehicle before it reached an altitude of 3,048 meters.

Ground egress procedures and support systems were being reviewed to determine their capability to ensure safe emergency evacuation from the orbiter at the pad or following a non-nominal landing. An egress slide,

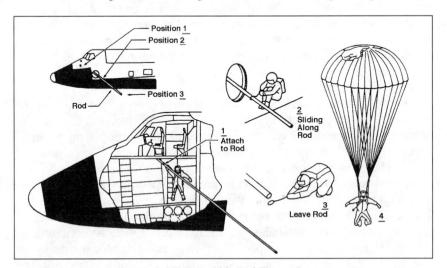

Figure 3–27. Extendible Rod Escape System
(In this system, the crew module hatch would be jettisoned and the rod would be extended through the hatch opening. The crew member would attach a lanyard to the rod, exit the vehicle in a tucked position, release at the end of the rod, and parachute to a ground or water landing.)

similar to that used on commercial aircraft, was being designed for use should an emergency escape be required after a runway landing. A study was initiated to evaluate future escape systems that would potentially expand the crew survival potential to include first-stage (solid rocket boosters thrusting) flight.

Recommendation VIII

The commission recommended that the nation not rely on a single launch vehicle capability for the future and that NASA establish a flight rate that is consistent with its resources.

Several major actions reduced the overall requirements for NSTS launches and provided for a mixed fleet of expendable launch vehicles and the Space Shuttle to ensure that the nation did not rely on a single launch vehicle for access to space. NASA and DOD worked together to identify DOD payloads for launch on expendable launch vehicles and to replan the overall launch strategy to reflect their launches on expendable launch vehicles. The presidential decision to limit the use of the NSTS for the launch of communications satellites to those with national security or foreign policy implications resulted in many commercial communications satellites, previously scheduled for launch on the NSTS, being reassigned to commercial expendable launch vehicles.

In March 1986, Admiral Truly directed that a "bottoms-up" Shuttle flight rate capability assessment be conducted. NASA established a flight rate capability working group with representatives from each Shuttle program element that affects flight rate. The working group developed ground rules to ensure that projected flight rates were realistic. These ground rules addressed such items as overall staffing of the work force, work shifts, overtime, crew training, and maintenance requirements for the orbiter, main engine, solid rocket motor, and other critical systems. The group identified enhancements required in the Shuttle mission simulator, the Orbiter Processing Facility, the Mission Control Center, and other areas, such as training aircraft and provisioning of spares. With these enhancements and the replacement orbiter, NASA projected a maximum flight rate capability of fourteen per year with four orbiters. This capacity, considering lead time constraints, "learning curves," and budget limitations, could be achieved no earlier than 1994 (Figure 3–28).

Controls were implemented to ensure that the Shuttle program elements were protected from pressures resulting from late manifest changes. While the manifest projects the payload assignments several years into the future, missions within eighteen months of launch were placed under the control of a formal change process controlled by the director, NSTS. Any manifest change not consistent with the defined capabilities of the Shuttle system would result in the rescheduling of the payload to another mission.

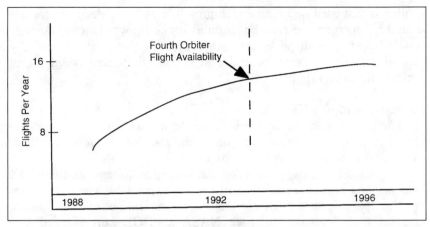

Figure 3–28. Availability of Fourth Orbiter
(With a fourth orbiter available, fourteen flights per year would be possible in 1994.)

Recommendation IX

The commission recommended that NASA develop and execute a maintenance inspection plan, perform structural inspections when scheduled, and restore the maintenance and spare parts program.

NASA updated the overall maintenance and flight readiness philosophy of the NSTS program to ensure that it was a rigorous and prominent part of the safety-of-flight process. A System Integrity Assurance Program was developed that encompassed the overall maintenance strategy, procedures, and test requirements for each element of flight hardware and software to ensure that each item was properly maintained and tested and was ready for launch. Figure 3–29 reflects the major capabilities of the System Integrity Assurance Program.

NASA alleviated the requirement for the routine removal of parts from one vehicle to supply another by expanding and accelerating various aspects of the NSTS logistics program. Procedures were being instituted to ensure that a sufficient rationale supported any future requirement for such removal of parts and that a decision to remove them underwent a formal review and approval process.

A vehicle checkout philosophy was defined that ensured that systems remain within performance limits and that their design redundancy features functioned properly before each launch. Requirements were established for identifying critical hardware items in the Operational Maintenance Requirements Specification Document (defines the work to be performed on the vehicle during each turnaround flow) and the Operations and Maintenance Instruction (lists procedures used in performing the work).

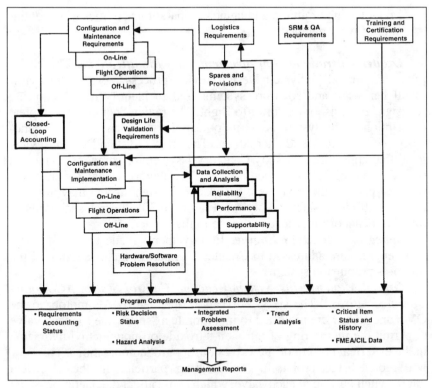

*Figure 3–29. System Integrity Assurance Program
(This program established the functional responsibilities and program requirements necessary to provide the proper configuration, operations, inspection, maintenance, logistics, and certified personnel to ensure that the NSTS was ready for flight.)*

Related Return-to-Flight Actions

At the time of the Rogers Commission report, NASA was engaged in several tasks in support of the return-to-flight activities that were not directly related to commission recommendations:

- A new launch target date and flight crew for the first flight were identified.
- The program requirements for flight and ground system hardware and software were being updated to provide a clear definition of the criteria that the project element designs must satisfy.
- The NSTS system designs were reviewed, and items requiring modification prior to flight were identified.
- Existing and modified hardware and software designs were being verified to ensure that they complied with the design requirements.
- The program and project documents, which implemented the redefined program requirements, were being reviewed and updated.

- Major testing, training, and launch preparation activities were continuing or were planned.

Orbiter Operational Improvements and Modifications. The NSTS program initiated the System Design Review process to ensure the review of all hardware and software systems and to identify items requiring redesign, analysis, or test prior to flight. The review included a complete description of the system issue, its potential consequences, recommended correction action, and alternatives. The orbiter System Design Review identified approximately sixty Category 1 system or component changes out of a total of 226 identified changes.[23] (Category 1 changes are those required prior to the next flight because the current design may not contain a sufficient safety margin.) Figure 3–30 illustrates the major improvements or modifications made to the orbiter.

Space Shuttle Main Engine. Improvements made to the Shuttle's main engines are addressed in Chapter 2 as part of the discussion of the Shuttle's propulsion system.

Orbital Maneuvering System/Reaction Control System AC-Motor-Operated Valves.[24] The sixty-four valves operated by AC motors in the OMS and RCS were modified to incorporate a "sniff" line for each valve to permit the monitoring of nitrogen tetroxide or monomethyl hydrazine in the electrical portion of the valves during ground operations. This new line reduced the probability of floating particles in the electrical microswitch portion of each valve, which could affect the operation of the

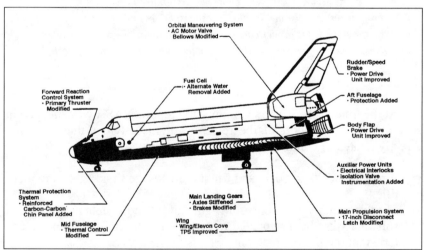

Figure 3–30. Major Orbiter Modifications

---

[23]*Aeronautics and Space Report of the President, 1988* (Washington, DC: U.S. Government Printing Office, 1989) p. 24.

[24]The information regarding additional changes presented from this point onward came from the *NSTS Shuttle Reference Manual* (1988), on-line from the Kennedy Space Center Home Page.

microswitch position indicators for on-board displays and telemetry. It also reduces the probability of nitrogen tetroxide or monomethyl hydrazine leakage into the bellows of each AC-motor-operated valve.

Primary RCS Modifications. The wiring of the fuel and oxidizer injector solenoid valves was wrapped around each of the thirty-eight primary RCS thrust chambers to remove electrical power from these valves in the event of a primary RCS thruster instability.

Fuel Cell Modifications. Modifications to the fuel cell included the deletion of end-cell heaters on each fuel cell power plant because of potential electrical failures and replacement with Freon coolant loop passages to maintain uniform temperature throughout the power plants; the improvement of the hydrogen pump and water separator of each fuel cell power plant to minimize excessive hydrogen gas entrained in the power plant product water; the addition of a current measurement detector to monitor the hydrogen pump of each fuel cell power plant and provide an early indication of hydrogen pump overload; the modification of the starting and sustaining heater system for each fuel cell power plant to prevent overheating and the loss of heater elements; and the addition of a stack inlet temperature measurement to each fuel cell power plant for full visibility of thermal conditions. Other improvements included the modification of the product water lines from all three fuel cell power plants to incorporate a parallel (redundant) path of product water to the Environmental Control and Life Support System's potable water tank B in the event of a freeze-up in the single water relief panel and the addition of a water purity sensor (pH) at the common product water outlet of the water relief panel to provide a redundant measurement of water purity.

Auxiliary Power Unit Modifications. The auxiliary power units that were used to date had a limited life. Each unit was refurbished after twenty-five hours of operation because of cracks in the turbine housing, degradation of the gas generator catalyst (which varied up to approximately thirty hours of operation), and operation of the gas generator valve module (which also varied up to approximately thirty hours of operation). The remaining parts of the auxiliary power unit were qualified for forty hours of operation.

Improved auxiliary power units were scheduled for delivery in late 1988. A new turbine housing would increase the life of the housing to seventy-five hours of operation (fifty missions); a new gas generator increased its life to seventy-five hours; a new standoff design of the gas generator valve module and fuel pump deleted the requirement for a water spray system that was required previously for each auxiliary power unit upon shutdown after the first OMS thrusting period or orbital checkout; and the addition of a third seal in the middle of the two existing seals for the shaft of the fuel pump/lube oil system (previously only two seals were located on the shaft, one on the fuel pump side and one on the gearbox lube oil side) reduced the probability of hydrazine leaking into the lube oil system. The deletion of the water spray system for the gas generator valve module and fuel pump for each auxiliary power unit resulted in a weight reduction of approximately sixty-eight kilograms for each orbiter. Upon the delivery of the improved units, the life-limited auxiliary power units would be refurbished to the upgraded design.

Main Landing Gear. The following modifications were made to improve the performance of the main landing gear elements:

1. An increase in the thickness of the main landing gear axle to provide a stiffer configuration that reduces brake-to-axle deflections, precludes brake damage experienced in previous landings, and minimizes tire wear
2. The addition of orifices to hydraulic passages in the brake's piston housing to prevent pressure surges and brake damage caused by a wobble/pump effect
3. The modification of the electronic brake control boxes to balance hydraulic pressure between adjacent brakes and equalize energy applications, with the removal of the anti-skid circuitry previously used to reduce brake pressure to the opposite wheel if a flat tire was detected
4. The replacement of the carbon-lined beryllium stator discs in each main landing gear brake with thicker discs to increase braking energy significantly
5. A long-term structural carbon brake program to replace the carbon-lined beryllium stator discs with a carbon configuration that provides higher braking capacity by increasing maximum energy absorption
6. The addition of strain gauges to each nose and main landing gear wheel to monitor tire pressure before launch, deorbit, and landing
7. Other studies involving arresting barriers at the end of landing site runways (except lake bed runways), the installation of a skid on the landing gear that could preclude the potential for a second blown tire on the same gear after the first tire has blown, the provision of "roll on rim" for a predictable roll if both tires are lost on a single or multiple gear, and the addition of a drag chute

Studies of landing gear tire improvements were conducted to determine how best to decrease tire wear observed after previous Kennedy Space Center landings and how to improve crosswind landing capability. Modifications were made to the Kennedy Space Center's Shuttle landing facility runway. The primary purpose of the modifications was to enhance safety by reducing tire wear during landing.

Nose Wheel Steering Modifications. The nose wheel steering system was modified on *Columbia* (OV-102) for the 61-C mission, and *Discovery* (OV-103) and *Atlantis* (OV-104) were being similarly modified before their return to flight. The modification allowed for a safe high-speed engagement of the nose wheel steering system and provided positive lateral directional control of the orbiter during rollout in the presence of high crosswinds and blown tires.

Thermal Protection System Modifications. The area aft of the reinforced carbon-carbon nose cap to the nose landing gear doors were damaged (tile slumping) during flight operations from impact during ascent and overheating during reentry. This area, which previously was covered with high-temperature reusable surface insulation tiles, would now be covered with reinforced carbon-carbon. The low-temperature thermal protection system tiles on *Columbia*'s mid-body, payload bay doors, and vertical tail were replaced with advanced flexible reusable surface insulation blankets. Because of evidence of plasma flow on the lower wing trailing edge and elevon landing edge tiles (wing/elevon cove) at the outboard elevon tip and inboard elevon, the low-temperature tiles were being replaced with fibrous refractory composite insulation and high-temperature tiles along with gap fillers on *Discovery* and *Atlantis.* On *Columbia,* only gap fillers were installed in this area.

Wing Modification. Before the wings for *Discovery* and *Atlantis* were manufactured, NASA instituted a weight reduction program that resulted in a redesign of certain areas of the wing structure. An assessment of wing air loads from actual flight data indicated greater loads on the wing structure than predicted. To maintain positive margins of safety during ascent, structural modifications were made.

Mid-Fuselage Modifications. Because of additional detailed analysis of actual flight data concerning descent-stress thermal-gradient loads, torsional straps were added to tie all the lower mid-fuselage stringers in bays 1 through 11 together in a manner similar to a box section. This eliminated rotational (torsional) capabilities to provide positive margins of safety. Also, because of the detailed analysis of actual descent flight data, room-temperature vulcanizing silicone rubber material was bonded to the lower mid-fuselage from bays 4 through 11 to act as a heat sink, distributing temperatures evenly across the bottom of the mid-fuselage, reducing thermal gradients, and ensuring positive margins of safety.

General Purpose Computers. NASA was to replace the existing general purpose computers aboard the Space Shuttle orbiters with new upgraded general purpose computers in late 1988 or early 1989. The upgraded computers allowed NASA to incorporate more capabilities into the orbiters and apply advanced computer technologies that were not available when the orbiter was first designed. The upgraded general purpose computers would provide two and a half times the existing memory capacity and up to three times the existing processor speed, with minimum impact on flight software. They would be half the size, weigh approximately half as much, and require less power to operate.

Inertial Measurement Unit Modifications. The new high-accuracy inertial navigation system were to be phased in to augment the KT-70 inertial measurement units in 1988–89. These new inertial measurement units would result in lower program costs over the next decade, ongoing production support, improved performance, lower failure rates, and reduced size and weight. The HAINS inertial measurement units also would contain an internal dedicated microprocessor with memory for processing and storing compensation and scale factor data from the vendor's calibration, thereby reducing the need for extensive initial load data for the orbiter's computers.

Crew Escape System. Hardware changes were made to the orbiter and to the software system to accommodate the crew escape system addressed in Recommendation VII.

Seventeen-Inch Orbiter/External Tank Disconnects. Each mated pair of seventeen-inch disconnects contained two flapper valves: one on the orbiter side and one on the external tank side. Both valves in each disconnect pair were opened to permit propellant flow between the orbiter and the external tank. Prior to separation from the external tank, both valves in each mated pair of disconnects were commanded closed by pneumatic (helium) pressure from the main propulsion system. The closure of both valves in each disconnect pair prevented propellant discharge from the external tank or orbiter at external tank separation. Valve closure on the orbiter side of each disconnect also prevented contamination of the orbiter main propulsion system during landing and ground operations.

Inadvertent closure of either valve in a seventeen-inch disconnect during main engine thrusting would stop propellant flow from the external tank to all three main engines. Catastrophic failure of the main engines and external tank feed lines would result. To prevent the inadvertent closure of the seventeen-inch disconnect valves during the Space Shuttle main engine thrusting period, a latch mechanism was added in each orbiter half of the disconnect. The latch mechanism provided a mechanical backup to the normal fluid-induced-open forces. The latch was mounted on a shaft in the flow stream so that it overlapped both flappers and obstructed closure for any reason.

In preparation for external tank separation, both valves in each seventeen-inch disconnect were commanded closed. Pneumatic pressure from the main propulsion system caused the latch actuator to rotate the shaft in each orbiter seventeen-inch disconnect ninety degrees, thus freeing the flapper valves to close as required for external tank separation. A backup mechanical separation capability was provided in case a latch pneumatic actuator malfunctioned. When the orbiter umbilical initially moved away from the external tank umbilical, the mechanical latch disengaged from the external tank flapper valve and permitted the orbiter disconnect flapper to toggle the latch. This action permitted both flappers to close.

Changes made to the Space Shuttle main engines as part of the Margin Improvement Program and solid rocket motor redesign were addressed in Chapter 2 as part of the discussion of launch systems.

Return to Flight

Preparation for STS-26

NASA selected *Discovery* as the Space Shuttle for the STS-26 mission in 1986. At the time of the STS 51-L accident, *Discovery* was in temporary storage in the Kennedy Space Center's Vehicle Assembly Building, awaiting transfer to the Orbiter Processing Facility for preparation for the first Shuttle flight from Vandenberg Air Force Base, California, scheduled for later that year. *Discovery* last flew in August 1985 on STS 51-I, the orbiter's sixth flight since it joined the fleet in November 1983.

In January 1986, *Atlantis* was in the Orbiter Processing Facility, prepared for the Galileo mission and ready to be mated to the boosters and tank in the Vehicle Assembly Building. *Columbia* had just completed the STS 61-C mission a few weeks prior to the *Challenger* accident and was also in the Orbiter Processing Facility undergoing postflight deconfiguration.

NASA was considering various Shuttle manifest options, and it was determined that *Atlantis* would be rolled out to Launch Pad 39-B for fit checks of new weather protection modifications and for an emergency egress exercise and a countdown demonstration test. During that year, NASA also decided that *Columbia* would be flown to Vandenberg for fit checks. *Discovery* was then selected for the STS-26 mission.

Discovery was moved from the Vehicle Assembly Building High Bay 2, where it was in temporary storage, into the Orbiter Processing Facility the last week of June 1986. Power-up modifications were active on the orbiter's systems until mid-September 1986, when *Discovery* was transferred to the Vehicle Assembly Building while technicians performed facility modifications in Bay 1 of Orbiter Processing Facility.

Discovery was moved back into the Orbiter Processing Facility's Bay 1 on October 30, 1987, a milestone that initiated an extensive modification and processing flow to ready the vehicle for flight. The hiatus in launching offered an opportunity to "tune up" and fully check out all of the orbiter's systems and treat the orbiter as if it was a new vehicle. Technicians removed most of the orbiter's major systems and components and sent them to the respective vendors for modifications or rebuilding.

After an extensive powered-down period of six months, which began in February 1987, *Discovery*'s systems were awakened when power surged through its electrical systems on August 3, 1987. *Discovery* remained in the Orbiter Processing Facility while workers implemented more than 200 modifications and outfitted the payload bay for the TDRS. Flight processing began in mid-September with the reinstallation and checkout of the major components of the vehicle, including the main engines, the right- and lefthand OMS pods, and the forward RCS.

In January 1988, *Discovery*'s three main engines arrived at the Kennedy Space Center and were installed. Engine 2019 arrived on January 6, 1988, and was installed in the number one position on January 10. Engine 2022 arrived on January 15 and was installed in the number two position on January 24. Engine 2028 arrived on January 21 and was installed in the number three position also on January 24.

The redesigned solid rocket motor segments began arriving at Kennedy on March 1, and the first segment, the left aft booster, was stacked on Mobile Launcher 2 in the Vehicle Assembly Building's High

Bay 3 on March 29. Technicians started with the left aft booster and continued stacking the four lefthand segments before beginning the righthand segments on May 5. They attached the forward assemblies/nose cones on May 27 and 28. The solid rocket boosters' field joints were closed out prior to mating the external tank to the boosters on June 10. An interface test between the boosters and tank was conducted a few days later to verify the connections.

The OASIS payload was installed in *Discovery*'s payload bay on April 19. TDRS arrived at the Orbiter Processing Facility on May 16, and its inertial upper stage arrived on May 24. The TDRS/inertial upper stage mechanical mating took place on May 31. *Discovery* was moved from the Orbiter Processing Facility to the Vehicle Assembly Building on June 21, where it was mated to the external tank and solid rocket boosters. A Shuttle interface test conducted shortly after the mate checked out the mechanical and electrical connections among the various elements of the Shuttle vehicle and the function of the on-board flight systems.

The assembled Space Shuttle vehicle aboard its mobile launcher platform was rolled out of the Vehicle Assembly Building on July 4. It traveled just over four miles to Launch Pad 39-B for a few major tests and final launch preparations.

A few days after *Discovery*'s OMS system pods were loaded with hypergolic propellants, a tiny leak was detected in the left pod (June 14). Through the use of a small, snake-like, fiber optics television camera, called a Cobra borescope, workers pinpointed the leak to a dynatube fitting in the vent line for the RCS nitrogen tetroxide storage tank, located in the top of the OMS pod. The tiny leak was stabilized and controlled by "pulse-purging" the tank with helium—an inert gas. Pulse-purge is an automated method of maintaining a certain amount of helium in the tank. In addition, console operators in the Launch Control Center firing room monitored the tank for any change that may have required immediate attention. It was determined that the leak would not affect the scheduled Wet Countdown Demonstration Test and the Flight Readiness Firing, and repair was delayed until after these tests.

The Wet Countdown Demonstration Test, in which the external tank was loaded with liquid oxygen and liquid hydrogen, was conducted on August 1. A few problems with ground support equipment resulted in unplanned holds during the course of the countdown. A leak in the hydrogen umbilical connection at the Shuttle tail service mast developed while liquid hydrogen was being loaded into the external tank. Engineers traced the leak to a pressure monitoring connector. During the Wet Countdown Demonstration Test, the leak developed again. The test was completed with the liquid hydrogen tank partially full, and the special tanking tests were deleted. Seals in the eight-inch fill line in the tail service mast were replaced and leak-checked prior to the Flight Readiness Firing. In addition, the loading pumps in the liquid oxygen storage farm were not functioning properly. The pumps and their associated motors were repaired.

After an aborted first attempt, the twenty-two-second Flight Readiness Firing of *Discovery*'s main engines took place on August 10. The first Flight Readiness Firing attempt was halted inside the T-ten-second mark because of a sluggish fuel bleed valve on the number two main engine. Technicians replaced this valve prior to the Flight Readiness Firing. This firing verified that the entire Shuttle system, including launch equipment, flight hardware, and the launch team, were ready for flight. With more than 700 pieces of instrumentation installed on the vehicle elements and launch pad, the test provided engineers with valuable data, including characteristics of the redesigned solid rocket boosters.

After the test, a team of Rockwell technicians began repairs to the OMS pod leak. They cut four holes into two bulkheads with an air-powered router on August 17 and bolted a metal "clamshell" device around the leaking dynatube fitting. The clamshell was filled with Furmanite—a dark thick material consisting of graphite, silicon, heavy grease, and glass fiber. After performing a successful initial leak check, covers were bolted over the holes on August 19, and the tank was pressurized to monitor any decay. No leakage or decay in pressure was noted, and the fix was deemed a success.

TDRS-C and its inertial upper stage were transferred from the Orbiter Processing Facility to Launch Pad 39-B on August 15. The payload was installed into *Discovery*'s payload bay on August 29. Then a Countdown Demonstration Test was conducted on September 8. Other launch preparations held prior to launch countdown included final vehicle ordinance activities, such as power-on stray-voltage checks and resistance checks of firing circuits, the loading of the fuel cell storage tanks, the pressurization of the hypergolic propellant tanks aboard the vehicle, final payload closeouts, and a final functional check of the range safety and solid rocket booster ignition, safe, and arm devices.

STS-26 Mission Overview

The Space Shuttle program returned to flight with the successful launch of *Discovery* on September 29, 1988. The Shuttle successfully deployed the TDRS, a 2,225-kilogram communications satellite attached to a 14,943-kilogram rocket. In addition, eleven scheduled scientific and technological experiments were carried out during the flight.

The STS-26 crew consisted of only experienced astronauts. Twenty months of preflight training emphasized crew safety. The crew members prepared for every conceivable mishap or malfunction.

Among the changes made in the Shuttle orbiter was a crew escape system for use if an engine should malfunction during ascent to orbit or if a controlled landing was risky or impossible. As part of this escape system, the crew wore newly designed partially pressurized flight suits during ascent, reentry, and landing. Each suit contained oxygen supplies, a parachute, a raft, and other survival equipment. The new escape system would permit astronauts to bail out of the spacecraft in an emergency during certain segments of their ascent toward orbit. To escape, the astronauts would blow off a hatch in the spacecraft cabin wall, extend a telescoping pole 3.65 meters beyond the spacecraft, and slide along the pole. From the pole, they would parachute to Earth.

The improved main engines were test-fired for a total of 100,000 seconds, which is equal to their use time in sixty-five Shuttle launches. The solid rocket boosters were tested with fourteen different flaws deliberately etched into critical components.

The launch was delayed for one hour and thirty-eight minutes because of unsuitable weather conditions in the upper atmosphere. Winds at altitudes between 9,144 and 12,192 meters were lighter than usual for that time of the year, and launch was prohibited because this condition had not been programmed into the spacecraft's computer. However, after specialists analyzed the situation, they judged that *Discovery* could withstand these upper-air conditions. Shuttle managers approved a waiver of the established flight rule and allowed the launch to proceed under the existing light wind conditions.

Upon the conclusion of the mission, *Discovery* began its return to Earth at 11:35 a.m., Eastern Daylight Time, on October 3. *Discovery* was traveling at about twenty-five times the speed of sound over the Indian Ocean when the astronauts fired the deorbit engines and started the hour-long descent. Touchdown was on a dry lake bed at Edwards Air Force Base.

Space Station

Overview and Background

The notion of a space station was not new or revolutionary when, in his State of the Union message of January 25, 1984, President Ronald Reagan directed NASA to develop a permanently occupied space station within the next ten years. Even before the idea of a Space Shuttle had been conceived in the late 1960s, NASA had envisioned a space station as a way to support high-priority science missions. Once the Shuttle's development was under way, a space station was considered as its natural complement—a destination for the orbiter and a base for its trip back to Earth. By 1984, NASA had already conducted preliminary planning efforts that sought the best space station concept to satisfy the requirements of potential users.

Reagan's space station directive underscored a national commitment to maintaining U.S. leadership in space. A space station would, NASA claimed, stimulate technology resulting in "spinoffs" that would improve the quality of life, create jobs, and maintain the U.S. skilled industrial base. It would improve the nation's competitive stance at a time when more and more high-technology products were being purchased in other countries. It offered the opportunity to add significantly to knowledge of

Earth and the universe.[25] The president followed up his directive with a request for $150 million for space station efforts in FY 1985. Congress approved this request and added $5.5 million in earlier year appropriations to total $155.5 million for the space station in FY 1985.[26]

From its start, international participation was a major objective of the Space Station program. Other governments would conduct their own definition and preliminary design programs in parallel with NASA and would provide funding. NASA anticipated international station partners who defined missions and used station capabilities, participated in the definition and development activities and who contributed to the station capabilities, and supported the operational activities of the station.

Events moved ahead, and on September 14, 1984, NASA issued a request for proposal (RFP) to U.S. industry for the station's preliminary design and definition. The RFP solicited proposals for four separate "work packages" that covered the definition and preliminary design of station elements:

1. Pressurized "common" modules with appropriate systems for use as laboratories, living areas, and logistics transport; environmental control and propulsive systems; plans for equipping one module as a laboratory and others as logistics modules; and plans for accommodations for orbital maneuvering and orbital transfer systems
2. The structural framework to which the various elements of the station would be attached; interface between the station and the Space Shuttle; mechanisms such as the RMS and attitude control, thermal control, communications, and data management systems; plans for equipping a module with sleeping quarters, wardroom, and galley; and plans for EVA
3. Automated free-flying platforms and provisions to service and repair the platforms and other free-flying spacecraft; provisions for instruments and payloads to be attached externally to the station; and plans for equipping a module for a laboratory
4. Electrical power generation, conducting, and storage systems.[27]

Proposals from industry were received in November 1984. Also in 1984, NASA designated the Johnson Space Center as the lead center for the Space Station program. In addition, NASA established seven inter-

---

[25]"Space Station," NASA Information Summaries, December 1986, p. 2.

[26]U.S. Congress, Conference Report, June 16, 1984, Chronological History, Fiscal Year 1985 Budget Submission authorized the initial $150 million. The Conference Committee authorized the additional $5 million from fiscal year 1984 appropriations as part of a supplemental appropriations bill, approved August 15, 1985.

[27]Space Station Definition and Preliminary Design, Request for Proposal, September 15, 1984.

center teams to conduct advanced development activities for high-potential technologies to be used in station design and development, and the agency assigned definition and preliminary design responsibilities to four field centers: the Marshall Space Flight Center, Johnson, the Goddard Space Flight Center, and the Lewis Research Center. The agency also established a Headquarters-based Space Station Program Office to provide overall policy and program direction.

Response to proposals for a space station was not uniformly favorable. In particular, the *New York Times* criticized the usefulness of the project. It called the proposed space station "an expensive yawn in space" (January 29, 1984) and "the ultimate junket" (November 9, 1984).[28] The *Times* claimed that unoccupied space platforms could accomplish anything that an occupied space platform could. Nevertheless, Reagan remained an enthusiastic proponent of the project, and NASA moved ahead.

NASA defined three categories of missions as the basis for space station design. Science and applications missions included astrophysics, Earth science and applications, solar system exploration, life sciences, materials science, and communications. Commercial missions included materials processing in space, Earth and ocean observations, communications, and industrial services. Technology development missions included materials and structures, energy conversion, computer science and electronics, propulsion, controls and human factors, station systems/operations, fluid and thermal physics, and automation and robotics.

NASA's 1984 plans called for the station to be operational in the early 1990s, with an original estimated U.S. investment of $8.0 billion (1984 dollars).[29] The station would be capable of growth both in size and capability and was intended to operate for several decades. It would be assembled at an altitude of about 500 kilometers at an inclination to the equator of twenty-eight and a half degrees. All elements of the station would be launched and tended by the Space Shuttle.[30]

On April 19, 1985, NASA's Space Station Program Office Manager Neil Hutchinson authorized the start of the definition phase contracts. Marshall, Johnson, Goddard, and Lewis each awarded competitive contracts on one of four work packages to eight industry teams (Table 3–50). These contracts extended for twenty-one months and defined the system requirements, developed supporting technologies and technology development plans, performed supporting systems and trade studies, developed preliminary designs and defined system interfaces, and developed plans, cost estimates, and schedules for the Phase C/D (design and development)

---

[28] *New York Times,* January 29, 1984; *New York Times,* November 9, 1984.

[29] Philip E. Culbertson, "Space Station: A Cooperative Endeavor," paper to 25th International Meeting on Space, Rome, Italy, March 26–28, 1985, p. 4, NASA Historical Reference Collection, NASA Headquarters, Washington, DC.

[30] Leonard David, *Space Station Freedom—A Foothold on the Future,* NASA pamphlet, Office of Space Science, 1986.

activities. In addition to the lead centers for each work package, the Kennedy Space Center was responsible for preflight and launch operations and would participate in logistics support activities. Other NASA centers would also support the definition and preliminary design activities.

Also during 1985, NASA signed memoranda of understanding (MOUs) with Canada, ESA, and Japan. The agreements provided a framework for cooperation during the definition and preliminary design phase (Phase B) of the program. Under the MOUs, the United States and its international partners would conduct and coordinate simultaneous Phase B studies. NASA also signed an MOU with Space Industries, Inc., of Houston, a privately funded venture to exchange information during Phase B. Space Industries planned to develop a pressurized laboratory that would be launched by the Space Shuttle and could be serviced from the station.

Progress on the station continued through 1986.[31] NASA issued a Technical and Management Information System (TMIS) RFP in July. The TMIS would be a computer-based system that would support the technical and management functions of the overall Space Station program. NASA also issued a Software Support Environment RFP for the "environment" that would be used for all computer software developed for the program. A draft RFP for the station's development phase (Phase C/D) was also issued in November 1986, with the definitive RFP released on April 24, 1987.[32]

In 1987, in accordance with a requirement in the Authorization Act for FY 1988, NASA began preparing a total cost plan spanning three years. Called the Capital Development Plan, it included the estimated cost of all direct research and development, spaceflight, control and data communications, construction of facilities, and resource and program management. This plan complemented the Space Station Development Plan submitted to Congress in November 1987.

Also during 1987, NASA awarded several station development contracts:

1. Boeing Computer Services Company was selected in May to develop the TMIS.
2. Lockheed Missiles and Space Company was chosen in June to develop the Software Support Environment contract.

---

[31]It is interesting to note that by 1986, the Soviet Union had already operated several versions of a space station. In February 1986, it placed into orbit a new space station called *Mir*, the Russian word for peace. The Soviets indicated they intended to occupy *Mir* permanently and make it the core of a busy complex of space-based factories, construction and repair facilities, and laboratories.

[32]"NASA Issues Requests for Proposals for Space Station Development," *NASA News,* Release 87-65, April 24, 1987.

3. Grumman Aerospace Corporation was picked in July to provide the Space Station Program Office with systems engineering and integration, in addition to a broad base of management support.

In addition, Grumman and Martin Marietta Astronautics Company were selected in November for definition and preliminary design of the Flight Telerobotic System, a space robot that would perform station assembly and spacecraft servicing tasks.

In December 1987, NASA selected the four work package contractors. These four aerospace firms were to design and build the orbital research base. Boeing Aerospace was selected to build the pressurized modules where the crews would work and live (Work Package 1). NASA chose McDonnell Douglas Astronautics Company to develop the structural framework for the station, as well as most of the major subsystems required to operate the facility (Work Package 2). GE Astro-Space Division was picked to develop the scientific platform that would operate above Earth's poles and the mounting points for instruments placed on the occupied base (Work Package 3). NASA selected the Rocketdyne Division of Rockwell International to develop the system that would furnish and distribute electricity throughout the station (Work Package 4).

The contracts included two program phases. Phase I covered the approximately ten-year period from contract start through one year after completion of station assembly. Phase II was a priced option that, if exercised, would enhance the capabilities of the station by adding an upper and lower truss structure, additional external payload attachment points, a solar dynamic power system, a free-flying co-orbiting platform, and a servicing facility. Contract negotiations with Boeing, McDonnell Douglas, GE Astro-Space, and Rocketdyne to design and build *Freedom*'s occupied base and polar platform were completed in September 1988. With these contracts in place, the definition and preliminary design (Phase B) ended and detailed design and development (Phase C/D) began. The award of these contracts followed approval by Congress and President Reagan of the overall federal funding bill that made available more than $500 million in FY 1988 for station development activities. This amount included funds remaining from the FY 1987 station appropriation as well as the new funding provided under the FY 1988 bill.[33]

In February 1988, the associate administrator for space station signed the Program Requirements Document. This top-level document contained requirements for station design, assembly, utilization, schedule, safety, evolution, management, and cost. In May, the Program Requirements Review began at the NASA Headquarters program office and was completed at the four work package centers by the end of the year. The Program Requirements Review provided a foundation to begin the

---

[33] "NASA Awards Contracts to Space Station Contractors," *NASA News*, Release 87-187, December 23, 1987.

detailed design and development process by verifying program requirements and ensuring that those requirements could be traced across all levels of the program and be met within the available technical and fiscal resources.

In July 1988, President Reagan named the international station *Freedom*. The U.S. international partners signed agreements to cooperate with the United States in developing, using, and operating the station. Government-level agreements between the United States and nine European nations, Japan, and Canada, and MOUs between NASA and ESA and between NASA and Canada were signed in September. The NASA-industry team proceeded to develop detailed requirements to guide design work beginning early in 1989.

Proposed Configurations

For the purpose of the 1984 RFP, NASA selected the "power tower" as the reference configuration for the station. NASA anticipated that this configuration could evolve over time. The power tower would consist of a girder 136 meters in length that would circle Earth in a gravity-gradient attitude. Pressurized laboratory modules, service sheds, and docking ports would be placed on the end always pointing downward; instruments for celestial observation would be mounted skyward; and the solar power arrays would be mounted on a perpendicular boom halfway up the tower.

After intensive reviews, NASA replaced the power tower configuration in 1985 with the "dual keel" configuration (Figure 3–31). This configuration featured two parallel 22.6-meter vertical keels, crossed by a single horizontal beam, which supported the solar-powered energy system by a double truss, rectangular-shaped arrangement that shortened the height of the station to ninety-one meters. This configuration made a stronger frame, thus better dampening the oscillations expected during operations. The design also moved the laboratory modules to the station's center of gravity to allow scientists and materials processing researchers to work near the quality microgravity zone within the station. Finally, the dual keel offered a far larger area for positioning facilities, attaching payloads, and storing supplies and parts. NASA formally adopted this design at its May 1986 Systems Requirements Review. Its Critical Evaluation Task Force modified the design in the fall of 1986 to increase the size of the nodes to accommodate avionics packages slated for attachment to the truss, thereby increasing pressurized volume available as well as decreasing the requirement for EVA.

In 1987, NASA and the administration, responding to significant increases in program costs, decided to take a phased approach to station development. In April 1987, the Space Station program was divided into Block I and Block II. Block I, the Revised Baseline Configuration, included the U.S. laboratory and habitat modules, the accommodation of attached payloads, polar platform(s), seventy-five kilowatts of photovoltaic power, European and Japanese modules, the Canadian Mobile

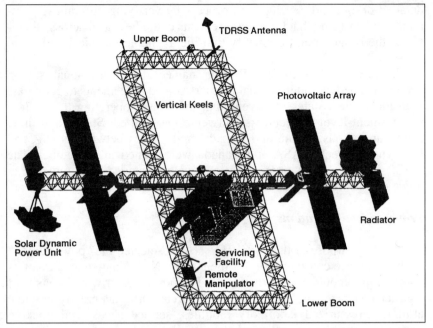

*Figure 3–31. Dual Keel Final Assembly Configuration
(adopted at May 1986 Systems Requirements Review)*

Servicing System, and provisions for evolution (Figure 3–32). The modules would be attached to a 110-meter boom. Block II, an Enhanced Configuration, would have an additional fifty kilowatts of power via a solar dynamic system, additional accommodation of attached payloads on dual keels and upper and lower booms, a servicing bay, and co-orbiting platforms (Figure 3–33).

Operations and Utilization Planning

NASA first formulated an operations concept for the space station in 1985 that considered preliminary launch, orbit, and logistics operational requirements, objectives such as reduced life-cycle costs, and international operations. It was determined that the station elements fulfill user requirements affordably and that NASA be able to afford the overall system infrastructure and logistics.

In 1985, the Space Station Utilization Data Base (later called the Mission Requirements Data Base) included more than 300 potential payloads from the commercial sector and from technology development, science, and applications communities. The information in this data base was used to evaluate potential designs of the station and associated platforms. Besides NASA, user sponsors included ESA, Canada, Japan, and the National Oceanic and Atmospheric Administration. In addition, a large number of private-sector users had requested accommodations on the station. Considerable interest was also expressed in using polar

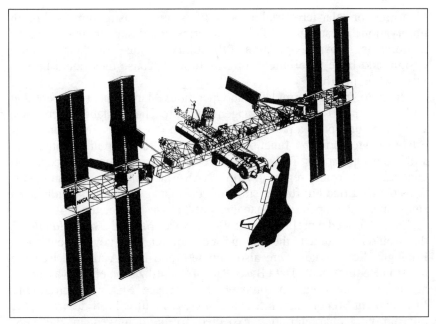

*Figure 3–32. Revised Baseline Configuration (1987), Block I
(This configuration would include the U.S. laboratory and habitat modules, accommodation of attached payloads, polar platform(s), seventy-five kilowatts of photovoltaic power, European and Japanese modules, the Canadian Mobile Servicing System, and provisions for evolution.)*

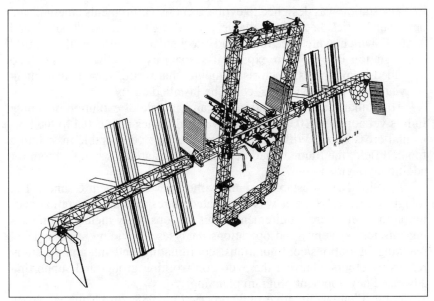

*Figure 3–33. Enhanced Configuration, Block II
(This would have an additional fifty kilowatts of power via a solar dynamic system, additional accommodation of attached payloads on dual keels and upper and lower booms, a servicing bay, and co-orbiting platforms.)*

platforms for solar-terrestrial physics, life sciences, astronomy, and Earth observation investigations. Polar platforms could support many related instruments, provide operational flexibility because of their modular design, and have indefinitely long lifetimes because they could be serviced while in orbit.

In 1986, NASA formulated an Operations Management Concept that outlined the philosophy and management approaches to station operations. Using the concept as a point of departure, an Operations Task Force was established to perform a functional analysis of future station operations. In 1987, the Operations Task Force developed an operations concept and concluded its formal report in April. NASA also implemented an operations plan, carried out further study of cost management, and conducted a study on science operations management that was completed in August.

NASA issued a preliminary draft of a *Space Station User's Handbook* that would be a guide to the station for commercial and government users. Pricing policy studies were also initiated, and NASA also revised the Mission Requirements Data Base. Part of the utilization effort was aimed at defining the user environment. The "Space Station Microgravity Environment" report submitted to Congress in July 1988 described the microgravity characteristics expected to be achieved in the U.S. Laboratory and compared these characteristics to baseline program operations and utilization requirements.

Evolution Planning

The station was designed to evolve as new requirements emerged and new capabilities became available. The design featured "hooks" and "scars," which were electronic and mechanical interfaces that would allow station designers to expand its capability. In this way, new and upgraded components, such as computer hardware, data management software, and power systems, could be installed easily.

The Enhanced Configuration was an example of evolution planning. In this version, two 103-meter-long vertical spines connected to the horizontal cross boom. With a near-rectangle shape comparable in size to a football field, the frame would be much stiffer and allow ample room for additional payloads.

In 1987, NASA established an Evolution Management Council. The Langley Research Center was designated as responsible for station evolution to meet future requirements. This responsibility included conducting mission, systems, and operations analyses, providing systems-level planning of options/configurations, coordinating and integrating study results by others, chairing the evolution working group, and supporting advanced development program planning.

A presidential directive of February 11, 1988, on "National Space Policy" stated that the "Space Station would allow evolution in keeping with the needs of station users and the long-term goals of the U.S."[34] This directive reaffirmed NASA's objective to design and build a station that could expand capabilities and incorporate improved technologies. Planning for evolution would occur in parallel with the design and development of the baseline station.

To support initiatives such as the Humans to Mars and Lunar Base projects, the station would serve as a facility for life science research and technology development and eventually as a transportation node for vehicle assembly and servicing. Another evolutionary path involved growth of the station as a multipurpose research and development facility. For these options, Langley conducted mission and systems analyses to determine primary resource requirements such as power, crew, and volume.

NASA Center Involvement

Marshall Space Flight Center

The Marshall Space Flight Center in Huntsville, Alabama, was designated as the Work Package 1 Center. Work Package 1 included the design and manufacture of the astronauts' living quarters, known as the habitation module (Figure 3–34); the U.S. Laboratory module; logistics elements, used for resupply and storage; node structures connecting the modules; the Environmental Control and Life Support System; and the thermal control and audio/video systems located within the pressurized modules.

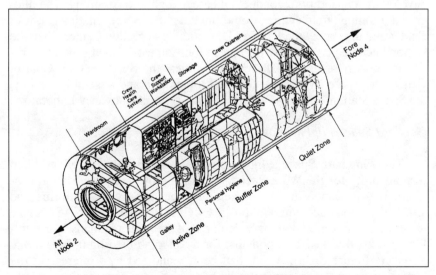

Figure 3–34. Habitation Module

---

[34]Office of the Press Secretary, "Fact Sheet: Presidential Directive on National Space Policy," February 11, 1988.

Marshall established the Space Station Freedom Projects Office to manage and direct the various design, development, and operational activities needed to successfully complete the Work Package 1 assignment, as well as several facilities to support its work package activities. These included the Payload Operations Integration Center, the Engineering Support Center, and the Payload Training Facility.

Johnson Space Center

The Johnson Space Center near Houston was responsible for the design, development, verification, assembly, and delivery of Work Package 2 flight elements and systems. This included the integrated truss assembly, propulsion assembly, mobile transporter, resource node design and outfitting, external thermal control, data management, operations management, communications and tracking, extravehicular systems, guidance, navigation, and control systems, and airlocks. Johnson was also responsible for the attachment systems, the STS for its periodic visits, the flight crews, crew training and crew emergency return definition, and operational capability development associated with operations planning. Johnson provided technical direction to the Work Package 1 contractor for the design and development of all station subsystems.

Johnson set up the Space Station Freedom Projects Office with the responsibility of managing and directing the various design, development, assembly, and training activities. This office reported to the Space Station Program Office in Reston, Virginia. The projects office at Johnson was to develop the capability to conduct all career flight crew training. The integrated training architecture would include the Space Station Control Center and ultimately the Payload Operations Integration Center when the station became permanently occupied. Johnson established several facilities in support of its various responsibilities: the Space Station Control Center, the Space Systems Automated Integration and Assembly Facility, the Space Station Training Facility, and the Neutral Buoyancy Laboratory.

Goddard Space Flight Center

The Goddard Space Flight Center in Greenbelt, Maryland, had responsibility for the Work Package 3 portion of the Space Station program. It was responsible for developing the free-flying platforms and attached payload accommodations, as well as for planning NASA's role in servicing accommodations in support of the user payloads and satellites. Goddard was also responsible for developing the Flight Telerobotic Servicer (Figure 3–35), which had been mandated by Congress in the conference report accompanying NASA's FY 1986 appropriations bill. The Flight Telerobotic Servicer was an outgrowth of the automation and robotics initiative of the station's definition and preliminary design phase.

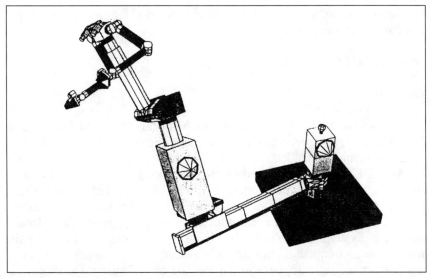

Figure 3–35. Flight Telerobotic Servicer

Lewis Research Center

The Lewis Research Center was responsible for the Work Package 4 portion of the Space Station program. Its station systems directorate was responsible for designing and developing the electric power system. This included responsibility for systems engineering and analysis for the overall electrical power system; all activities associated with the design, development, test, and implementation of the photovoltaic systems (Figure 3–36); hooks and scars activities in solar dynamics and in support of Work Package 2 in resistojet propulsion technology; power management and distribution system development; and activities associated with

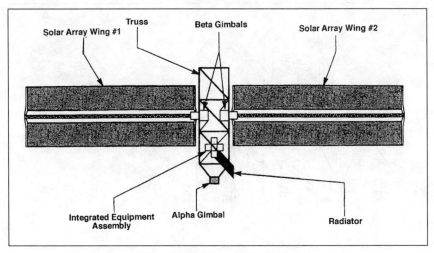

Figure 3–36. Photovoltaic Module

the Lewis station power system facilities and in planning electric power system mission operations.

International Cooperation

Canada

In March 1986, Canadian Prime Minister Brian Mulroney and President Reagan agreed to Canadian participation in the Space Station program. Canada intended to commit $1.2 billion to the program through the year 2000. Canada planned to provide the Mobile Servicing Center for Space Station *Freedom*. Together with a U.S.-provided, rail-mounted, mobile transporter, which would move along the truss, the Mobile Servicing Center and the transporter would comprise the Mobile Servicing System. The Mobile Servicing System was to play the main role in the accomplishing the station's assembly and maintenance, moving equipment and supplies around the station, releasing and capturing satellites, supporting EVAs, and servicing instruments and other payloads attached to the station. It would also be used for docking the Space Shuttle orbiter to the station and then loading and unloading materials from its cargo bay.

NASA considered the Mobile Servicing Center as part of the station's critical path: an indispensable component in the assembly, performance, and operation of the station. In space, Canada would supply the RMS, the Mobile Servicing Center and Maintenance Depot, the special purpose dexterous manipulator, Mobile Servicing System work and control stations, a power management and distribution system, and a data management system (Figure 3–37). On the ground, Canada would build a manipulator development and simulation facility and a mission operations facility. The Canadian Space Agency would provide project management.

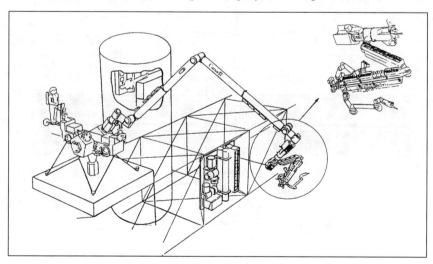

Figure 3–37. Mobile Servicing System and Special Purpose Dexterous Manipulator

European Space Agency

ESA gave the name "Columbus" to its program to develop the three elements that Europe was to contribute to the station: the Columbus Attached Laboratory, the Columbus Free-Flying Laboratory, and the Columbus Polar Platform. Columbus would provide an in-orbit and ground infrastructure compatible with European and international user needs from the mid-1990s onward. The program would also provide Europe with expertise in human, human-assisted, and fully automatic space operations as a basis for future autonomous missions. The program aimed to ensure that Europe establish the key technologies required for these various types of spaceflight.

The concept of Columbus was studied in the early 1980s as a follow-up to the Spacelab. The design, definition, and technology preparation phase was completed at the end of 1987. The development phase was planned to cover 1988–98 and would be completed by the initial launch of Columbus's three elements

Columbus Attached Laboratory. This laboratory would be permanently attached to the station's base. It would have a diameter of approximately four meters and would be used primarily for materials sciences, fluid physics, and compatible life sciences missions (Figure 3–38). The attached laboratory would be launched from the Kennedy Space Center on a dedicated Space Shuttle flight, removed from the Shuttle's payload bay, and berthed at the station's base.

Figure 3–38. Columbus Attached Laboratory

Figure 3–39. Columbus Free-Flying Laboratory

Columbus Free-Flying Laboratory. This free-flying laboratory (the "Free Flyer") would operate in a microgravity optimized orbit with a twenty-eight-and-a-half-degree inclination, centered on the altitude of the station (Figure 3–39). It would accommodate automatic and remotely controlled payloads, primarily from the materials sciences and technology disciplines, together with its initial payload, and would be launched by an Ariane 5 from the Centre Spatial Guyanais in Kourou, French Guiana. The laboratory would be routinely serviced in orbit by a Hermes at approximately six-month intervals. Initially, this servicing would be performed at Space Station *Freedom,* which the Free Flyer would also visit every three to four years for major external maintenance events.

Columbus Polar Platform. This platform would be stationed in a highly inclined Sun-synchronous polar orbit with a morning descending node (Figure 3–40). It would be used primarily for Earth observation missions. The platform was planned to operate in conjunction with one or more additional platforms provided by NASA and/or other international partners and would accommodate European and internationally provided payloads. The platform would not be serviceable and would be designed to operate for a minimum of four years. The platform would accommodate between 1,700 and 2,300 kilograms of ESA and internationally provided payloads.

Japan

Japan initiated its space program in 1985 in response to the U.S. invitation to join the Space Station program. The Space Activities Commission's Ad Hoc Committee on the Space Station concluded that Japan should participate in the Phase B (definition) study of the program with its own experimental module. On the basis of the committee's conclusion, the Science and Technology Agency concluded a Phase B MOU with NASA. Under the supervision of the Science and Technology Agency, the National Space Development Agency of Japan, a quasi-

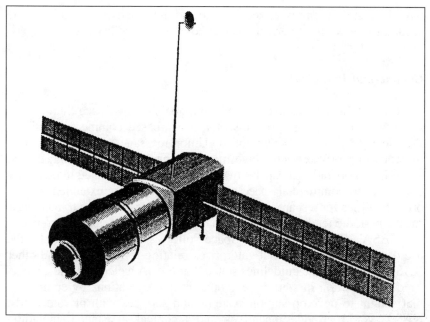

Figure 3–40. Columbus Polar Platform

governmental organization responsible for developing and implementing Japanese space activities, began the detailed definition and the preliminary design of the Japanese Experiment Module (JEM), which is shown in Figure 3–41 and would be attached to the Space Station. The JEM would be a multipurpose laboratory consisting of a pressurized module, an exposed facility, and an experiment logistics module (Table 3–51). The JEM would be launched on two Space Shuttle flights. The first flight

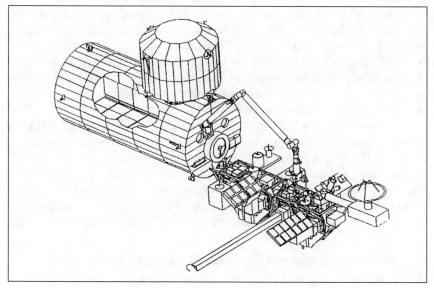

Figure 3–41. Japanese Experiment Module

would transport the pressurized module and the first exposed facility. The second flight would transport the second exposed facility and the experiment logistics module.

Commercial Participation

From its inception, one of the prime goals of the Space Station program was to encourage private-sector, space-based commercial activity. President Reagan's 1984 State of the Union message stated the objective of promoting private-sector investment in space through enhanced U.S. space-based operational capabilities. The station was planned to be highly conducive to commercial space activities by providing extended time in orbit, facilities for research and testing, and the presence of a trained crew for the periodic tending, repair, and handling of unexpected occurrences.

NASA's 1985 "Commercial Space Policy" set forth guidelines for the use of space for commercial enterprises relating to the station and other NASA activities. The guidelines stated that NASA welcomed and encouraged participation in station development and operations by companies that sought to develop station systems and services with private funds. NASA would provide incentives and technical assistance, including access to NASA data and facilities, where appropriate. NASA would protect proprietary rights and would request privately owned data only when necessary to carry outs its responsibilities.[35]

NASA expected the private sector to be a principal user of station capabilities. It also expected the private sector to participate in the program by providing services, both on the ground and in orbit. The private sector would participate in the program through procurements to design and build elements of the station and its related systems. In 1986, NASA's Commercial Advocacy Group conducted workshops to identify and encourage potential commercial use of the station, particularly in the areas of materials processing, Earth and ocean remote sensing, communications satellite delivery, and industrial services. In August 1986, NASA established "Guidelines for United States Commercial Enterprises for Space Station Development and Operations." These guidelines were to encourage U.S. private-sector investment and involvement in developing and operating station systems and services.

In November 1987, NASA issued a series of new program initiatives designed to expand the opportunities for pioneering commercial ventures in space. The initiatives built on earlier commercial development policies and provided for the continued encouragement of private space activities. The 1988 National Space Policy mandated the provision for commercial participation in the Space Station program. Commercial participation would be possible through commercial utilization and commercial

---

[35]"NASA Guidelines for United States Commercial Enterprises for Space Station Development and Operations," Office of Space Station, NASA, 1985, NASA Historical Reference Collection, Washington, DC.

infrastructure activities. Commercial utilization activities would involve commercial users of the station who would conduct space-based research and development activities. Commercial infrastructure activities would involve provisions for selected station-related systems and services on a commercial basis to NASA and station users.

In October 1988, NASA published revised policy guidelines for proposals from commercial entities to provide the infrastructure for the station. These guidelines, revised in response to President Reagan's Commercial Space Initiatives, issued in February 1988, were intended to provide a framework to encourage U.S. commercial investment and involvement in the development and operation of Space Station *Freedom*. NASA would use these guidelines to evaluate proposals from industry for participating in the Space Station program.[36]

---

[36] "NASA Issues Draft Guidelines on Station Commercial Infrastructure," *NASA News*, Release 88-144, October 25, 1988.

Table 3–1. Total Human Spaceflight Funding History (in thousands of dollars)

Year and Budget Item	Request	Authorization	Appropriation	Programmed (Actual)
1979				
Space Shuttle Program	1,439,300	1,443,300	a	1,638,300
Suppl. Appr.	185,000	185,000	185,000	
Space Flight Operations	311,900	315,900	b	299,700
1980			c	
Space Shuttle Program	1,586,000	1,586,000		1,871,000
Suppl. Appr.	300,000	—	285,000	
Space Flight Operations	446,604 d	463,300	e	446,600
1981				
Space Shuttle	1,873,000	1,873,000	1,995,000 f	1,995,000
Space Flight Operations	683,700 g	779,500	679,200 h	679,200
1982				
Space Shuttle	2,194,000	2,189,000	2,194,000	2,638,350
Space Flight Operations	895,900 j	907,900	k	466,500
1983				
Space Shuttle	1,718,000	1,798,000	1,769,000	2,144,700
Space Flight Operations	1,452,000 l	1,699,000	1,796,000	1,421,700
1984				
Space Transportation Capability Dev. (R&D)	1,927,400	2,009,400	427,400 m	484,795
Space Transportation Capability Dev. (SFC&DC)			1,545,000 n	1,569,303
Space Transportation Ops. (R&D)	1,452,000 o	1,545,600	p	41,669
Space Transportation Ops. (SFC&DC)			1,570,600	1,397,638

Table 3–1 continued

Year and Budget Item	Request	Authorization	Appropriation	Programmed (Actual)
1985				
Space Transportation Capability Dev.	361,400	351,400	407,400 *q*	391,400
Space Production & Operational Capability (SFC&DC) *r*	1,465,600	1,470,600	1,510,600	1,484,500
Space Transportation Operations	1,339,000	1,319,000	1,339,000	1,314,000
Space Station	150,000	150,000	155,000 *s*	155,500
1986				
Space Transportation Capability Dev.	459,300	437,300	439,300	402,383
Shuttle Production & Operational Capability (SFC&DC) *t*	976,500	961,500	976,500	1,329,390
Space Transportation Operations	1,725,000 *u*	1,710,100	1,725,100	1,573,412
Space Station	205,000 *v*	205,000	205,000	184,702
1987				
Space Transportation Capability Dev.	507,500 *w*	515,500	507,500	491,100
Shuttle Production & Operational Capability	1,134,400 *x*	1,156,400	2,984,400	3,408,100
Space Transportation Operations	1,847,000 *y*	1,881,700	1,867,700	1,746,000
Space Station	410,000	410,000	410,000	420,000
1988				
Space Transportation Capability Dev.	568,600	553,600	549,600	598,564
Shuttle Production & Operational Capability	1,229,600	1,174,600	1,100,609	1,815,517
Space Transportation Operations	1,838,000 *z*	1,885,800	1,885,800	1,836,352
Space Station	392,300 *aa*	767,000	425,000 *bb*	489,509

a Undistributed. Total 1979 R&D appropriation = $3,477,200,000.
b Undistributed. Total 1979 R&D appropriation = $3,477,200,000.
c Undistributed. Total 1980 R&D appropriation = $4,091,086,000.

Table 3–1 continued

d Amended budget submission. Original budget submission = $467,300,000.
e Undistributed. Total 1980 R&D appropriation = $4,091,086,000.
f Reflects recission.
g Amended budget submission. Original budget submission = $809,500,000.
h Reflects recission.
i Amended budget submission. Original budget submission = $2,230,000,000.
j Amended budget submission. Original budget submission = $1,043,000,000.
k Undistributed. Total 1982 R&D appropriation = $4,973,100,000.
l Amended budget submission. Original budget submission = $1,707,000,000.
m In 1984, funding for most transportation activities moved from the R&D appropriation to the SFC&DC appropriation. The amount of $427,400,000 remained in R&D for upper stages ($143,200,000), Spacelab ($119,600,000), engineering and technology base ($93,100,000), advanced programs ($53,200,000), tethered satellite system ($15,000,000), and teleoperator maneuvering system ($3,300,000) activities.
n Space Transportation & Capability Development included under the SFC&DC appropriation funded Shuttle Production and Capability Development ($1,500,000,000) and a reserve of $45,000,000, which included an allocation of $50,000,000 for orbiter/engine spares and reduced the amount earmarked for Shuttle training aircraft advance payment by $5,000,000.
o Amended budget submission. Original budget submission = $1,570,600,000.
p In 1984, all funds for Space Transportation Operations activities moved to the SFC&DC appropriation.
q Reflects supplemental legislation, which increased appropriation from $367,400,000 to $407,400,000. Accompanying note states that increased appropriation was designated for upper stages and would not be available until March 1, 1986.
r New budget category includes funding for orbiter, launch and mission support, propulsion systems, and changes and systems upgrading.
s Reflects supplemental legislation, which increased appropriation from $150,000,000 to $155,500,000.
t New budget category includes funding for orbiter, launch and mission support, propulsion systems, and changes and systems upgrading.
u Amended budget submission. Original budget submission = $1,725,000,000.
v Amended budget submission. Original budget submission = $230,000,000.
w Amended budget submission. Original budget submission = $465,500,000.
x Amended budget submission. Original budget submission = $745,400,000.
y Amended budget submission. Original budget submission = $1,524,700,000.
z Amended budget submission. Original budget submission = $1,885,800,000.
aa Amended budget submission. Original budget submission = $767,000,000.
bb Reflects transfer of $100,000,000 to FY 1988 for Space Station from FY 1987.

Table 3–2. Programmed Budget by Budget Category (in thousands of dollars)

Budget Category/Year	1979	1980	1981	1982	1983
Space Shuttle Program	1,638,300	1,871,000	1,995,000	2,638,350	2,144,700
Space Flight Operations Program	299,700	446,600	679,200	466,500	1,421,700
Shuttle Production & Operational Capability		Not a budget category until 1985			
Space Transportation (System) Operations		Not a budget category until 1984			
Space Transportation Capability Development		Not a budget category until 1984			
Space Station Program		Not a budget category until 1984			

Budget Category/Year	1984	1985	1986	1987	1988
Space Shuttle Program		Not a budget category after 1983			
Space Flight Operations Program		Not a budget category after 1983			
Shuttle Production & Operational Capability	—	1,484,500	1,329,390	3,408,100	1,815,517
Space Transportation (System) Operations	1,439,307 (R&D + SFC&DC)	1,314,000	1,573,412	1,746,000	1,836,352
Space Transportation Capability Development	2,048,098 (R&D + SFC&DC)	391,400	402,383	491,100	598,564
Space Station Program	—	155,500	184,702	420,000	489,509

Table 3–3. Orbiter Funding History (in thousands of dollars)

Year (Fiscal)	Submission	Authorization	Appropriation	Programmed
1979	DDT&E 536,500	DDT&E 536,500	b	DDT&E 727,800
	Production 397,000	Production a		Production 264,500
Suppl Appr.	c	61,500 d	e	
1980	DDT&E 560,900 f	DDT&E 420,800	h	DDT&E 641,900
	Production 572,600 g	Production 570,600		Production 572,600
Suppl Appr.	i	— j		
1981	DDT&E 521,000 l	DDT&E 320,900	DDT&E 521,000 n	DDT&E 510,500
	Production 727,500 m	Production 768,200	Production 727,500 o	Production 779,000
1982	DDT&E 422,000 p	DDT&E 372,000	DDT&E 372,000	Production 916,850
	Production 860,000 q	Production 832,000	Production 837,000	
1983	904,400 r	1,018,500	984,500	903,910
1984	716,300 s	t		724,900
1985	655,300 u	651,800	651,800	674,200
1986	333,600	333,600	333,600	396,400
1987	373,000 v	211,000	211,000	448,100 w
1988	328,600 x	348,200	403,200	321,300

a Undistributed. Total 1979 Production authorization = $458,000,000.
b Undistributed. Total 1979 R&D appropriation = $3,477,200,000.
c Supplemental appropriation submission of $185,000,000 undistributed among functions.
d Distribution for supplemental appropriation not specified in budget figures. Distribution indicated in supporting congressional committee documentation.
e Supplemental appropriation of $185,000,000 undistributed among functions.
f Amended budget submission. Original submission = $420,800,000.
g Amended budget submission. Original submission = $570,600,000.
h Undistributed. Total 1980 R&D appropriation = $4,091,086,000.
i Specific functions for supplemental appropriation not specified. Supporting committee documentation states that the administration requested additional funds for design, development, test, and evaluation of orbiter ($140,100,000), external tank ($11,000,000), solid rocket booster ($3,700,000), launch and landing ($45,200,000), and changes/systems upgrading ($100,000,000) for a total of $300,000,000.

Table 3–3 continued

j   House committee introduced and passed supplemental appropriation for $300,000,000, but no further action was taken by the whole House. Senate committee authorized supplemental appropriation of $300,000,000, which was passed by the whole Senate. However, there was no conference committee authorization for a supplemental appropriation.

k   Undistributed supplemental appropriation of $285,000,000 approved by conference committee and Congress.

l   Amended budget submission. Original submission = $320,900,000. The increase reflects technical problems that delayed the scheduled rollout of the orbiter *Columbia* from the Orbiter Processing Facility and led to the delay in the Orbital Flight Test program and the initial operational capability date.

m   Amended budget submission. Original submission = $768,200,000.

n   Reflects recission.

o   Reflects recission.

p   Amended budget submission. Original submission = $372,000,000.

q   Amended budget submission. Original submission = $837,000,000.

r   Amended budget submission. Original submission = $933,500,000.

s   Amended budget submission. Original submission = $729,600,000.

t   No authorization or appropriation for orbiter budget category.

u   Amended budget submission. Original submission = $606,800,000.

v   Amended budget submission. Original budget submission = $211,000,000.

w   Amount is for Orbiter Operational Capability.

x   Amended budget submission. Original budget submission = $403,200,000.

Table 3–4. Orbiter Replacement Funding History (in thousands of dollars)

Year (Fiscal)	Submission	Authorization	Appropriation	Programmed
1987	250,000	272,000	2,100,000	2,000,000

262 NASA HISTORICAL DATA BOOK

Table 3–5. Launch and Mission Support Funding History (in thousands of dollars) [a]

Year (Fiscal)	Submission	Authorization	Appropriation	Programmed
1979	DDT&E 128,100 Production 11,000	DDT&E 128,000 Production [b]	[c]	DDT&E 149,600 Production 7,000
Suppl. Appr.	[d]	19,500 [e]	[f]	
1980	DDT&E 188,400 [g] Production 16,400 [h]	DDT&E 143,200 Production 20,000	[i]	DDT&E 188,400 Production 16,400
Suppl. Appr.	[j]	[k]	[l]	
1981	DDT&E 204,000 [m] Production 34,000 [n]	DDT&E 154,400 Production 40,400	DDT&E 204,000 [o] Production 34,000 [p]	DDT&E 214,500 Production 33,000
1982	DDT&E 260,000 [q] Production 63,000 [r]	DDT&E 199,000 Production 57,000	DDT&E 199,000 Production 57,000	Operations 6,400 Production and Capability Dev. 134,900
1983	246,600	67,000 [s]	67,000	246,300
1984	No budget item [t]			277,700
1985	229,800 [u]	219,800	234,800	218,100
1986	169,000 [v]	158,900	163,900	180,000
1987	148,200 [w]	161,000	161,000	151,200
1988	164,800 [x]	249,300	552,100 [y]	167,400

[a] Budget category titled Launch and Landing (Design, Development, Test, and Evaluation) in 1979–1983.
[b] Undistributed. Total 1979 Production authorization = $458,000,000.
[c] Undistributed. Total 1979 R&D appropriation = $3,477,200,000.
[d] Supplemental appropriation submission undistributed functions.
[e] Distribution for supplemental appropriation not specified in budget figures. Distribution indicated in supporting congressional committee documentation.
[f] Supplemental appropriation undistributed among functions.
[g] Amended budget submission. Original submission = $143,200,000.
[h] Amended budget submission. Original submission = $20,000,000.

Table 3–5 continued

i   Undistributed. Total 1980 R&D appropriation = $4,091,086,000.
j   Specific functions for supplemental appropriation not specified. Supporting committee documentation states that the administration requested additional funds for design, development, test, and evaluation of orbiter ($140,100,000), external tank ($11,000,000), solid rocket booster ($3,700,000), launch and landing ($45,200,000), and changes/systems upgrading ($100,000,000), for a total of $300,000,000.
k   House committee introduced and passed supplemental appropriation for $300,000,000, but no further action was taken by the whole House. Senate committee authorized supplemental appropriation of $300,000,000, which was passed by the whole Senate. However, there was no conference committee authorization for a supplemental appropriation.
l   Undistributed supplemental appropriation of $285,000,000 approved by conference committee and Congress.
m Amended budget submission. Original submission = $154,400,000. Increase reflects the compressed effort required to process the Shuttle vehicle for launch, which increased the development and support contractor funding requirements substantially.
n   Amended budget submission. Original submission = $40,400,000.
o   Reflects recission.
p   Reflects recission.
q   Amended budget submission. Original submission = $199,000,000. Increase reflects growth in requirements related to the support for launch processing operations and for facility and equipment modifications needed to resolve problems experienced during the first two Shuttle launches.
r   Amended budget submission. Original submission = $57,000,000.
s   Authorization and appropriation figures reflect amounts for Space Shuttle program only. Mission support figures not provided for Space Flight Operations program.
t   Amount for Launch and Mission Support included as part of Space Transportation System Capability Development. Request = $1,927,400,000; authorization = $2,009,400,000; and appropriation (SFC&DC) = $1,570,000. Budget submission for Launch and Mission Support per budget submission submission according to NASA Comptroller's Office = $277,700,000.
u   Amended budget submission. Original budget submission = $234,800,000.
v   Amended budget submission. Original budget submission = $163,900,000.
w   Amended budget submission. Original budget submission = $161,000,000.
x   Amended budget submission. Original budget submission = $249,300,000.
y   Reflects $302,000,000 transferred from Propulsion Systems to Launch and Mission Support.

Table 3–6. Launch and Landing Operations Funding History (in thousands of dollars)

Year (Fiscal)	Submission	Authorization	Appropriation	Programmed
1984	357,200	a	b	329,000
1985	275,900 c	265,000	265,000	275,500
1986	332,400 d	e		315,100
1987	353,300 f	285,000	300,500	359,800
1988	449,800 g	401,600	401,600	452,200

a Undistributed. Total Shuttle Operations authorization = $,495,600,000.
b No budget item. Shuttle Operations transferred to SFC&DC appropriation.
c Amended budget submission. Original budget submission = $265,000,000.
d Amended budget submission. Initial budget submission = $335,900,000.
e No authorization or appropriation for Launch and Landing Operations.
f Amended budget submission. Initial budget submission = $285,000,000.
g Amended budget submission. Initial budget submission = $401,600,000.

Table 3–7. Spaceflight Operations Program Funding History (in thousands of dollars)

Year (Fiscal)	Submission	Authorization	Appropriation	Programmed
1979	311,900	315,900	a	299,700
1980	446,600 b	463,300	c	446,600
1981	683,700 d	779,500	679,200 e	679,200
1982	895,900 f	907,900	g	466,500 h
1983	1,453,700 i	1,699,000	1,796,000	1,421,700
1984 j	1,452,000 k	1,545,600	1,570,600	1,452,000
1985	1,339,000	1,319,000	1,339,000	1,314,000
1986	1,725,000 l	1,710,000	1,725,000	1,640,200
1987	1,847,000 m	1,881,700	1,867,700	1,746,000
1988	1,838,000 n	1,885,800	1,885,800	1,833,600

a   Undistributed. Total 1979 R&D appropriation = $3,477,200,000.
b   Amended budget submission. Original submission = $467,300,000.
c   Undistributed. Total 1980 R&D appropriation = $4,091,086,000.
d   Reflects amended budget submission. Initial budget submission = $809,500,000.
e   Amended budget submission. Original submission = $767,500,000. Primary reflects rephasing of Shuttle and Spacelab operations. Also reflects recission.
f   Amended budget submission. Initial budget submission = $1,043,000,000.
g   Undistributed. Total 1982 R&D appropriation = $4,740,900,000 (reflects effect of Gen. Prov. Sec. 412).
h   Reduction reflects reordering of budget categories included in Space Flight Operations.
i   Amended budget submission. Initial budget submission = $1,707,000.
j   Became Space Transportation Operations in 1984. Appropriation transferred to SFC&DC.
k   Budget category reconfigured as Space Transportation Operations. Includes Shuttle Operations and Expendable Launch Vehicles programs. Amended budget submission. Original budget submission = $1,570,600,000.
l   Amended budget submission. Initial budget submission = $1,725,100,000.
m   Amended budget submission. Initial budget submission = $1,524,700,000.
n   Amended budget submission. Initial budget submission = $1,885,800,000.

Table 3–8. Flight Operations Funding History (in thousands of dollars)

Year (Fiscal)	Submission	Authorization	Appropriation	Programmed
1984	315,000	a	b	333,900
1985	316,000 c	316,000	316,000	315,600
1986	435,000 d	e		434,200
1987	557,700 f	360,600	399,000	515,000
1988	583,600 g	561,100	561,100	597,100

a   Undistributed. Total Shuttle Operations authorization = $1,495,600,000.
b   No budget category for Flight Operations in appropriations activity.
c   Amended budget submission. Initial budget submission = $315,200,000.
d   Amended budget submission. Initial budget submission = $425,200,000.
e   No authorization or appropriation for Flight Operations budget category.
f   Amended budget submission. Initial budget submission = $360,600,000. Increase reflects increased funding to replace lost reimbursable income and an increase in program requirements that have resulted in increases in support to system design reviews and safety/reliability oversight.
g   Amended budget submission. Initial budget submission = $561,100,000.

Table 3–9. Spacelab Funding History (in thousands of dollars)

Year (Fiscal)	Submission	Authorization	Appropriation	Programmed
1979 a				26,700
1980 b	58,800			58,800
1981	139,700 c	144,700	139,700 d	138,800
1982	100,800 e	f	110,700	100,800
1983	121,200 g	h	113,200	121,200
1984	112,500 i	119,600	119,600	111,000
1985	58,300 j	69,300	69,300	55,700
1986	92,900 k	96,700	96,700	78,000
1987	73,900 l	66,700	68,800	72,000
1988	66,500 m	73,500	73,500	66,500

a No budget item for Spacelab indicated in 1979 budget submission. Spacelab is mentioned in Space Flight Operations budget category narrative for FY 1979 Senate authorization bill (legislative day: April 24, 1978), which states that Spacelab is being developed and paid for by the European Space Agency (ESA) and that NASA supports ESA's Spacelab development effort. "This support includes developing a crew transfer tunnel and procurement of necessary mockups, trainers and ground support equipment not provided by ESA. Other activities include procurement of flight and ground hardware, and system activation activities to assure Spacelab compatibility with the orbiter and an operational capability." Actual programmed amount for Spacelab for FY 1979 = $26,700,000.
b No budget item for Spacelab indicated in FY 1980 budget submission. Only mention of Spacelab in FY 1980 budget submission occurs in narrative accompanying Senate authorization bill, which says that the "principal areas of [Space Flight Operations] activity include the Spacelab, the space transportation upper stages…."
c Amended budget submission. Initial budget submission = $151,700,000.
d Amended budget submission. Initial submission = $149,700,000. Reflects recission.
e Amended budget submission. Initial budget submission = $140,700,000.
f Undistributed. Total 1982 R&D authorization = $4,973,100,000.
g Amended budget submission. Original budget submission = $113,200,000.
h Undistributed. Total Space Flight Operations authorization = $1,699,000,000.
i Amended budget submission. Initial budget submission = $119,600,000.
j Amended budget submission. Initial budget submission = $69,300,000.
k Amended budget submission. Initial budget submission = $96,700,000.
l Amended budget submission. Initial budget submission = $89,700,000.
m Amended budget submission. Initial budget submission = $73,500,000.

Table 3–10. Space Station Funding History (in thousands of dollars)

Year (Fiscal)	Submission	Authorization	Appropriation	Programmed
1985	150,000	150,000	155,500	155,500
1986	205,000 a	205,000	205,000	184,702
1987	410,000	410,000	410,000	420,000
1988	392,300 b	767,000	425,000 c	489,509

a Amended budget submission. Initial budget submission = $230,000,000.
b Amended budget submission. Initial budget submission = $767,300,000. The reduction reflects deferring the buildup of the prime contractor personnel, limiting supporting development personnel hiring and equipment purchases and constraining supporting engineering development capabilities.
c Reflects transfer of $100,000,000 to FY 1988 for Space Station from FY 1987.

Table 3–11. Orbiter Characteristics

Component	Characteristics
Length	37.24
Height	17.25
Vertical Stabilizer	8.01
Wingspan	23.79
Body Flap	
Area (sq m)	12.6
Width	6.1
Aft Fuselage	
Length	5.5
Width	6.7
Height	6.1
Mid-Fuselage	
Length	18.3
Width	5.2
Height	4.0
Airlock (cm)	
Inside Diameter	160
Length	211
Minimum Clearance	91.4
Opening Capacity	46 x 46 x 127
Forward Fuselage Crew Cabin (cu m)	71.5
Payload Bay Doors	
Length	18.3
Diameter	4.6
Surface Area (sq m)	148.6
Weight (kg)	1,480
Wing	
Length	18.3
Maximum Thickness	1.5
Elevons	4.2 and 3.8
Tread Width	6.91
Structure Type	Semimonocoque
Structure Material	Aluminum
Gross Takeoff Weight	Variable
Gross Landing Weight	Variable
Inert Weight (kg) (approx.)	74,844
Main Engines	
Number	3
Average Thrust	1.67M newtons at sea level
	2.10M newtons in vacuum
Nominal Burn Time	522 seconds

Table 3–11 continued

Component	Characteristics
OMS Engines	
Number	2
Average Thrust	26,688 newtons
Dry Weight (kg)	117.9
Propellant	Monomethyl hydrazine and nitrogen tetroxide
RCS Engines	
Number	38 primary (4 forward, 12 per aft pod)
	6 vernier (2 forward, 4 aft)
Average Thrust	3,870 newtons in each primary engine
	111.2 newtons in each vernier engine
Propellant	Monomethyl hydrazine and nitrogen tetroxide
Major Systems	Propulsion; Power Generation; Environmental Control and Life Support; Thermal Protection; Communications; Avionics; Data Processing; Purge, Vent, and Drain; Guidance, Navigation, and Control; Dedicated Display; Crew Escape

All measurements are in meters unless otherwise noted.

Table 3–12. Typical Launch Processing/Terminal Count Sequence

Time	Event
T-11 hr	Start retraction of rotating service structure (completed by T-7 hr 30 min)
T-5 hr 30 min	Enter 6-hr built-in hold, followed by clearing of pad
T-5 hr	Start countdown, begin chill down of liquid oxygen/liquid hydrogen transfer system
T-4 hr 30 min	Begin liquid oxygen fill of external tank
T-2 hr 50 min	Begin liquid hydrogen fill of external tank
T-2 hr 4 min	1-hr built-in hold, followed by crew entry operations
T-1 hr 5 min	Crew entry complete; cabin hatch closed; start cabin leak check (completed by T-25 min)
T-30 min	Secure white room; ground crew retires to fallback area by T-10 min; range safety activation/Mission Control Center guidance update
T-25 min	Mission Control Center/crew communications checks; crew given landing weather information for contingencies of return-to-abort or abort once around
T-20 min	Load flight program; beginning of terminal count
T-9 min	10-min built-in hold (also a 5-min hold capability between T-9 and T-2 min and a 2-min hold capability between T-2 min and T-27 sec)
T-9 min	Go for launch/start launch processing system ground launch sequencer (automatic sequence)
T-7 min	Start crew access arm retraction
T-5 min	Activate orbiter hydraulic auxiliary power units (APUs)
T-4 min 30 sec	Orbiter goes to internal power
T-3 min	Gimbal main engines to start position
T-2 min 55 sec	External tank oxygen to flight pressure
T-1 min 57 sec	External tank hydrogen to flight pressure
T-31 sec	Onboard computers' automatic launch sequence software enabled by launch processing system command
T-30 sec	Last opportunity for crew to exit by slidewire
T-27 sec	Latest hold point if needed (following any hold below the T-2 min mark, the countdown will be automatically recycled to T-9 min)
T-25 sec	Activate solid rocket booster hydraulic power units; initiative for management of countdown sequence assumed by onboard computers; ground launch sequencer remains on line
T-18 sec	Solid rocket booster nozzle profile conducted
T-3.6 sec	Main propulsion system start commands issued by the onboard GPCs
T-3.46 sec to 3.22 sec	Main engines start
T-0	Main engines at 90 percent thrust
T+2.64 sec	Solid rocket booster fire command/holddown bolts triggered
T+3 sec	LIFTOFF

Table 3–13. Space Shuttle Launch Elements

Event	Time min:sec	Geodetic altitude (km)	Inertial velocity (km/hr)
SSME ignition	–00:03.46	.056	0
Solid rocket booster ignition	00:03	.056	0
Begin pitchover	00:07	.166	0
Maximum dynamic pressure reached	01:09	13.4	6.4
Solid rocket booster separation	02:04	47.3	38.1
Main engine cutoff	08:38	117.5	1,335
External tank separation	08:50	118.3	1,427
OMS-1 ignition	10:39	126	2,221
OMS-1 cutoff	12:24	133.9	2,993
OMS-2 ignition	43:58	279.4	15,731
OMS-2 cutoff	45:34	280.3	16,526

Table 3–14. Mission Command and Control Positions and Responsibilities

Position	Function
Flight Director	Leads the flight control team. The Flight Director is responsible for mission and payload operations and decisions relating to safety and flight conduct.
Spacecraft Communicator	Primary communicator between Mission Command and Control and the Shuttle crew.
Flight Dynamics Office	Plans orbiter maneuvers and follows the Shuttle's flight trajectory along with the Guidance Officer.
Guidance Officer	Responsible for monitoring the orbiter navigation and guidance computer software.
Data Processing Systems Engineer	Keeps track of the orbiter's data processing systems, including the five on-board general purpose computers, the flight-critical and launch data lines, the malfunction display system, mass memories, and systems software.
Flight Surgeon	Monitors crew activities and is for the medical operations flight control team, providing medical consultations with the crew, as required, and keeping the Flight Director informed on the state of the crew's health.
Booster Systems Engineer	Responsible for monitoring and evaluating the main engine, solid rocket booster, and external tank performance before launch and during the ascent phases of a mission.
Propulsion Systems Engineer	Monitors and evaluates performance of the reaction control and orbital maneuvering systems during all flight phases and is charged with management of propellants and other consumables for various orbiter maneuvers.
Guidance, Navigation, and Control Systems Engineer	Monitors all Shuttle guidance, navigation, and control systems. Also keeps the Flight Director and crew notified of possible abort situations, and keeps the crew informed of any guidance problems.
Electrical, Environmental, and Consumables Systems Engineer (EECON)	Responsible for monitoring the cryogenic supplies available for the fuel cells, avionics and cabin cooling systems, and electrical distribution, cabin pressure, and orbiter lighting systems.
Instrumentation and Communication Systems Engineer	Plans and monitors in-flight communications and instrumentation systems.
Ground Control	Responsible for maintenance and operation of Mission Command and Control hardware, software, and support facilities. Also coordinates tracking and data activities with the Goddard Space Flight Center, Greenbelt, Maryland.
Flight Activities Officer	Plans and supports crew activities, checklists, procedures, and schedules.
Payloads Officer	Coordinates the ground and on-board system interfaces between the flight control team and the payload user. Also monitors Spacelab and upper stage systems and their interfaces with payloads.
Maintenance, Mechanical Arm, and Crew Systems Engineer	Monitors operation of the remote manipulator arm and the orbiter's structural and mechanical systems. May also observe crew hardware and in-flight equipment maintenance.
Public Affairs Officer	Provides mission commentary and augments and explains air-to-ground conversations and flight control operations for the news media and public.

Table 3–15. Shuttle Extravehicular Activity

Mission	Date	Astronaut	Duration (Hr: Min)
STS-6	April 8, 1983	Musgrave	3:54
		Peterson	3:54
STS 41-B	February 8, 1984	McCandless	11:37
		Stewart	11:37
STS 41-C	April 11, 1984	Nelson	10:06
		van Hoften	10:06
STS 41-G	October 12, 1984	Leestma	3:29
		Sullivan	3:29
STS 51-A	November 21, 1984	Allen	12:14
		Gardner	12:14
STS 51-D	April 17, 1985	Griggs	3:10
		Hoffman	3:10
STS 51-I	September 1, 1985	van Hoften	4:31
		W. Fisher	4:31
STS 61-B	November 30, 1985	Spring	12:12
	December 1, 1985	Ross	12:12

Table 3–16. STS-1–STS-4 Mission Summary

Mission	Dates	Crew	Payload and Experiments
STS-1	Apr. 12–14, 1981	Cmdr: John W. Young Pilot: Robert L. Crippen	Aerodynamic Coefficient Identification Package Data Flight Instrumentation Package Passive Optical Sample Assembly
STS-2	Nov. 12–14, 1981	Cmdr: Joe H. Engle Pilot: Richard H. Truly	Aerodynamic Coefficient Identification Package Catalytic Surface Experiment Data Flight Instrumentation Dynamic, Acoustic and Thermal Experiment Induced Environment Contamination Monitor Tile Gap Heating Effects Experiment OSTA-1 Payload (Office of Space and Terrestrial Applications) • Feature Identification and Location Experiment • Heflex Bioengineering Test • Measurement of Air Pollution From Satellites • Night-Day Optical Survey of Lightning • Ocean Color Experiment • Shuttle Imaging Radar-A • Shuttle Multispectral Infrared Radiometer
STS-3	Mar. 22–30, 1982	Cmdr: Jack R. Lousma Pilot: C. Gordon Fullerton	Data Flight Instrumentation Aerodynamic Coefficient Identification Package Induced Environment Contamination Monitor Tile Gap Heating Effects Experiment Catalytic Surface Experiment Dynamic, Acoustic and Thermal Experiment Monodisperse Latex Reactor Electrophoresis Test Heflex Bioengineering Test Infrared Imagery of Shuttle OSS-1 Payload (Office of Space Science) • Contamination Monitor • Microabrasion Foil Experiment • Plant Growth Unit • Plasma Diagnostics Package • Shuttle-Spacelab Induced Atmosphere • Solar Flare X-Ray Polarimeter • Solar Ultraviolet Spectral Irradiance Monitor • Thermal Canister Experiment • Vehicle Charging and Potential Experiment

Table 3–16 continued

Mission	Dates	Crew	Payload and Experiments
STS-3 continued			Get-Away Special Canister • Flight Verification Shuttle Student Involvement Project • Insects in Flight
STS-4	June 27, 1982– July 4, 1982	Cmdr: Thomas K. Mattingly Pilot: Henry W. Hartsfield, Jr.	Aerodynamic Coefficient Identification Package Catalytic Surface Experiment Continuous Flow Electrophoresis System Data Flight Instrumentation Department of Defense Payload DOD-82-1 Dynamic, Acoustic and Thermal Experiment Induced Environment Contamination Monitor Infrared Imagery of Shuttle Monodisperse Latex Reactor Night/Day Optical Survey of Lightning Tile Gap Heating Effects Experiment Get-Away Special • G-001 Utah State University Shuttle Student Involvement Project • Effects of Diet, Exercise, Zero Gravity on Lipoprotein Profiles • Effects of Space Travel on Trivalent Chromium in the Body

Table 3–17. STS-1 Mission Characteristics

Crew	Cmdr: John W. Young Pilot Robert L. Crippen
Launch	7:00:03 a.m., EST, April 12, 1981, Kennedy Space Center The launch followed a scrubbed attempt on April 10. The countdown on April 10 proceeded normally until T-20 minutes when the orbiter general purpose computers (GPCs) were scheduled for transition from the vehicle checkout mode to the vehicle flight configuration mode. The launch was held for the maximum time and scrubbed when the four primary GPCs would not provide the correct timing of the backup flight system GPC. Analysis and testing indicated the primary set of GPCs provided incorrect timing to the backup flight system at initialization and caused the launch scrub. The problem resulted from a Primary Ascent Software System (PASS) skew during initialization. The PASS GPCs were reinitialized and dumped to verify that the timing skew problem had cleared. During the second final countdown attempt on April 12, transition of the primary set of orbiter GPCs and the backup flight system GPC occurred normally at T-20 minutes. The Shuttle cleared its 106-meter launch tower in six seconds and reached Earth orbit in about 12 minutes.
Orbital Altitude & Inclination	237 km/40 degrees The crew changed their orbit from its original elliptical 106 km x 245 km by firing their orbital maneuvering system on apogee.
Total Weight in Payload Bay	4,870 kg
Landing & Post-landing Operations	10:27:57 a.m., PST, April 14, 1981, Dry Lakebed Runway 23, Edwards AFB Orbiter was returned to Kennedy April 28, 1981.
Rollout Distance	2,741 m
Rollout Time	60 seconds
Mission Duration	2 days, 6 hours, 20 minutes, and 53 seconds
Landed Revolution No.	37
Mission Support	Spacecraft Tracking and Data Network (STDN)
Deployed Satellites	None
Get-Away Specials	None
Experiments	Data Flight Instrumentation (DFI). This subsystem included special-purpose sensors required to monitor spacecraft conditions and performance parameters not already covered by critical operational systems. The subsystem consisted of transducers, signal conditioning equipment, pulse-code modulation (PCM) encoding equipment, frequency multiplex equipment, PCM recorders, analog recorders, timing equipment, and checkout equipment.

Table 3–17 continued

	Passive Optical Sample Assembly. This assembly consisted of an array of passive samples with various types of surfaces exposed to all STS-1 mission phases. The array was mounted on the DFI pallet in the orbiter payload bay. Ground-based assessments were to evaluate contamination constraints to sensitive payloads to be flown on future missions.

Aerodynamic Coefficient Identification Package (ACIP). This package consisted of three linear accelerometers, three angular accelerometers, three rate gyros, and signal conditioning and PCM equipment mounted on the wing box carry-through structure near the longitudinal center-of-gravity. The instruments sensed vehicle motions during flight from entry initiation to touchdown to provide data for postflight determination of aerodynamic coefficients, aerocoefficient derivatives, and vehicle-handling qualities. |
| **Mission Success** | Successful |

Table 3–18. STS-2 Mission Characteristics

Crew	Cmdr: Joe H. Engle Pilot: Richard H. Truly
Launch	10:09:59 a.m., EST, Nov. 12, 1981, Kennedy Space Center Launch set for October 9 was rescheduled when a nitrogen tetroxide spill occurred during loading of forward reaction control system. Launch on November 4 was delayed and then scrubbed when countdown computer called for a hold in count because of an apparent low reading on fuel cell oxygen tank pressures. During hold, high oil pressures were discovered in two of three auxiliary power units (APUs) that operated hydraulic system. APU gear boxes were flushed and filters replaced, forcing launch reschedule. Launch on November 12 was delayed 2 hours, 40 minutes to replace multiplexer/demultiplexer and additional 9 minutes, 59 seconds to review systems status. Modifications to launch platform to overcome solid rocket booster overpressure problem were effective.
Orbital Altitude & Inclination	222 x 230 km/38 degrees
Total Weight in Payload Bay	8,900 kg
Landing & Post-landing Operations	8:40 a.m., PST, November 14, 1981, Dry Lakebed Runway 23, Edwards AFB Orbiter was returned to Kennedy November 25, 1981.
Rollout Distance	2,350 m
Rollout Time	53 seconds
Mission Duration	2 days, 6 hours, 13 minutes, 12 seconds Mission was shortened by approximately 3 days because of number one fuel cell failure.
Landed Revolution No.	36
Mission Support	Spacecraft Tracking and Data Network (STDN)
Deployed Satellites	None
Get-Away Specials	None
Experiments	Data Flight Instrumentation (see STS-1) Aerodynamic Coefficient Identification Package (see STS-1) Induced Environment Contamination Monitor (IECM). This monitor measured and recorded concentration levels of gaseous and particulate contamination near the payload bay during flight. During ascent and entry, the IECM obtained data on relative humidity and temperature, dewpoint temperature, trace quantities of various compounds, and airborne particulate concentration.

Table 3–18 continued

	Tile Gap Heating Effects Experiment. Analysis and ground tests have indicated that the gap between thermal protection system (TPS) tiles will generate turbulent airflow, resulting in increased heating during entry. Analysis and ground tests also showed that this may be reduced significantly by reconfiguring the tiles with a larger edge radius. To test this effect under actual orbiter entry conditions, a panel with various tile gaps and edge radii was carried.
	Catalytic Surface Experiment. Various orbiter tiles were coated with a highly efficient catalytic overlay. The coating was applied to standard instrumented tiles. This experiment provided a better understanding of the effects of catalytic reaction on convective heat transfer, perhaps permitting a weight reduction in the TPS of future orbiters and other reentry vehicles.
	Dynamic, Acoustic and Thermal Experiment (DATE). The DATE program was to develop improved techniques for predicting the dynamic, acoustic, and thermal environments and associated payload response in cargo areas of large reusable vehicles. The first step was to obtain baseline data of the orbiter environment using existing sensors and data systems. These data served as the basis for developing better prediction methods, which would be confirmed and refined on subsequent flights and used to develop payload design criteria and assess flight performance.
	OSTA-1 Payload (Office of Space and Terrestrial Applications) (see Table 5–55)
Mission Success	Successful

Table 3–19. STS-3 Mission Characteristics

Crew	Cmdr: Jack R. Lousma Pilot: C. Gordon Fullerton
Launch	11:00 a.m., EST, March 22, 1982, Kennedy Space Center The launch was delayed by 1 hour because of the failure of a heater on a nitrogen gas ground support line.
Orbital Altitude & Inclination	208 km/38 degrees
Total Weight in Payload Bay	10,220 kg
Landing & Post-landing Operations	9:04:46 a.m., MST, March 30, 1982, Northrup Strip, White Sands, New Mexico Landing site was changed from Edwards AFB to White Sands because of wet conditions on Edwards dry lakebed landing site. High winds at White Sands resulted in a 1-day extension of mission. Some brake damage upon landing and dust storm caused extensive contamination of orbiter. Orbiter was returned to Kennedy April 6, 1982.
Rollout Distance	4,186 m
Rollout Time	83 seconds
Mission Duration	8 days, 0 hours, 4 minutes, 465 seconds
Landed Revolution No.	130
Mission Support	Spacecraft Tracking and Data Network (STDN)
Deployed Satellites	None
Get-Away Specials	Get-Away Special Verification Payload. This test payload, a cylindrical canister 61 centimeters in diameter and 91 centimeters deep, measured the environment in the canister during the flight. Those data were recorded and analyzed for use by Get-Away Special experimenters on future Shuttle missions.
Experiments	Data Flight Instrumentation (see STS-1) Aerodynamic Coefficient Identification Package (see STS-1) Induced Environment Contamination Monitor (see STS-2) Tile Gap Heating Effects Experiment (see STS-2) Catalytic Surface Experiment (see STS-2) Dynamic, Acoustic and Thermal Experiment (see STS-2) Monodisperse Latex Reactor (MLR). This experiment studied the feasibility of making monodisperse (identically sized) polystyrene latex microspheres, which may have major medical and industrial research applications.

Table 3–19 continued

	Electrophoresis Test. This test evaluated the feasibility of separating cells according to their surface electrical charge. It was a forerunner to planned experiments with other equipment that would purify biological materials in the low gravity environment of space.
	Heflex Bioengineering Test. This preliminary test supported an experiment called Heflex, part of the Spacelab 1 mission. The Heflex experiment would depend on plants grown to a particular height range. The relationship between initial soil moisture content and final height of the plants needed to be determined to maximize the plant growth during the Spacelab mission.
	Infrared Imagery of Shuttle. This experiment obtained high-resolution infrared imagery of the orbiter lower and side surfaces during reentry from which surface temperatures and hence aerodynamic heating may be inferred. The imagery was obtained using a 91.5 cm telescope mounted in the NASA C-141 Gerard P. Kuiper Airborne Observatory positioned at an altitude of 13,700 m along the entry ground track of the orbiter.
	OSS-1 Payload (see Table 4–49)
	Shuttle Student Involvement Project Insects in Flight Motion Study. Investigated two species of insects under uniform conditions of light, temperature, and pressure, the variable being the absence of gravity in space.
Mission Success	Successful

Table 3–20. STS-4 Mission Characteristics

Crew	Cmdr: Thomas K. Mattingly Pilot: Henry W. Hartsfield, Jr.
Launch	June 27, 1982, Kennedy Space Center This was the first Shuttle launch with no delays in schedule. Two solid rocket booster casings were lost when main parachutes failed and they hit the water and sank. Some rainwater penetrated the protective coating of several tiles while the orbiter was on the pad. On orbit, the affected area turned toward the Sun and water evaporation prevented further tile damage from freezing water.
Orbital Altitude & Inclination	258 km/28.5 degrees
Total Weight in Payload Bay	11,021 kg
Landing & Post-landing Operations	July 4, 1982, Runway 22, Edwards AFB This was the first landing on the 15,000-foot-long concrete runway at Edwards AFB. Orbiter was returned to Kennedy July 15, 1982.
Rollout Distance	3,011 m
Rollout Time	73 seconds
Mission Duration	7 days, 1 hour, 9 minutes, 31 seconds
Landed Revolution No.	113
Mission Support	Spacecraft Tracking and Data Network (STDN)
Deployed Satellites	None
Get-Away Specials	G-001 Customer: R. Gilbert Moore Moore, a Morton Thiokol Corporation executive, donated this Get-Away Special to Utah State University. It consisted of 10 experiments dealing with the effects of microgravity on various processes.
Experiments	Aerodynamic Coefficient Identification Package (see STS-1) Catalytic Surface Experiment (see STS-2) Data Flight Investigation (see STS-1) Dynamic, Acoustic and Thermal Experiment (see STS-2) Induced Environment Contamination Monitor (see STS-2) Infrared Imagery of Shuttle (see STS-3) Monodisperse Latex Reactor (see STS-3) Night/Day Optical Survey of Lightning (see STS-2) Tile Gap Heating Experiment (see STS-2)

	Table 3–20 continued
	Continuous Flow Electrophoresis System. This experiment obtained flight data on system performance. During operation, a sample of biological material was continuously injected into a flowing medium, which carried the sample through a separating column where it was under the influence of an electric field. The force exerted by the field separated the sample into its constituent types at the point of exit from the column where samples were collected.
	Department of Defense DOD-82-1 (Classified)
	Shuttle Student Involvement Project • Effects of Diet, Exercise, and Zero Gravity on Lipoprotein Profiles. This project documented the diet and exercise program for the astronauts preflight and postflight. The goal of the research was to determine whether any changes occurred in lipoprotein profiles during spaceflight. • Effects of Space Travel on Trivalent Chromium in the Body. This project was to determine whether any changes occurred in chromium metabolism during spaceflight.
Mission Success	Successful

Table 3–21. STS-5–STS-27 Mission Summary

Mission/ Orbiter	Dates	Crew	Payload and Experiments
STS-5 *Columbia*	Nov. 11–16, 1982	Cmdr: Vance D. Brand Plt: Robert F. Overmyer MS: Joseph P. Allen, William B. Lenoir	**Commercial Payloads** • Satellite Business Systems Satellite (SBS-C)/PAM-D • Telesat-E (Anik C-3)/PAM-D **Experiments and Equipment** • Tile Gap Heating Effects Experiment • Catalytic Surface Effects Experiment • Dynamic, Acoustic and Thermal Environment Experiment (DATE) • Oxygen Atom Interaction With Materials Test • Atmospheric Luminosities Investigation (Glow Experiment) • Development Flight Instrumentation (DFI) • Aerodynamic Coefficient Identification Package (ACIP) **Get-Away Special** • G-026 (DFVLR, West Germany) **Shuttle Student Involvement Program** • Formation of Crystals in Weightlessness • Growth of Porifera in Zero-Gravity • Convection in Zero-Gravity
STS-6 *Challenger*	Apr. 4–9, 1983	Cmdr: Paul J. Weitz Plt: Karol J. Bobko MS: F. Story Musgrave, Donald H. Peterson	**NASA Payload** • Tracking and Data Relay Satellite (TDRS-1)/IUS **Experiments and Equipment** • Continuous Flow Electrophoresis System • Monodisperse Latex Reactor • Nighttime/Daytime Optical Survey of Lightning • ACIP **Get-Away Specials** • G-005 (Asahi Shimbun, Japan) • G-049 (Air Force Academy) • G-381 (Park Seed Company, South Carolina)

STS-7 *Challenger*	June 18–24, 1983	Cmdr: Robert L. Crippen
		Plt: Frederick H. Hauck
		MS: John M. Fabian, Sally K. Ride, Norman E. Thagard
			Commercial Payloads
			• Telesat-F (Anik C-2)/PAM-D
			• Palapa-B1/PAM-D
			NASA Payload
			• OSTA-2 (Office of Space and Terrestrial Applications)
			  – Mission Peculiar Equipment Support Structure (MPESS)
			  – Materials Experiment Assembly (MEA)
			  – Liquid Phase Miscibility Gap Materials
			  – Vapor Growth of Alloy-Type Semiconductor Crystals

Table 3–21 continued

Mission/Orbiter	Dates	Crew	Payload and Experiments
STS-7 continued			– Containerless Processing of Glass Forming Melts – Stability of Metallic Dispersions – Particles at a Solid/Liquid Interface **Detachable Payload** • Shuttle Pallet Satellite (SPAS)-01 **Experiments and Equipment** • Continuous Flow Electrophoresis System (CFES) • Monodisperse Latex Reactor **Get-Away Specials** • G-002 (Kayser Threde, West Germany) • G-009 (Purdue University) • G-012 (RCA/Camden, New Jersey, Schools) • G-033 (California Institute of Technology, Steven Spielberg) • G-088 (Edsyn, Inc.) • G-305 (Air Force/Naval Research Laboratory (NRL), Department of Defense Space Test Program) • G-345 (Goddard Space Flight Center/NRL)
STS-8 *Challenger*	Aug. 30– Sept. 5, 1983	Cmdr: Richard H. Truly Plt: Daniel C. Brandenstein MS: Dale A. Gardner, Guion S. Bluford, Jr., William E. Thornton	**International Payload** • Insat-1B/PAM-D **Detachable Payload** • Payload Flight Test Article **Experiments and Equipment** • Radiation Monitoring Experiment • Development Flight Instrumentation Pallet – Heat Pipe – Oxygen Interaction on Materials • Investigation of STS Atmospheric Luminosities • Animal Enclosure • Continuous Flow Electrophoresis System • Modular Auxiliary Data System (MADS) • ACIP **Get-Away Specials** • G-346 (GSFC/Neupert) • G-347 (GSFC/Adolphsen) • G-348 (GSFC/McIntosh) • G-475 (Asahi/Shimbun, Japan) **Shuttle Student Involvement Program** • SE 81-1 (Biofeedback Mediated Behavioral Training in Physiological Self Regulator: Application in a Near Zero Gravity Environment) **Other** • Postal Covers

Table 3–21 continued

Mission/ Orbiter	Dates	Crew	Payload and Experiments
STS-9 *Columbia*	Nov. 28– Dec. 8, 1983	Cmdr: John W. Young Plt: Brewster H. Shaw MS: Owen K. Garriott, Robert A.R. Parker PS: Byron K. Lichtenberg, Ulf Merbold (ESA)	**International Payload (NASA/ESA)** • Spacelab-1 (long module and pallet —ESA) • Spacelab Attach Hardware, TK, set, Misc.
STS 41-B *Challenger*	Feb. 3–11 1984	Cmdr: Vance D. Brand Plt: Robert L. Gibson MS: Robert L. Stewart, Bruce McCandless, II, Ronald E. McNair	**Commercial Payloads** • Westar VI/PAM-D • Palapa-B2/PAM-D **Attached Payload** • Shuttle Pallet Satellite (SPAS)-01A **Experiments and Equipment** • Integrated Rendezvous Target • Acoustic Containerless Experiment System • Isoelectric Focusing • Radiation Monitoring Experiment • Monodisperse Latex Reactor • Cinema 360 • Manned Maneuvering Unit (MMU) • Manipulation Foot Restraint • Cargo Bay Storage Assembly **Get-Away Specials** • G-004 (Utah State Univ./Aberdeen Univ.) • G-008 (AIAA/Utah State Univ./Brighton High School) • G-051 (GTE Laboratories, Inc.) • G-309 (Air Force Space Test Program) • G-349 (Goddard Space Flight Center) **Shuttle Student Involvement Program** • SE 81-40 (Arthritis, Dan Weber-Pfizer/GD)
STS 41-C *Challenger*	April 6–13, 1984	Cmdr: Robert L. Crippen Plt: Francis R. Scobee MS: Terry J. Hart, James D.A. van Hoften, George D. Nelson	**NASA Payloads** • Long Duration Exposure Facility (LDEF) • Solar Max Mission Flight Support System **Experiments and Equipment:** • Manned Maneuvering Unit Flight Support System • Manned Foot Restraint • Cinema 360 • IMAX • Radiation Monitoring Experiment **Shuttle Student Involvement Program** • Honeycomb construction by bee colony
STS 41-D *Discovery*	Aug. 30– Sept. 5, 1984	Cmdr: Henry W. Hartsfield, Jr. Plt: Michael L. Coats MS: Richard M. Mullane, Steven A. Hawley,	**Commercial Payload** • SBS-4/PAM-D • Syncom IV-2/Unique Upper Stage (Leasat-2) • Telstar 3-C/PAM-D

Table 3–21 continued

Mission/ Orbiter	Dates	Crew	Payload and Experiments
STS 41-D continued		Judith A. Resnik PS: Charles D. Walker	**NASA Payload** • OAST-1/MPESS **Experiments and Equipment** • CFES III • IMAX • Radiation Monitoring Experiment • Clouds Logic to Optimize Use of Defense Systems (CLOUDS) • Vehicle Glow Experiment **Shuttle Student Involvement Program** • SE 82-14 (Purification and Growth of Single Crystal Gallium by the Float Zone Technique in a Zero Gravity Environment, Shawn Murphy/Rockwell International)
STS 41-G *Challenger*	Oct. 5–13, 1984	Cmdr: Robert L Crippen Plt: Jon A. McBride MS: Sally K. Ride, Kathryn D. Sullivan, David C. Leestma PS: Marc D. Garneau, Paul D. Scully-Power	**NASA Payloads** • Earth Radiation Budget Satellite (ERBS) **Experiments and Equipment:** • OSTA-3/Pallet • Large Format Camera (LFC)/CRS/ MPESS • IMAX • Radiation Monitoring Experiment • Auroral Photography Experiment • Thermoluminescent Dosimeter • Canadian Experiment (CANEX) **Get-Away Specials** • G-007 (Student Experiment, Radio Transmission Experiment, Alabama Space and Rocket Center) • G-013 (Halogen Lamp Experiment (HALEX), Kayser-Threde/ESA) • G-032 (Physics of Solids and Liquids, International Space Corp., Asahi Nat. Broadcasting Corp., Japan) • G-038 (Vapor Deposition, McShane/Marshall Space Flight Center) • G-074 (Fuel System Test, MDAC) • G-306 (Trapped Ions in Space, NRL/Navy) • G-469 (Cosmic Ray Upset Experiment, NASA/Goddard/IBM) • G-518 (Physics and Materials Processing, Utah State Univ.)
STS 51-A *Discovery*	Nov. 8–16, 1984	Cmdr: Frederick H. Hauck Plt: David M. Walker MS: Joseph P. Allen, Anna L. Fisher, Dale A. Gardner	**Commercial Payloads** • Telesat-H/PAM-D (Anik D2) • Syncom IV-1/Unique Upper Stage (Leasat-1) • Satellite Retrieval Pallets (2) (Palapa B-2, Westar-6)

Table 3–21 continued

Mission/ Orbiter	Dates	Crew	Payload and Experiments
STS 51-A continued			**Experiments and Equipment** • MMU/Fixed Service Structure (FSS) (2) • Diffuse Mixing of Organic Solids • Radiation Monitoring Experiment • Manual Foot Restraint
STS 51-C *Discovery*	Jan. 24–27, 1985	Cmdr: Thomas K. Mattingly, II Plt: Loren J. Shriver MS: Ellison S. Onizuka, James F. Buchli PS: Gary E. Payton	**NASA Payloads** • DOD 85-1/IUS **Experiments and Equipment** • Aggregation of Red Blood Cells, Middeck Experiment—University of Sydney
STS 51-D *Discovery*	April 12–19, 1985	Cmdr: Karol J. Bobko Plt: Donald E. Williams MS: M. Rhea Seddon, S. David Griggs, Jeffrey A. Hoffman PS: Charles D. Walker, Sen. E.J. Garn	**Commercial Payloads** • Telesat-I/PAM-D (Anik C-1) • Syncom IV-3/Unique Upper Stage (UUS) (Leasat-3) **Experiments and Equipment** • Office of Space Science and Applications Middeck Experiments: – American Flight Echocardiograph – Phase Partitioning Experiment – Protein Crystal Growth (PCG) • CFES III • Image Intensifier Investigation • Informal Science Study (Toys in Space) • Medical Experiments **Get Away Specials** • G-035 (Physics of Solids and Liquids, Asahi, Japan) • G-471 (Capillary Pumped Loop Experiment, Goddard Space Flight Center) **Shuttle Student Involvement Program** • SE 82-03 (Statoliths in Corn Root Caps-Amberg/Martin Marietta) • SE 83-03 (Effect of Weightlessness on Aging of Brain Cells-A. Fras/ USC/Los Angeles Orthopedic Hospital) **Other** • Statue of Liberty Replicas (2)
STS 51-B *Challenger*	April 29–May 6, 1985	Cmdr: R.F. Overmyer Plt: F.D. Gregory MS: Don L. Lind, Norman E Thagard, William Thornton PS: Lodewijk van den Berg, Taylor Wang	**International Payload (NASA/ESA)** • Spacelab 3 (long module and MPESS) **Get-Away Specials (Deployable)** • NUSAT • GLOMR (not deployed)
STS 51-G *Discovery*	June 17–24, 1985	Cmdr: Daniel Brandenstein Plt: John O. Creighton MS: John M. Fabian, Steven R. Nagel, Shannon W. Lucid	**Commercial Payloads** • Morelos-A/PAM-D • Arabsat-A/PAM-D • Telstar 3-D/PAM-D

Table 3–21 continued

Mission/ Orbiter	Dates	Crew	Payload and Experiments
STS 51-G continued		PS: Patrick Baudry (CNES), Prince Sultan Salman Al-Saud	**Deployable** • Spartan-1/MPESS **Experiments and Equipment** • French Echocardiograph Experiment • French Postural Experiment • Automated Directional Solidification Furnace • High-Precision Tracking Experiment **Get-Away Specials** • G-025 (Dynamic Behavior of Liquid Properties, ERNO, West Germany) • G-027 (Slipcasting Under Microgravity, DFVLR, West Germany) • G-028 (Functional Study of MnBi, DFVLR, West Germany) • G-034 (Biological/Physical Science Experiment, El Paso/ Dickshire Coors, Ysleta, Texas) • G-314 (Space Ultraviolet Radiation Environment (SURE), Air Force/NRL) • G-471 (Capillary Pumped Loop Experiment, Goddard)
STS 51-F *Challenger*	July 29– Aug. 6, 1985	Cmdr: C. Gordon Fullerton Plt: Roy Bridges, Jr. MS: F. Story Musgrave, Anthony W. England, Karl G. Henize PS: Loren W. Acton, John-David Bartoe	**International Payload (NASA/ESA)** • Spacelab 2 **Experiments and Equipment** • Shuttle Amateur Radio Experiment • Protein Crystal Growth in a Microgravity Environment **Deployable** • Plasma Diagnostics Package (part of Spacelab 2)
STS 51-I *Discovery*	Aug. 27– Sept. 3, 1985	Cmdr: Joe H. Engle Plt: Richard O. Covey MS: James D.A. van Hoften, John M. Lounge, William F. Fisher	**Commercial Payload** • Aussat-1/PAM-D • ASC-1/PAM-D • Syncom IV-4/Unique Upper Stage (Leasat-4) **Experiments and Equipment** • Physical Vapor Transport of Organic Solids • Syncom IV-3 Repair Equipment
STS 51-J *Atlantis*	Oct. 3–7, 1985	Cmdr: Karl Bobko Plt: Ronald J. Grabe MS: Robert L. Stewart, David C. Hilmers PS: William A. Pailes	DOD Mission

Table 3–21 continued

Mission/ Orbiter	Dates	Crew	Payload and Experiments
STS 61-A *Challenger*	Oct. 30– Nov. 6, 1985	Cmdr: Henry Hartsfield, Jr. Plt: Steven Nagel MS: Bonnie Dunbar, James Buchli, Guion Bluford PS: Ernst Messerschmid, Reinhard Furrer, Wubbo Ockels (ESA)	**International Payload (Germany)** • German Spacelab D-1 (Long Module + Unique Support Structure) **Get-Away Special (Deployed)** • G-308 (GLOMR—DOD) **Experiments and Equipment** • MEA
STS 61-B *Atlantis*	Nov. 26– Dec. 3, 1985	Cmdr: Brewster H. Shaw, Jr. Plt: Bryan D. O'Connor MS: Mary L. Cleave, Sherwood C. Spring, Jerry L. Ross PS: Rodolfo Neri Vela, Charles Walker	**Commercial Payloads** • Morelos B/PAM-D • Aussat-2/PAM-D • Satcom KU-2/PAM-DII **Experiments and Equipment** • EASE/ACCESS/MPESS • IMAX Payload Bay Camera • CFES III • Diffusive Mixing of Organic Solutions • Protein Crystal Growth (PCG) • Morelos Payload Specialist Experiments **Get-Away Special** • G-479 (Primary Surface Mirrors and Metallic Crystals, Telesat, Canada)
STS 61-C *Columbia*	Jan. 12–18, 1986	Cmdr: Robert L. Gibson Plt: C.F. Bolden, Jr. MS: F.R. Chang-Diaz, George D. Nelson, Steven A. Hawley PS: Robert J. Cenker, Congressman Bill Nelson	**Commercial Payloads** • Satcom KU-1/PAM-D2 **Experiments and Equipment** • Materials Science Lab (MSL-2) • Hitchhiker G-1 • Infrared Imaging Experiment • Initial Blood Storage Experiment • Comet Halley Active Monitoring Program • GAS Bridge Assembly (includes 12 GAS cans) **Get-Away Specials** • G-007 (Alabama Space and Rocket Center) • G-062 (Pennsylvania State Univ./ General Electric Co. Space Div.) • G-310 (Air Force Academy/DOD Space Test Program) • G-332 (Booker T. Washington High School, Houston, Texas) • G-446 (High Performance Liquid Chromatography/Alltech Associates Inc.) • G-449 (Joint Utilization of Laser Integrated Experiments/St. Mary's Hospital, Milwaukee)

Table 3–21 continued

Mission/Orbiter	Dates	Crew	Payload and Experiments
STS 61-C continued			Experiment, NASA OSSA) • G-470 (Dept. of Agriculture/Goddard) • G-481 (Vertical Horizons) • G-494 (Photometric Thermospheric Oxygen Nightglow Study/National Research Council of Canada) • Unnumbered (Environmental Monitoring Package, Goddard) **Shuttle Student Involvement Program** • Argon Injection as an Alternative to Honeycombing • Formation of Paper in Microgravity • Measurement of Auxin Levels and Starch Grains in Plant Roots
STS 51-L *Challenger*	Jan. 28–28, 1986	Cmdr: Francis R. Scobee Plt: Michael J. Smith MS: Judith A. Resnik, Ellison S. Onizuka, Ronald E. McNair PS: Gregory Jarvis, S. Christa McAuliffe	**NASA Payload (Planned)** • TDRS-B/IUS-NASA/Spacecom **Experiments and Equipment (Planned)** • Spartan-Halley/MPESS • Comet Halley Active Monitoring Program • Fluid Dynamics Experiment • Radiation Monitoring Experiment • Phase Partitioning Experiment • Teacher in Space Project **Shuttle Student Involvement Program (Planned)** • Utilizing a Semi-Permeable Membrane to Direct Crystal Growth • Effects of Weightlessness on Grain Formation and Strength in Metals • Chicken Embryo Development in Space
STS-26 *Discovery*	Sept. 29– Oct. 3, 1988	Cmdr: Frederick H. Hauck Plt: Richard O. Covey MS: John M. Lounge, David C. Hilmers, George D. Nelson	**NASA Payload** • TDRS-3/IUS **Experiments and Equipment** • Orbiter Experiments Autonomous Supporting Instrumentation System (OASIS) • Automated Directional Solidification Furnace • Aggregation of Red Blood Cells • Earth Limb Radiance Experiment • Isoelectric Focusing Experiment • Infrared Communication Flight Experiment • Mesoscale Lightning Experiment • Protein Crystal Growth (PCG) • Phased Partitioning Experiment • Physical Vapor Transport of Organic Solids

Table 3–21 continued

Mission/Orbiter	Dates	Crew	Payload and Experiments
STS-26 continued			**Shuttle Student Involvement Program** • 82-4 (Utilizing a Semi-Permeable Membrane to Direct Crystal Growth, MDAC/Lloyd Bruce) • 82-5 (Effects of Weightlessness on Grain Formation and Strengthening Metals, Union College/R. Caboli)
STS-27 *Atlantis*	Dec. 2–6, 1988	Cmdr: Robert L. Gibson Plt: Guy S. Gardner MS: Jerry L. Ross, Richard M. Mullane, William M. Shepherd	DOD Payload

Table 3–22. STS-5 Mission Characteristics

Vehicle	*Columbia* (OV-102)
Crew	Cmdr: Vance D. Brand
	Pilot: Robert F. Overmyer
	MS: Joseph P. Allen, William B. Lenoir
Launch	November 11, 1982, 7:19:00 a.m., EST, Kennedy Space Center
	The launch proceeded as scheduled with no delays.
Orbital Altitude & Inclination	298.172 km/28.5 degrees
Launch Weight a	112,090.4 kg
Landing & Post-landing Operations	November 16, 1982, 6:33:26 am PST, Runway 22, Edwards AFB
	Orbiter was returned to Kennedy November 22, 1982.
Rollout Distance	2,911.8 m
Rollout Time	63 seconds
Mission Duration	5 days, 2 hours, 14 minutes, 26 seconds
Landed Revolution No.	82
Mission Support	Spaceflight Tracking and Data Network (STDN)
Deployed Satellites	SBS-C/PAM-D
	Telesat-E 3/PAM-D (Anik C-3)
Get-Away Specials	G-026
	Customer: DFVLR, the German Aerospace Research Establishment
	This GAS was the first in a series of 25 GAS payloads managed by DFVLR. It was part of the German material science program, Project MAUS. Investigators used their knowledge that several combinations of two metals can be dissolved together in their liquid state above a certain temperature (consolute temperature), but not below this temperature. They used a combination of gallium and mercury to investigate the dissolution process above the consolute temperature. X-ray recordings provided real-time data of the different states of the experiment sequence.
Experiments	Tile Gap Heating Effects Experiment. This investigated the heat generated by gaps between the tiles of the thermal protection system on the Shuttle.
	Catalytic Surface Effects Experiment. This investigated the chemical reaction caused by impingement of atomic oxygen on the Shuttle thermal protection system, which was designed with the assumption that the atomic oxygen would recombine at the thermal protection system wall.
	Dynamic, Acoustic and Thermal Environment (DATE) Experiment. This collected data for use in making credible predictions of cargo bay environments. These environments were neither constant nor consistent throughout the bay and were influenced by interactions between cargo elements.

Table 3–22 continued

Atmospheric Luminosities Investigation (Glow Experiment). This experiment was to determine the spectral content of the STS-induced atmospheric luminosities that had relevance for scientific and engineering aspects of payload operations.

Oxygen Atom Interaction With Materials Test. This was conducted to obtain quantitative reaction rates of low-Earth orbit oxygen atoms with various materials used on payloads. Data obtained on STS-2 through 4 indicated that some payloads might be severely limited in life because of oxygen effect. The STS-5 test provided data for assessment of oxygen effects and possible fixes.

Development Flight Instrumentation. This was a data collection and recording package, located in the aft areas of the payload bay, consisting of three magnetic tape recorders, wideband frequency division multiplexers, a pulse code modulation master unit, and signal conditioners.

Aerodynamic Coefficient Identification Package (ACIP). This package, which has flown on STS-1 through 4, continued to collect aerodynamic data during the launch, entry, and landing phases of the Shuttle; establish an extensive aerodynamic database for verification of the Shuttle's aerodynamic performance and the verification and correlation with ground-based data, including assessments of the uncertainties of such data; and provide flight dynamics data in support of other technology areas, such as aerothermal and structural dynamics.

Shuttle Student Involvement Program:
1. Growth of Porifera in Zero-Gravity studied the effect of zero gravity on sponge, Porifera, in relation to its regeneration of structure, shape, and spicule formation following separation of the sponge.
2. Convection in Zero-Gravity studied surface tension convection in zero gravity and the effects of boundary layer conditions and geometries on the onset and character of the convection.
3. Formation of Crystals in Weightlessness compared crystal growth in zero gravity to that in one-g to determine whether weightlessness eliminates the causes of malformation of crystals.

Mission Success	Successful

a   Weight includes all cargo but does not include consumables.

Table 3–23. STS-6 Mission Characteristics

Vehicle	*Challenger* (OV-099)
Crew	Cmdr: Paul J. Weitz
	Pilot: Karol J. Bobko
	MS: Donald H. Peterson, F. Story Musgrave
Launch	April 4, 1983, 1:30:00 p.m., EST, Kennedy Space Center
	The launch set for January 20, 1983, was postponed because of a hydrogen leak into the number one main engine aft compartment, which was discovered during the 20-second Flight Readiness Firing (FRF) on December 18. Cracks in the number one main engine were confirmed to be the cause of the leak during the second FRF performed January 25, 1983. All three main engines were removed while the Shuttle was on the pad, and fuel line cracks were repaired. Main engines two and three were reinstalled following extensive failure analysis and testing. The number one main engine was replaced. An additional delay was caused by contamination to the Tracking and Data Relay Satellite (TDRS-1) during a severe storm. The launch on April 4 proceeded as scheduled.
Orbital Altitude & Inclination	284.5 km/28.45 degrees
Launch Weight	116,459 kg
Landing & Post-landing Operations	April 9, 1983, 10:53:42 a.m., PST, Runway 22, Edwards AFB Orbiter was returned to Kennedy April 16, 1983.
Rollout Distance	2,208 m
Rollout Time	49 seconds
Mission Duration	5 days, 0 hours, 23 minutes, 42 seconds
Landed Revolution No.	81
Mission Support	Spaceflight Tracking and Data Network (STDN)
Deployed Satellites	Tracking and Data Relay Satellite-1/IUS
Get-Away Specials	G-005
	Customer: The Asahi Shimbun
	This experiment was proposed by two Japanese high school students to make artificial snowflakes in the weightlessness of space. The experiment was to contribute to crystallography, especially the crystal growth of semiconductors or other materials from a vapor source.

Table 3–23 continued

	G-049 Customer: Air Force Academy Academy cadets conducted six experiments: 1. Metal Beam Joiner demonstrated that soldering of beams can be accomplished in space. 2. Metal Alloy determined whether tin and lead will combine more uniformly in a zero-gravity environment. 3. Foam Metal generated foam metal in zero-gravity forming a metallic sponge. 4. Metal Purification tested the effectiveness of the zone-refining methods of purification in a zero-gravity environment. 5. Electroplating determined how evenly a copper rod can be plated in a zero-gravity environment. 6. Microbiology tested the effects of weightlessness and space radiation on microorganism development. G-381 Customer: George W. Park Seed Company, Inc. This payload consisted of 46 varieties of flower, herb, and vegetable seeds. It studied the impact of temperature fluctuations, vacuum, gravity forces, and radiation on germination rate, seed vigor, induced dormancy, and varietal purity. An objective was to determine how seeds should be packaged to withstand spaceflight.
Experiments	Continuous Flow Electrophoresis System (CFES). A sample of biological material was continuously injected into a flowing medium, which carried the sample through a separating column where it was under the influences of an electric field. The force exerted by the field separated the sample into its constituent types at the point of exit from the column where samples were collected. Monodisperse Latex Reactor. This materials processing experiment continued the development of uniformly sized (monodisperse) latex beads in a low-gravity environment, where the effects of buoyancy and sedimentation were minimized. The particles may have major medical and industrial research applications. Night/Day Optical Survey of Lightning. This studied lightning and thunderstorms from orbit for a better understanding of the evolution of lightning in severe storms. Aerodynamic Coefficient Identification Package (ACIP) (see STS-5)
Mission Success	Successful

Table 3–24. STS-7 Mission Characteristics

Vehicle	*Challenger* (OV-099)
Crew	Cmdr: Robert L. Crippen
	Pilot: Frederick H. Hauck
	MS: John M. Fabian, Sally K. Ride, Norman E. Thagard
Launch	June 18, 1983, 7:33:00 a.m., EDT, Kennedy Space Center
	The launch proceeded as scheduled.
Orbital Altitude & Inclination	296.3 km/28.45 degrees
Launch Weight	113,027.1 kg
Landing & Post-landing Operations	June 24, 1983, 6:56:59 a.m., PDT, Runway 15, Edwards AFB
	The planned landing at Kennedy was scrubbed because of poor weather conditions, and the mission was extended two revolutions to facilitate landing at Edwards. Orbiter was returned to Kennedy June 29, 1983.
Rollout Distance	3,185 m
Rollout Time	75 seconds
Mission Duration	6 days, 2 hours, 23 minutes, 59 seconds
Landed Revolution No.	98
Mission Support	Spaceflight Tracking and Data Network (STDN)
Deployed Satellites	Telesat-F/PAM-D (Anik C-2), Palapa-B1/PAM-D
Get-Away Specials	G-002
	Customer: Kayser-Threde GMBH
	German high school students provided the experiments for this GAS. Their five experiments studied crystal growth, nickel catalysts, plant contamination by heavy metals, microprocessor controlled sequencers, and a biostack studying the influence of cosmic radiation on plant seeds.
	G-305
	Customer: Department of Defense Space Test Program
	The Space Ultraviolet Radiation Environment (SURE) instrument, developed by the U.S. Naval Research Laboratory (NRL) Space Science Division, marked the debut of the GAS motorized door assembly (MDA). The MDA allowed the payload's spectrometer to measure the natural radiation in the upper atmosphere at extreme ultraviolet wavelengths. SURE was the first in a series of experiments planned by the NRL that ultimately would provide global pictures of "ionospheric weather."
	G-033
	Customer: Steven Speilberg
	Movie director Steven Speilberg donated this GAS to the California Institute of Technology after receiving the payload as a gift. Caltech students designed and built one experiment, which examined oil and water separation in microgravity, and a second, which grew radish seeds, testing the theory that roots grow downward because gravity forces dense structures (amyloplasts) to settle to the bottom of root cells.

Table 3–24 continued

	G-009 Customer: Purdue University Purdue University students conducted three experiments: 1. Seeds were germinated in microgravity on a spinning disk. 2. Nuclear Particle Detection Experiment traced and recorded the paths of nuclear particles encountered in the near-Earth space environment. 3. Fluid Dynamics Experiment measured the bulk oscillations of a drop of mercury immersed in a clear liquid. G-088 Customer: Edsyn, Inc. Edsyn ran more than 60 experiments on soldering and de-soldering equipment. Passive experiments determined how soldering gear would function in space. Powered experiments investigated the physics of soldering in microgravity and a vacuum. G-345 Customer: Goddard Space Flight Center The Ultraviolet Photographic Test Package exposed film samples to the space environment. G-012 Customer: RCA High school students from Camden, New Jersey, with the backing of RCA Corporation and Temple University, investigated whether weightlessness would affect the social structure of an ant colony.
Detachable Payload	Shuttle Pallet Satellite (SPAS)-01. Ten experiments mounted on SPAS-01 performed research in forming metal alloys in microgravity and using a remote-sensing scanner. The orbiter's small control rockets fired while SPAS-01 was held by the RMS to test movement on the extended arm.
Experiments	OSTA-2 Payload (see Chapter 5, "Space Applications") Continuous Flow Electrophoresis System (CFES) (see STS-6) Monodisperse Latex Reaction (see STS-6)
Mission Success	Successful

Table 3–25. STS-8 Mission Characteristics

Vehicle	*Challenger* (OV-99)
Crew	Cmdr: Richard H. Truly Pilot: Daniel C. Brandenstein MS: Dale A. Gardner, Guion S. Bluford, Jr., William E. Thornton
Launch	August 30, 1983, 2:32:00 a.m., EDT, Kennedy Space Center Launch was delayed 17 minutes because of weather.
Orbital Altitude & Inclination	296.3 km/28.45 degrees
Launch Weight	110,107.8 kg
Landing & Post-landing Operations	September 5, 1983, 12:40:43 a.m. PDT, Runway 22, Edwards AFB Orbiter was returned to Kennedy September 9, 1983.
Rollout Distance	2,856.3 m
Rollout Time	50 seconds
Mission Duration	6 days, 1 hour, 8 minutes, 43 seconds
Landed Revolution No.	98
Mission Support	Spaceflight Tracking and Data Network (STDN)
Deployed Satellites	Insat 1B/PAM-D
Get-Away Specials	G-346 Customer: Goddard Space Flight Center The Cosmic Ray Upset Experiment attempted to resolve many of the questions concerning upsets caused by single particles. An upset, or change in logic state, of a memory cell can result from a single, highly energetic particle passing through a sensitive volume in a memory cell. G-347 Customer: Goddard Space Flight Center The Ultraviolet-Sensitive Photographic Emulsion Experiment evaluated the effect of the orbiter's gaseous environment on ultraviolet-sensitive photographic emulsions. G-475 Customer: The Asahi Shimbun The Japanese Snow Crystal Experiment attempted to create the first snowflakes in space, which had been attempted unsuccessfully on STS-6. G-348 Customer: Goddard Space Flight Center The Contamination Monitor Package measured the changes in outer coatings and thermal blanket coverings on the Shuttle that were caused by atomic oxygen erosion.
Experiments	Development Flight Instrumentation Pallet (DFI Pallet): • High Capacity Heat Pipe Demonstration (DSO 0101) provided an in-orbit demonstration of the thermal performance of a high-capacity heat pipe designed for future spacecraft heat rejection systems. • Evaluation of Oxygen Interaction with Materials (DSO 0301) obtained quantitative rates of oxygen interaction with materials used on the orbiter and advanced payloads.

SPACE TRANSPORTATION/HUMAN SPACEFLIGHT

Table 3–25 continued

	Biofeedback Experiments. Six rats were flown in the Animal Enclosure Module to observe animal reactions in space and to demonstrate that the module was capable of supporting six healthy rats in orbit without compromising the health and comfort of either the astronaut crew or the rats. Continuous Flow Electrophoresis System (CFES) (see STS-6) Aerodynamic Coefficient Identification Package (ACIP) (see STS-5) Radiation Monitoring Experiment. This consisted of hand-held and pocket-sized monitors, which measured the level of background radiation present at various times in orbit. The two devices were self-contained and powered by 9-volt batteries. At appointed times, the crew took and recorded measurements of any radiation that penetrated the cabin. Investigation of STS Atmospheric Luminosities (see STS-5) Shuttle Student Involvement Program: Biofeedback Mediated Behavioral Training in Physiological Self Regulator: Application in Near Zero Gravity Environment. This aimed to determine whether biofeedback training learned in a one-g environment can be successfully implemented at zero-g.
Mission Success	Successful

Table 3–26. STS-9 Mission Characteristics

Vehicle	*Columbia* (OV-102)
Crew	Cmdr: John W. Young
	Pilot: Brewster H. Shaw
	MS: Owen K. Garriott, Robert A.R. Parker
	PS: Byron K. Lichtenberg, Ulf Merbold (ESA)
Launch	November 28, 1983, 11:00:00 a.m., EST, Kennedy Space Center Launch set for September 30, 1983, was delayed 28 days because of a suspect exhaust nozzle on the right solid rocket booster. The problem was discovered while the Shuttle was on the launch pad. The Shuttle was returned to the Vehicle Assembly Building and demated. The suspect nozzle was replaced, and the vehicle was restacked. The countdown on November 28 proceeded as scheduled. During launch and ascent, verification flight instrumentation (VFI) operated the Spacelab and the Spacelab interfaces with the orbiter. This instrumentation monitored Spacelab subsystem performance and Spacelab-to-orbiter interfaces. Data were recorded during launch and ascent on the VFI tape recorder and played back to receiving stations on Earth during acquisition of signal periods using the Tracking and Data Relay Satellite System (TDRSS).
Orbital Altitude & Inclination	287.1 km/57.0 degrees
Launch Weight	112,320 kg
Landing & Post-landing Operations	December 8, 1983, 3:47:24 p.m., PST, Runway 17, Edwards AFB Landing was delayed approximately 8 hours to analyze problems when general purpose computers one and two failed and inertial measurement unit one failed. During landing, two of the three auxiliary power units caught fire. During descent and landing, the VFI continued to monitor and record selected Spacelab parameters within the payload bay. One hour after touchdown, power to the induced environment contamination monitor was removed. Orbiter was returned to Kennedy December 15, 1983.
Rollout Distance	2,577.4 m
Rollout Time	53 seconds
Mission Duration	10 days, 7 hours, 47 minutes, 24 seconds
Landed Revolution No.	167
Mission Support	Spaceflight Tracking and Data Network (STDN)/Tracking and Data Relay Satellite System (TDRSS)
Deployed Satellites	INSAT-1B/PAM-D
Get-Away Specials	None
Experiments	See Table 4–45, Spacelab 1 Experiments
Mission Success	Successful

Table 3–27. STS 41-B Mission Characteristics

Vehicle	*Challenger* (OV-099)
Crew	Cmdr: Vance D. Brand
	Pilot: Robert L. Gibson
	MS: Bruce McCandless II, Ronald E. McNair, Robert L. Stewart
Launch	February 3, 1984, 8:00:00 a.m., EST, Kennedy Space Center
	The launch, set for January 29, was postponed for 5 days while the orbiter was still in the Orbiter Processing Facility to allow changeout of all three auxiliary power units (APUs), a precautionary measure in response to APU failures on the STS-9 mission.
Orbital Altitude & Inclination	350 km/28.5 degrees
Launch Weight	113,605 kg
Landing & Post-landing Operations	February 11, 1984, 7:15:55 a.m., EST, Runway 15, Kennedy
	This was the first end-of-mission landing at Kennedy.
Rollout Distance	3,294 m
Rollout Time	67 seconds
Mission Duration	7 days, 23 hours, 15 minutes, 55 seconds
Landed Revolution No.	128
Mission Support	Spaceflight Tracking and Data Network (STDN)
Deployed Satellites	Westar-VI/PAM-D, Palapa-B2/PAM-D
Get-Away Specials	G-004
	Customer: Utah State University
	Students at the University of Aberdeen in Scotland used one of Utah State's spacepaks on this payload. Aberdeen students flew experiments on spore growth, three-dimensional Brownian motion, and dimensional stability. Two other spacepaks contained experiments on capillary action in the absence of gravity.
	G-008
	Customer: Utah State University
	This payload was purchased by the Utah Section of the American Institute of Aeronautics and Astronautics and donated to Utah State University:
	1. In the experiment conducted by students from Brighton High School, Salt Lake City, radish seeds sprouted in a zero-g environment. About one-half of the germinated seeds had flown earlier in an STS-6 experiment.
	2. Students from Utah State University attempted to crystallize proteins in a controlled-temperature environment under zero-g conditions. The crystallization of proteins was necessary for studies in x-ray crystallography.
	3. Two Utah State students devised this payload. The first experiment reran a soldering experiment flown on GAS G-001. The second tested an experimental concept for creating a flow system for electophoresis experiments.

Table 3–27 continued

Experiments	G-349 Customer: Goddard Space Flight Center Contamination Monitor Package (flown on STS-8) measured the flow of atomic oxygen by determining the mass loss of carbon and osmium, known to readily oxidize. The mass loss indicated the atomic oxygen flux as a function of time, which was correlated to altitude, attitude, and direction. This experiment exposed the Shuttle's outer coatings and thermal blanket coverings to normal orbit conditions. G-051 Customer: GTE Laboratories, Inc. Arc Lamp Research studied the configuration of an arc lamp in gravity-free surroundings. Scientists hoped the experiment would pave the way for the development of a more energy-efficient commercial lamp. G-309 Customer: U.S. Air Force Cosmic Ray Upset Experiment (CRUX) was a repeat of G-346 initially flown by Goddard on STS-8. This experiment investigated upsets or changes in the logic state of a memory cell caused by highly active energetic particles passing through a sensitive volume in the memory cell. Acoustic Containerless Experiment (ACES). This materials processing furnace experiment was enclosed in two airtight canisters in the orbiter middeck. Activated at 23 hours mission elapsed time, ACES ran a preprogrammed sequence of operations and shut itself off after 2 hours. Monodisperse Latex Reaction (see STS-6) Radiation Monitoring Experiment (see STS-8) Isoelectric Focusing Experiment. This self-contained experiment package in the middeck lockers was activated by the crew at the same time as ACES. It evaluated the effect of electro-osmosis on an array of eight columns of electrolyte solutions as DC power was applied and pH levels between anodes and cathodes increased.

Table 3–27 continued

	Cinema 360 Camera. Two Cinema 360 cameras were carried on board to provide a test for motion picture photography in a unique format designed especially for planetarium viewing. One camera was located in the crew cabin area and the other in a GAS canister in the payload bay. The primary objective was to test the equipment and concept. Film footage taken by the two systems was also of considerable value. Arriflex 35mm Type 3 motion picture cameras with an 8mm/f2.8 "fisheye" lens were used. The Cinema 360 camera, including an accessory handle and lens guard/support, weighed about 5 kilograms. A system power supply weighed an additional 7.7 kilograms. Filming inside the orbiter focused on activities on the flight deck. The camera system located in the GAS canister in the payload bay provided film on exterior activities, including EVA/MMU operations, satellite deployment, and RMS operations. Lens focus, diaphragm setting, and frame speed were preset, thus requiring no light level readings or exposure calculations by the crew. Each camera carried a 122-meter film magazine. Filming done on this flight and subsequent missions was used in the production of a motion picture about the Space Shuttle program. Shuttle Student Involvement Program: This experiment tested the hypothesis that arthritis may be affected by gravity.
Mission Success	Successful

Table 3–28. STS 41-C Mission Characteristics

Vehicle	*Challenger* (OV-099)
Crew	Cmdr: Robert L. Crippen
	Pilot: Francis R. Scobee
	MS: George D. Nelson, James D. A. van Hoften, Terry J. Hart
Launch	April 6, 1984 8:58:00 a.m., EST, Kennedy Space Center
	The launch proceeded as scheduled with no delays.
Orbital Altitude & Inclination	579.7 km/28.5 degrees
Launch Weight	115,329.6 kg
Landing & Post-landing Operations	April 13, 1984, 5:38:07 a.m., PST, Runway 17, Edwards AFB
	The mission was extended 1 day when astronauts were initially unable to grapple the Solar Maximum Mission spacecraft. The planned landing at Kennedy was scrubbed and the mission extended one revolution to facilitate landing at Edwards. Orbiter was returned to Kennedy April 18, 1984.
Rollout Distance	2,656.6 m
Rollout Time	49 seconds
Mission Duration	6 days, 23 hours, 40 minutes, 7 seconds
Landed Revolution No.	108
Mission Support	Spaceflight Tracking and Data Network (STDN)
Deployed Satellites	Long Duration Exposure Facility-1 (LDEF-1)
Get-Away Specials	None
Experiments	The experiments carried aboard the reusable LDEF fell into four major groups: material structures, power and propulsion, electronics and optics, and science. The 57 separate experiments involved more that 200 investigators from the United States and eight other countries and were furnished by government laboratories, private companies, and universities. They are described in Chapter 4, "Space Science."
	Radiation Monitoring Experiment (see STS-8)
	Cinema 360 (see STS 41-B)
	IMAX. The IMAX camera made the first of three scheduled trips into space on this mission. Footage from the flight was assembled into a film called *The Dream Is Alive*. The IMAX camera was part of a joint project among NASA, the National Air and Space Museum, IMAX Systems Corporation of Toronto, Canada, and the Lockheed Corporation.
	Shuttle Student Involvement Program: This experiment studied the honeycomb structure built by bees in zero gravity, compared to the structure built by bees on Earth.
Mission Success	Successful

Table 3–29. STS 41-D Mission Characteristics

Vehicle	*Discovery* (OV-103)
Crew	Cmdr: Henry W. Hartsfield, Jr.
	Pilot: Michael L. Coats
	MS: Judith A. Resnik, Richard M. Mullane, Steven A. Hawley
	PS: Charles D. Walker
Launch	August 30, 1984, 8:41:50 a.m., EDT, Kennedy Space Center
	The launch attempt on June 25 was scrubbed during a T-9 minute hold because of failure of the orbiter's back-up general purpose computer (GPC). The launch attempt on June 26 aborted at T-4 seconds when the GPC detected an anomaly in the orbiter's number three main engine. *Discovery* was returned to the Orbiter Processing Facility and the number three main engine replaced. (To preserve the launch schedule of future missions, the 41-D cargo was remanifested to include payload elements from both the 41-D and 41-F flights, and the 41-F mission was canceled.) After replacement of the engine, the Shuttle was restacked and returned to the pad. The third launch attempt on August 29 was delayed when a discrepancy was noted in flight software of *Discovery*'s master events controller relating to solid rocket booster fire commands. A software patch was verified and implemented to assure all three booster fire commands were issued in the proper time interval. The launch on August 30 was delayed 6 minutes, 50 seconds when a private aircraft intruded into the warning area off the coast of Cape Canaveral.
Orbital Altitude & Inclination	340.8 km/28.5 degrees
Launch Weight	119,513.2 kg
Landing & Post-landing Operations	September 5, 1984, 6:37:54 a.m. PDT, Runway 17, Edwards AFB Orbiter was returned to Kennedy September 10, 1984.
Rollout Distance	3,131.8 m
Rollout Time	60 seconds
Mission Duration	6 days, 0 hours, 56 minutes, 4 seconds
Landed Revolution No.	97
Mission Support	Spaceflight Tracking and Data Network (STDN)
Deployed Satellites	SBS-4/PAM-D, Syncom IV-2/UUS (Leasat-2), and Telstar 3-C/PAM-D
Get-Away Specials	None
Experiments	Cloud Logic to Optimize Use of Defense Systems (CLOUDS). Sponsored by the Air Force, this payload consisted of two 250-exposure camera assemblies with battery-powered motor drives, which were used at the aft flight deck station for cloud photography data collection.

Table 3–29 continued

	Vehicle Glow Experiment. This experiment characterized surface-originated vehicle glow on strips of material that were attached to the robot arm. Observations made during previous Shuttle flights indicated that optical emissions originated on spacecraft surfaces facing the direction of orbital motion. These emissions showed differing spectral distribution and intensity of the glow for different materials and spacecraft altitude. These results had significance for observations made from the space telescope and space station. CFES-III (see STS-6) Radiation Monitoring Experiment (see STS-8) IMAX (see STS 41-C) Shuttle Student Involvement Program: Purification and Growth of Single Crystal Gallium by the Float Zone Technique in a Zero Gravity Environment, Shawn Murphy/Rockwell International. This experiment compared a crystal grown by the "Flat Zone" technique in a low-gravity environment with one grown in an identical manner on Earth.
Mission Success	Successful

Table 3–30. STS 41-G Mission Characteristics

Vehicle	*Challenger* (OV-099)
Crew	Cmdr: Robert L. Crippen
	Pilot: Jon A. McBride
	MS: David C. Leestma, Sally K. Ride, Kathryn D. Sullivan
	PS: Paul D. Scully-Power, Marc Garneau (Canadian Space Agency)
Launch	October 5, 1984 7:03:00 a.m., EDT, Kennedy Space Center
	Launch proceeded as scheduled with no delays.
Orbital Altitude & Inclination	403.7 km/57.0 degrees
Launch Weight	110,125 kg
Landing & Post-landing Operations	October 13, 1984, 12:26:38 p.m., EDT, Runway 33, Kennedy
Rollout Distance	3,220 m
Rollout Time	54 seconds
Mission Duration	8 days, 5 hours, 23 minutes, 38 seconds
Landed Revolution No.	133
Mission Support	Spaceflight Tracking and Data Network (STDN)/Tracking and Data Relay Satellite System (TDRSS)
Deployed Satellites	Earth Radiation Budget Satellite (ERBS)
Get-Away Specials	G-013
	Customer: Kayser-Threde GMBH
	The Halogen Lamp Experiment (HALEX) tested the performance of halogen lamps during periods of microgravity. The flight was financed by ESA.
	G-007
	Customer: Alabama Space and Rocket Center
	Project Explorer Payload:
	1. This experiment attempted to transmit radio-frequency measurements to ground-based radio hams around the world. This experiment was built by the Marshall Space Flight Center Amateur Radio Club.
	2. Alabama university students investigated the growth of a complex inorganic compound with exceptional conductive properties, the solidification of an alloy with superplastic properties, and the germination and growth of radish seeds in space.
	The payload did not operate, and a reflight was scheduled for STS 61-C.
	G-032
	Customer: International Space Corp.
	This experiment studied the strength of surface tension in the absence of gravity by firing BBs at free-standing spheres of water in microgravity. A second experiment on this GAS used five small electrical furnaces to produce new materials.

Table 3–30 continued

G-306

Customer: Department of Defense Space Test Program
The Trapped Ions in Space experiment recorded the tiny radiation damage tracks left by heavy ions as they passed through a stack of track-detecting plastic sheets during flight. Upon return to Earth, the tracks were etched chemically, revealing cone-shaped pits where particles had passed. Investigators then studied the pits to deduce the energies and arrival direction of the different types of ions that were collected.

G-038

Customer: Marshall—McShane
The investigator used vacuum deposition techniques to coat glass spheres with gold, platinum, and other metals to create lustrous space sculptures. The process was similar to that used on Earth to coat lenses, glass, and mirrors, but the vacuum and weightlessness of space allowed a highly uniform coating only a few microns thick. A control sphere was evacuated to the natural vacuum level of space and sealed. Back on Earth, the investigator took measurements of it to determine the vacuum level at which the depositions had occurred.

G-518

Customer: Utah State University
Four experiments flown on STS 41-B were reflown. The experiments explored basic physical processes in a microgravity environment: capillary waves caused when water is excited, separation of flux and solder, thermocapillary convection, and a fluid flow system in a heat pipe.

G-074

Customer: McDonnell Douglas Astronautics Co.
This experiment demonstrated two methods of delivering partially full tanks of liquid fuel, free of gas bubbles, to engines that control and direct orbiting spacecraft.

G-469

Customer: Goddard Space Flight Center
The Cosmic Ray Upset Experiment (CRUX) III evolved from experiments flown on STS-8 and STS 41-B. It tested fur types of advanced, state-of-the-art microcircuits, totaling more than 12 megabytes. This environment for this experiment was harsher by orders of magnitude than for previous CRUX payloads carried at lower latitudes.

Table 3–30 continued

Experiments	Aurora Photography Experiment. This was conducted for the U.S. Air Force.
	Orbital Refueling System. This developed and demonstrated the equipment and techniques for refueling existing satellites in orbit. Four fuel transfers, controlled by the crew from within the crew cabin, were performed during the mission, in addition to a spacewalk designed to connect a servicing tool to valves that simulated existing satellites not originally designed for on-orbit refueling.
	Radiation Monitoring Experiment (see STS-8)
	IMAX (see STS 41-C)
	Canadian Experiment (CANEX). Mark Garneau, the Canadian payload specialist, conducted ten experiments for the National Research Council of Canada. They fell into the categories of space technology, space science, and life sciences.
	Thermoluminescent Dosimeter (TLD). The Central Research Institute for Physics in Budapest, Hungary, developed a small portable dosimetry system that was carried in a cabin locker. It received doses of cosmic radiation during spaceflight for comparison with the currently used dosimetry systems.
Mission Success	Successful, with the exception of the Shuttle Imaging Radar (SIR)-B. *Challenger*'s Ku-band antenna problems severely affected the SIR-B. A reflight of SIR-B was requested and manifested on STS 72-A, at that time scheduled for launch in February 1987.

Table 3–31. STS 51-A Mission Characteristics

Vehicle	*Discovery* (OV-103)
Crew	Cmdr: Frederick H. Hauck
	Pilot: David M. Walker
	MS: Anna L. Fisher, Dale A. Gardner, Joseph P. Allen
Launch	November 8, 1984, 7:15:00 a.m., EST, Kennedy Space Center Launch attempt on November 7 was scrubbed during a built-in hold at T-20 minutes because of wind shears in the upper atmosphere. The countdown on November 8 proceeded as scheduled.
Orbital Altitude & Inclination	342.6 km/28.5 degrees
Launch Weight	119,443.7 kg
Landing & Post-landing Operations	November 16, 1984, 6:59:56 a.m., EST, Runway 15, Kennedy
Rollout Distance	2,881.6 m
Rollout Time	58 seconds
Mission Duration	7 days, 23 hours, 44 minutes, 56 seconds
Landed Revolution No.	127
Mission Support	Spaceflight Tracking and Data Network (STDN)/ Tracking and Data Relay Satellite System (TDRSS)
Deployed Satellites	Telesat-H/PAM-D (Anik D2), Syncom IV-1/PAM-D (Leasat-1)
Get-Away Specials	None
Experiments	The Diffusive Mixing of Organic Solutions (DMOS) experiment, a collaboration between 3M and NASA, was the first attempt to grow organic crystals in the microgravity environment of the orbiter. The program's ultimate goal was to produce commercially valuable products in the fields of organic and polymer chemistry. The experiment studied the physical processes that govern the crystal growth and evaluated the diffusive mixing method of crystal growth. It also evaluated the type of apparatus used for its suitability for crystal growth in the weightless environment of low-Earth orbit.
	Radiation Monitoring Experiment (see STS-8)
Mission Success	Successful

Table 3–32. STS 51-C Mission Characteristics

Vehicle	*Discovery* (OV-103)
Crew	Cmdr: Thomas K. Mattingly II Pilot: Loren J. Shriver MS: James F. Buchli, Ellison S. Onizuka PS: Gary E. Payton
Launch	January 24, 1985, 2:50:00 p.m., EST, Kennedy Space Center The January 23 launch was scrubbed because of freezing weather conditions. (*Challenger* was scheduled for STS 51-C, but thermal tile problems forced the substitution of *Discovery*.)
Orbital Altitude & Inclination	407.4 km/28.5 degrees
Launch Weight	113,804.2 kg
Landing & Post-landing Operations	January 27, 1985, 4:23:23 p.m., EST, Runway 15, Kennedy
Rollout Distance	2,240.9 m
Rollout Time	50 seconds
Mission Duration	3 days, 1 hour, 33 minutes, 23 seconds
Landed Revolution No.	49
Mission Support	Spaceflight Tracking and Data Network (STDN)/ Tracking and Data Relay Satellite System (TDRSS)
Deployed Satellites	DOD 85-1/IUS
Get-Away Specials	None
Experiments	Aggregation of Red Blood Cells. This tested the capability of the apparatus to study in weightlessness some of the various characteristics of blood, such as viscosity, and their disease dependencies. Preliminary results indicated that: • It was possible to obtain perfect microphotographs of blood cells in space under conditions of heavy vibration. • Cells form aggregates that grow with time, analogous to patterns on Earth. • The internal organization and structure of aggregates seem to be different under zero gravity. • Individual red cells do not show abnormal shapes under zero gravity; notwithstanding the origin of the blood samples, they looked normal. • Because there was no sludging under weightlessness, studies on interactions between cells should be much easier. • Changes in shape of red cells in astronauts (as reported by Johnson Space Center) must be caused by a change of calcium metabolism.
Mission Success	Successful

Table 3–33. STS 51-D Mission Characteristics

Vehicle	*Discovery* (OV-103)
Crew	Cmdr: Karol J. Bobko
	Pilot: Donald E. Williams
	MS: M. Rhea Seddon, S. David Griggs, Jeffrey A. Hoffman
	PS: Charles D. Walker, Sen. E.J. Garn
Launch	April 12, 1985, 8:59:05 a.m., EST, Kennedy Space Center
	The launch set for March 19 was rescheduled to March 28 because of remanifesting of payloads from canceled mission 51-E. The launch was delayed further because of damage to the orbiter's payload bay door when the facility access platform dropped. The April 12 launch was delayed 55 minutes when a ship entered the restricted solid rocket booster recovery area.
Orbital Altitude & Inclination	527.8 km/28.5 degrees
Launch Weight	113,804.2 kg
Landing & Post-landing Operations	April 19, 1985, 8:54:28 a.m. EST, Runway 33, Kennedy
	Extensive brake damage and a blown tire during landing prompted the landing of future flights at Edwards AFB until the implementation of nose wheel steering.
Rollout Distance	3,138.8 m
Rollout Time	63 seconds
Mission Duration	6 days, 23 hours, 55 minutes, 23 seconds
Landed Revolution No.	110
Mission Support	Spaceflight Tracking and Data Network (STDN)/Tracking and Data Relay Satellite System (TDRSS)
Deployed Satellites	Telesat-I/PAM-D (Anik C-1), Syncom IV-3/UUS (Leasat-3)
Get-Away Specials	G-035
	Customer: The Asahi Shimbun
	Physics of Solids and Liquids in Zero Gravity was designed to determine what happened when a metal or plastic (solid) was allowed to collide with a water ball (liquid) in weightlessness. The behavior of the metal or plastic ball and the water ball after collision was observed on video systems.
	G-471
	Customer: Goddard Space Flight Center
	Capillary Pump Loop Experiment investigated whether a capillary pump system could transfer waste heat from a spacecraft out into space. The experiment consisted of two capillary pump evaporators with heaters and was designed to demonstrate that such a system can be used under zero-gravity conditions of spaceflight to provide thermal control of scientific instruments, advanced orbiting spacecraft, and space station components.

Table 3–33 continued

Experiments	Phase Partitioning Experiment. Phase partitioning is a selective, yet gentle and inexpensive technique used to separate biomedical materials, such as cells and proteins. It establishes a two-phase system by adding various polymers to a water solution containing the materials to be separated. Theoretically, phase partitioning should separate cells with significantly higher resolution than was presently obtained in the laboratory. Investigators believed that when the phases are emulsified on Earth, the rapid, gravity-driven fluid movements occurring as the phases coalesce tended to randomize the separation process. They expected that the theoretical capabilities of phase partitioning systems could be more closely approached in the weightlessness of orbital spaceflight, where gravitational effects of buoyancy and sedimentation were minimized.
	American Flight Echocardiograph. This experiment studied the effects of weightlessness on the cardiovascular system of astronauts, which was important for both personal and operational safety reasons. The newly available instrument gathered in-flight data on these effects during space adaptation to develop optimal countermeasures to crew cardiovascular changes (particularly during reentry) and to ensure long-term safety to people living in weightlessness.
	Protein Crystal Growth (PCG). This experiment studied the composition and structure of proteins, extremely important to the understanding of their nature and chemistry and the ability to manufacture them for medical purposes. However, for most complex proteins, it had not been possible to grow, on Earth, crystals large enough to permit x-ray or neutron diffraction analyses to obtain this information. A key objective of the overall PCG program was to enable drug design without the present empirical approach to enzyme engineering and the manufacture of chemotherapeutic agents.
	Toys in Space. The crew demonstrated the behavior of simple toys in a weightless environment. The results, recorded and videotaped, became part of a curriculum package for elementary and junior high students through the Houston Museum of Natural Science. Studies showed that students could learn physics concepts by watching mechanical systems in action. In an Earth-based classroom, the gravitational field has a constant value of 1-g. Although the gravity force varied greatly throughout the universe and in noninertial reference frames, students could only experiment in a constant 1-g environment. The filming of simple generic-motion toys in the zero-g environment enabled students to discover how the different toy mechanical systems work without gravity.

Table 3–33 continued

	Image Intensifier Investigation. This tested low-light-level photographic equipment, in preparation for the visit by Halley's Comet. Astronaut Hoffman examined an image intensifier coupled with a Nikon camera, a combination that intensified usable light by a factor of about 10,000. It was believed that the equipment could be used to observe objects of astronomical interest through the Shuttle's window, including Halley's Comet. Hoffman photographed objects at various distances from the Sun when it was below the horizon, similar to lighting conditions when the comet appeared. Continuous Flow Electrophoresis System (CFES) III (see STS-8) Shuttle Student Involvement Program: 1. Statoliths in Corn Root Caps examined the effect of weightlessness on the formation of statoliths (gravity-sensing organs) in plants and was tested by exposing plants with capped and uncapped roots to spaceflight. The root caps of the flight and control plants were examined postflight by an electron microscope for statolith changes. 2. Effect of Weightlessness on the Aging of Brain Cells used houseflies and was expected to show accelerated aging in their brain cells based on an increased accumulation of age pigment in, and deterioration of, the neurons.
Medical Experiments	Utah Senator E.J. "Jake" Garn was the first public official to fly aboard the Space Shuttle. Garn was a payload specialist and congressional observer. As payload specialist, he conducted medical physiological tests and measurements. Tests on Garn detected and recorded changes the body underwent in weightlessness, an ongoing program that began with astronauts on the fourth Shuttle flight. Garn accomplished the following: • During launch, Garn wore a waist belt with two stethoscope microphones fastened to an elastic bandage. At main engine cutoff, about 8.5 minutes into the flight, the belt was plugged into a portable tape recorder stored in the seat flight bag and began recording bowel sounds to evaluate early in-flight changes in gastric mobility. • An electrocardiogram recorded electrical heart rhythm in the event of space motion sickness in orbit. • Garn was wore a leg plethysmography stocking to measure leg volume. It recorded the shifting of fluids during adaptation to weightlessness. • Blood pressure and heart rate were recorded in orbit and during reentry. • Another test measured Garn's height and girth in space to determine the amount of growth and change in body shape associated with weightlessness. Space travelers may grow up to 2 inches while weightless. • Tests determined whether a medication dosage on Earth was adequate in space with acetaminophen. Garn's saliva was collected for analysis after each dose.
Mission Success	Successful

Table 3–34. STS 51-B Mission Characteristics

Vehicle	*Challenger* (OS-099)
Crew	Cmdr: Robert F. Overmyer
	Pilot: Frederick D. Gregory
	MS: Don L. Lind, Norman E. Thagard, William E. Thornton
	PS: Lodewijk van den Berg, Taylor G. Wang
Launch	April 29, 1985, 12:02:18 p.m., EDT, Kennedy Space Center
	This flight was first manifested as 51-E. It was rolled back from the pad because of a timing problem with the TDRS-B payload. Mission 51-E was canceled, and the orbiter was remanifested with 51-B payloads. The launch on April 29 was delayed 2 minutes, 18 seconds because of a launch processing system failure.
Orbital Altitude & Inclination	411.1 km/57.0 degrees
Launch Weight	111,984.8 kg
Landing & Post-landing Operations	May 6, 1985, 9:11:04 a.m. PDT, Runway 17, Edwards AFB Orbiter was returned to Kennedy May 11, 1985.
Rollout Distance	2,535 m
Rollout Time	59 seconds
Mission Duration	7 days, 0 hours, 8 minutes, 46 seconds
Landed Revolution No.	111
Mission Support	Spaceflight Tracking and Data Network (STDN)/ Tracking and Data Relay Satellite System (TDRSS)
Deployed Satellites	NUSAT (Get-Away Special); GLOMR was scheduled for deployment but was rescheduled on STS 61-A
Get-Away Specials	G-010
	Customer: R. Gilbert Moore
	Northern Utah Satellite (NUSAT) was a cooperative effort among the Federal Aviation Administration (FAA), Weber State College, Utah State University, New Mexico State University, Goddard, the U.S. Air Force, and more than 26 private corporations. It was deployed into a 20-month orbit. It was an air traffic control radar calibration system that measured antenna patterns for ground-based radar operated in the United States and in member countries of the International Civil Aviation Organization.
	G-308
	Customer: Department of Defense Space Test Program
	The Global Low Orbiting Message Relay (GLOMR) satellite was planned to pick up digital data streams from ground users, store the data, and deliver the messages in these data streams to customers' computer terminals upon command. It was designed to remain in orbit for 1 year. However, because of a malfunction in the Motorized Door Assembly, GLOMR was not deployed on this mission.
Experiments	See Table 4–46, Spacelab 3 Experiments
Mission Success	Successful

Table 3–35. STS 51-G Mission Characteristics

Vehicle	*Discovery* (OV-103)
Crew	Cmdr: Daniel C. Brandenstein
	Pilot: John O. Creighton
	MS: Shannon W. Lucid, Steven R. Nagel, John M. Fabian
	PS: Patrick Baudry (CNES), Sultan Salman Al-Saud
Launch	June 17, 1985, 7:33:00 a.m., EDT, Kennedy Space Center
	The launch proceeded as scheduled with no delays.
Orbital Altitude & Inclination	405.6 km/28.5 degrees
Launch Weight	116,363.8 kg
Landing & Post-landing Operations	June 24, 1985, 6:11:52 a.m., PDT, Runway 23, Edwards AFB
	Orbiter was returned to Kennedy June 28, 1985.
Rollout Distance	2,265.6 m
Rollout Time	42 seconds
Mission Duration	7 days, 1 hour, 38 minutes, 52 seconds
Landed Revolution No.	112
Mission Support	Spaceflight Tracking and Data Network (STDN)/Tracking and Data Relay Satellite System (TDRSS)
Deployed Satellites	Morelos-A/PAM-D, Telstar-3D/PAM-D, Arabsat-A/PAM-D, Spartan-1 (deployed and retrieved)
Get-Away Specials	G-025
	Customer: ERNO-Raumfahrttechnik GMBH
	Liquid Sloshing Behavior in Microgravity examined the behavior of liquid in a tank in microgravity. It was representative of phenomena occurring in satellite tanks with liquid propellants. The results were expected to validate and refine mathematical models describing the dynamic characteristics of tank-fluid systems. This in turn would support the development of future spacecraft tanks, in particular the design of propellant management devices for surface-tension tanks.
	G-027
	Customer: DFVLR
	Slipcasting Under Microgravity Conditions was performed by Germany's material research project, MAUS. Its goal was to demonstrate with model materials the possibility of slipcasting in microgravity, even with unstabilized suspensions using mixtures of powders with different density, grain size, and concentration.
	G-028
	Customer: DFVLR
	Fundamental Studies in Manganese-Bismuth produced manganese-bismuth specimens with possibly better magnetic properties than currently was possible under Earth gravity.

Table 3–35 continued

	G-034 Customer: Dickshire Coors Texas Student Experiments featured twelve different biological and physical science experiments designed by high school students from El Paso and Ysleta, Texas. The microgravity experiments studied the growth of lettuce seeds, barley seed germination, the growth of brine shrimp, germination of turnip seeds, the regeneration of the flat work planeria, the wicking of fuels, the effectiveness of antibiotics on bacteria, the growth of soil mold, crystallization in zero gravity, the symbiotic growth of the unicellular algae chlorella and the milk product kefir, the operation of liquid lasers, and the effectiveness of dynamic random access memory computer chips without ozone protection. G-314 Customer: DOD Space Test Program Space Ultraviolet Radiation Environment (SURE) measured the natural radiation field in the upper atmosphere at extreme ultraviolet wavelengths, between 50 and 100 nanometers. These measurements provided a way of remotely sensing the ionosphere and upper atmosphere. G-471 Customer: Goddard Space Flight Center Capillary Pumped Loop investigated the thermal control capability of a capillary-pumped system under zero-gravity conditions for ultimate use in large scientific instruments, advanced orbiting spacecraft, and space station components.
Experiments	Spartan 1. This was the first in a series of Shuttle-launched, short-duration free-flyers designed to extend the capabilities of sounding rocket class experiments. Its primary mission was to perform medium-resolution mapping of the x-ray emission from extended sources and regions, specifically the hot (10,000 degrees Celsius) gas pervading a large cluster of galaxies in the constellation Perseus and in the galactic center and Scorpius-X-2. In addition, it mapped the x-ray emissions from the nuclear region of the Milky Way galaxy. French Echocardiograph Experiment. This measured and studied the evolution of the fundamental parameters that characterized cardiac function, blood vessel circulation, and cardiovascular adaptation. After reviewing the data, the principal investigator observed a decrease of cardiac volume, stroke volume, and left ventricular diastolic volume, a decrease in cerebral circulatory resistance, and noted variations in peripheral resistance and vascular stiffness of the lower limbs.

Table 3–35 continued

	French Posture Experiment. This had five general objectives: a study of the adaptive mechanisms of postural control, the influence of vision in adaptations, the role of the otoliths in the oculomotor stabilization reflexes, their role in the coordination of eye and head movements, and mental representation of space. After reviewing the data, the principal investigator observed a change in vertical optokinetic nystagmus, which included an asymmetry reversal and a downward shift in beating field of the nystagmus, as well as a decrease in the gain of the vestibular ocular reflex.
	Automated Directional Solidification Furnace. Experiments carried out in the furnace demonstrated the capability of the furnace equipment and provided preliminary scientific results on magnetic composites. Future missions would demonstrate the feasibility of producing improved magnetic composite materials for commercial use. These materials could eventually lead to smaller, lighter, stronger and longer lasting magnets for electrical motors used in aircraft and guidance systems, surgical instruments, and transponders.
	High-Precision Tracking Experiment. Flown by the Strategic Defense Initiative Organization, this tested the ability of a ground laser beam director to accurately track an object in low-Earth orbit.
Mission Success	Successful

Table 3–36. STS 51-F Mission Characteristics

Vehicle	*Challenger* (OV-099)
Crew	Cmdr: C. Gordon Fullerton
	Pilot: Roy D. Bridges, Jr.
	MS: F. Story Musgrave, Karl G. Henize, Anthony W. England
	PS: Loren W. Acton, John-David F. Bartoe
Launch	July 29, 1985, 5:00:00 p.m., EDT, Kennedy Space Center
	The launch countdown on July 12 was halted at T-3 seconds when a malfunction of the number two main engine coolant valve caused a shutdown of all three main engines. Launch countdown was initiated on July 27 and continued to about T-9 minutes on July 29. At that time, launch was delayed 1 hour, 37 minutes because of a problem with the table maintenance block update uplink. In addition, ascent was hampered when at 5 minutes, 45 seconds into ascent, the number one main engine shut down prematurely, resulting in an abort-to-orbit trajectory. The abort-to-orbit trajectory resulted in the orbiter's insertion orbit altitude being approximately 108 x 143 nautical miles. A final orbit if 314.84 x 316.69 kilometers was achieved to meet science payload requirements. During launch and ascent, verification flight instrumentation (VFI) operated. The VFI was strategically located throughout Spacelab and at the Spacelab interfaces with the orbiter. The VFI monitored Spacelab subsystem performance and Spacelab/orbiter interfaces. Data were recorded during launch and ascent on the VFI tape recorder and played back to ground receiving stations during acquisition of signal periods utilizing the Tracking and Data Relay Satellite System (TDRSS).
Orbital Altitude & Inclination	314.84 km/49.5 degrees
Launch Weight	114,695 kg
Landing & Post-landing Operations	August 6, 1985, 12:45:26 p.m., PDT, Runway 23, Edwards AFB
	The VFI continued to monitor and record selected Spacelab parameters on the VFI tape recorder and the orbiter payload recorder during descent and landing. Approximately 25 minutes after landing, orbiter power was removed from Spacelab. The mission was extended 17 revolutions for additional payload activities because of the abort-to-orbit. Orbiter was returned to Kennedy August 11, 1985.
Rollout Distance	2,611.8 m
Rollout Time	55 seconds
Mission Duration	7 days, 22 hours, 45 minutes, 26 seconds
Landed Revolution No.	127
Mission Support	Spaceflight Tracking and Data Network (STDN)/ Tracking and Data Relay Satellite System (TDRSS)
Deployed Satellites	Plasma Diagnostics Package (PDP) (see experiments below)
Get-Away Specials	None

322 NASA HISTORICAL DATA BOOK

Table 3–36 continued

Experiments	Spacelab 2 (see Table 4–47, Spacelab 2 Experiments)
	Plasma Diagnostics Package. The instrument package was extended and released by the RMS to take measurements after the orbiter maneuvered to selected attitudes. After taking measurements, the manipulator arm recaptured the PDP and returned it to the vicinity of the payload bay. Before landing, it was locked back in place on the aft pallet. Instruments mounted within the PDP included a quadrispherical low-energy proton and electron differential analyzer, a plasma wave analyzer and electric dipole and magnetic search coil sensors, a direct current electric field meter, a triaxial flux-gate magnetometer, a Langmuir probe, a retarding potential analyzer and differential flux analyzer, an ion mass spectrometer, and a cold cathode vacuum gauge. (See Chapter 4 for further data on the PDP.)
	Protein Crystal Growth in a Microgravity Environment. The purpose was to develop hardware and procedures for growing proteins and other organic crystals by two methods in the orbiter during the low-gravity portion of the mission. Generally, hardware for both methods worked as planned. Postflight analysis showed that minor modification in the flight hardware was needed and a means of holding the hardware during activation, crystal growth, deactivation, and photography was desirable. The dialysis method produced three large tetragonal lysozyme crystals with average dimensions of 1.3 mm x 0.65 mm x 0.65 mm. The solution growth methods produced small crystals of lysozyme, alpha-2 interferon, and bacterial purine nucleoside phosphorylase. (See also STS 51-D.)
	Gravity-Influenced Lignification in Higher Plants/Plant Growth Unit. Mung beans and pine seedlings, planted in the Plant Growth Unit before flight, were flown to monitor the production of lignin, a structural rigidity tissue found in plants.
	Shuttle Amateur Radio Experiment (SAREX). Astronauts England and Bartoe conversed from *Challenger* with amateur radio operators through a handheld radio.
Mission Success	Successful

Table 3–37. STS 51-I Mission Characteristics

Vehicle	*Discovery* (OV-103)
Crew	Cmdr: Joseph H. Engle
	Pilot: Richard O. Covey
	MS: James D.A. van Hoften, John M. Lounge, William F. Fisher
Launch	August 27, 1985, 6:58:01 a.m., EDT, Kennedy Space Center
	The August 24 launch was scrubbed at T-5 minutes because of thunderstorms in the vicinity. The launch on August 25 was delayed when the orbiter's number five on-board general purpose computer failed. The launch on August 27 was delayed 3 minutes, 1 second because of a combination of weather and an unauthorized ship entering the restricted solid rocket booster recovery area.
Orbital Altitude & Inclination	514.9 km/28.5 degrees
Launch Weight	118,983.4 kg
Landing & Post-landing Operations	September 3, 1985, 6:15:43 a.m., PDT, Runway 23, Edwards AFB
	The mission was shortened 1 day when the Aussat sunshield hung up on the Remote Manipulator System camera and Aussat had to be deployed before scheduled. Orbiter was returned to Kennedy September 8, 1985.
Rollout Distance	1,859.3 m
Rollout Time	47 seconds
Mission Duration	7 days, 2 hours, 17 minutes, 42 seconds
Landed Revolution No.	112
Mission Support	Spaceflight Tracking and Data Network (STDN)/Tracking and Data Relay Satellite System (TDRSS)
Deployed Satellites	ASC-1/PAM-D; Aussat-1/PAM-D, Syncom IV-4/UUS (Leasat-4); Syncom IV-4 failed to function after reaching correct geosynchronous orbit
Get-Away Specials	None
Experiments	Physical Vapor Transport Organic Solid Experiment (PVTOS). In this second of some 70 experiments the 3M Corporation planned to conduct by 1995, solid materials were vaporized into a gaseous state to form thick crystalline films on selected substrates of sublimable organics. 3M researchers studied the crystals produced by PVTOS for their optical properties and other characteristics that might ultimately have important applications to 3M's businesses in the areas of electronics, imaging, and health care.
Mission Success	Successful

Table 3–38. STS 51-J Mission Characteristics

Vehicle	*Atlantis* (OV-104)
Crew	Cmdr: Karol J. Bobko
	Pilot: Ronald J. Grabe
	MS: Robert L. Stewart, David C. Hilmers
	PS: William A. Pailes
Launch	October 3, 1985, 11:15:30 a.m., EDT, Kennedy Space Center
	The launch was delayed 22 minutes, 30 seconds because of the main engine liquid hydrogen prevalve close remote power controller showing a faulty "on" indication.
Orbital Altitude & Inclination	590.8 km/28.5 degrees
Launch Weight	classified
Landing & Post-landing Operations	October 7, 1985, 10:00:08 a.m., PDT, Runway 23, Edwards AFB Orbiter returned to Kennedy October 11, 1985.
Rollout Distance	2,455.5 m
Rollout Time	65 seconds
Mission Duration	4 days, 1 hour, 44 minutes, 38 seconds
Landed Revolution No.	64
Mission Support	Spaceflight Tracking and Data Network (STDN)
Deployed Satellites	Not available
Get-Away Specials	None
Experiments	Not available
Mission Success	Successful

Table 3–39. STS 61-A Mission Characteristics

Vehicle	*Challenger* (OV-099)
Crew	Cmdr: Henry W. Hartsfield, Jr.
	Pilot: Steven R. Nagel
	MS: James F. Buchli, Guion S. Bluford, Jr., Bonnie J. Dunbar
	PS: Reinhard Furrer, Ernst Messerschmid, Wubbo J. Ockels (ESA)
Launch	October 30, 1985, 12:00:00 noon, EST, Kennedy Space Center
	Launch proceeded as scheduled with no delays.
Orbital Altitude & Inclination	383.4 km/57.0 degrees
Launch Weight	110,570.4 kg
Landing & Post-landing Operations	November 6, 1985, 9:44:53 a.m., PST, Runway 17, Edwards AFB Orbiter was returned to Kennedy November 11, 1985.
Rollout Distance	2,531.1 m
Rollout Time	45 seconds
Mission Duration	7 days, 0 hours, 44 minutes, 53 seconds
Landed Revolution No.	112
Mission Support	Spaceflight Tracking and Data Network (STDN)/Tracking and Data Relay Satellite System (TDRSS)
Deployed Satellites	Global Low Orbiting Message Relay (GLOMR) deployed from G-308
Get-Away Specials	G-308
	Customer: Department of Defense Space Test Program
	GLOMR, originally planned for deployment on STS 51-B, was successfully deployed and remained in orbit for 14 months. The GLOMR satellite, a 68-kilogram, 62-side polyhedron, was a data-relay, communications spacecraft. Its purpose was to demonstrate the ability to read signals and command oceanographic sensors, locate oceanographic and other ground sensors, and relay data from them to customers. The satellite could pick up digital data streams from ground users, store the data, and deliver the streams to customers' computer terminals upon command.
Experiments	Spacelab D-1 (see Table 4–48, Spacelab D-1 Experiments)
Mission Success	Successful

Table 3–40. STS 61-B Mission Characteristics

Vehicle	*Atlantis* (OV-104)
Crew	Cmdr: Brewster H. Shaw, Jr.
	Pilot: Bryan D. O'Connor
	MS: Mary L. Cleave, Sherwood C. Spring, Jerry L. Ross
	PS: Rodolfo Neri Vela, Charles D. Walker
Launch	November 26, 1985, 7:29:00 p.m., EST, Kennedy Space Center
	The launch proceeded as scheduled with no delays.
Orbital Altitude & Inclination	416.7 km/28.5 degrees
Launch Weight	118,596 kg
Landing & Post-landing Operations	December 3, 1985, 1:33:49 p.m., PST, Runway 22, Edwards AFB
	The mission was shortened one revolution because of lightning conditions at Edwards. *Atlantis* landed on a concrete runway because the lakebed was wet. Orbiter was returned to Kennedy December 7, 1985.
Rollout Distance	3,279.3 m
Rollout Time	78 seconds
Mission Duration	6 days, 21 hours, 4 minutes, 49 seconds
Landed Revolution No.	109
Mission Support	Spaceflight Tracking and Data Network (STDN)/ Tracking and Data Relay Satellite System (TDRSS)
Deployed Satellites	Morelos-B/PAM-D, AUSSAT-2/PAM-D, Satcom Ku-2/PAM-DII
Get-Away Specials	G-479
	Customer: Telesat Canada
	Telesat, Canada's domestic satellite carrier, sponsored a national competition soliciting science experiments from Canadian high school students. The selected experiment, called Towards a Better Mirror, proposed to fabricate mirrors in space that would provide higher performance than similar mirrors made on Earth.
Experiments	Orbiter Experiments (OEX). An onboard experimental digital autopilot software package was tested. The autopilot software could be used with the orbiter, another space vehicle, such as the Orbital Transfer Vehicle, which was under development, or the space station. OEX was designed to provide precise stationkeeping capabilities between various vehicles operating in space.
	Protein Crystal Growth Experiment (PCG). This experiment studied the possibility of crystallizing biological materials, such as hormones, enzymes, and other proteins. Successful crystallization of these materials, which were very difficult to crystallize on Earth, would allow their three-dimensional atomic structure to be determined by x-ray crystallography.
	IMAX Cargo Bay Camera. The IMAX camera was used to document payload bay activities associated with the EASE/ACCESS assembly during the two spacewalks.
	Experimental Assembly of Structures in Extravehicular Activity (EASE). This experiment studied EVA dynamics and human factors in the construction of structures in space.

Table 3–40 continued

Assembly Concept for Construction of Erectable Space Structures (ACCESS). This experiment validated ground-based timelines based on the neutral buoyancy water simulator at the Marshall Space Flight Center, Huntsville, Alabama. Crew members manually assembled and disassembled a 45-foot truss to evaluate concepts for assembling larger structures in space.

Morelos Payload Specialist Experiments. Rodolfo Neri Vela performed a series of mid-deck cabin experiments and took photographs of Mexico:
1. Effects of Spatial Environment on the Reproduction and Growing of Bacteria. Cultures of *Escherichia coli* B-strain were mixed in orbit with different bacteriophages that attack the *E. coli* and were observed for possible changes and photographed as required.
2. Transportation of Nutrients in a Weightless Environment. Ten plant specimens were planted in containers that allowed a radioactive tracer to be released in orbit for absorption by the plants. At selected intervals, each plant was sectioned and the segments retained for postflight analysis to determine the rate and extent of absorption.
3. Electropuncture and Biocybernetics in Space. This experiment validated electropuncture theories, which stated that disequilibrium in the behavior of human organs could be monitored and stimulated using electric direct current in specified zones. The experiment was performed by measuring the conductance of electricity in a predetermined zone. If a disequilibrium was detected, exercises or stimulus would be applied for a certain period until the value of the conductance fell into the normal range.
4. Effects of Weightlessness and Light on Seed Germination. Seed specimens of amaranth, lentil, and wheat were planted in orbit in two identical containers. One container was exposed to illumination and the other to constant darkness. Photographs of the resulting sprouts were taken every 24 hours. One day prior to landing, the sprouts were submitted to a metabolical detection process for subsequent histological examination on Earth to determine the presence and localization of starch granules.
5. Photography of Mexico. Postearthquake photos were taken of Mexico and Mexico City.

Diffusive Mixing of Organic Solutions. This experiment grew organic crystals in near-zero gravity. 3M scientists hoped to produce single crystals that are more pure and larger than those available on Earth to study their optical and electrical properties.

Continuous Flow Electrophoresis System (CFES) (see STS-6)

Mission Success	Successful

Table 3–41. STS 61-C Mission Characteristics

Vehicle	Columbia (OV-102)
Crew	Cmdr: Robert L. Gibson
	Pilot: Charles F. Bolden, Jr.
	MS: Franklin R. Chang-Diaz, Steven A. Hawley, George D. Nelson
	PS: Robert J. Cenker, Congressman Bill Nelson
Launch	January 12, 1986, 6:55:00 a.m., EST, Kennedy Space Center. The launch set for December 18, 1985, was delayed 1 day when additional time was needed to close out the orbiter aft compartment. The December 19 launch attempt was scrubbed at T-14 seconds because of an indication that the right solid rocket booster hydraulic power unit was exceeding RPM redline speed limits. (This was later determined to be a false reading.) After an 18-day delay, a launch attempt on January 6, 1986, was halted at T-31 seconds because of the accidental draining of liquid oxygen from the external tank. The January 7 launch attempt was scrubbed at T-9 minutes because of bad weather at both transoceanic landing sites (Moron, Spain, and Dakar, Senegal). After a 2-day delay, the launch set for January 9 was delayed because of a launch pad liquid oxygen sensor breaking off and lodging in the number two main engine prevalve. The launch set for January 10 was delayed for 2 days because of heavy rains. The launch countdown on January 12 proceeded with no delays.
Orbital Altitude & Inclination	392.6 km/28.5 degrees
Launch Weight	116,123 kg
Landing & Post-landing Operations	January 18, 1986, 5:58:51 a.m. PST, Runway 22, Edwards AFB. The planned landing at Kennedy, originally scheduled for January 17, was moved to January 16 to save orbiter turnaround time. The landing attempts on January 16 and 17 were abandoned because of unacceptable weather at Kennedy. A landing was set for January 18 at Kennedy, but persisting bad weather forced a one-revolution extension of mission and landing at Edwards AFB. Orbiter was returned to Kennedy January 23, 1986.
Rollout Distance	3,110 m
Rollout Time	59 seconds
Mission Duration	6 days, 2 hours, 3 minutes, 51 seconds
Landed Revolution No.	98
Mission Support	Spaceflight Tracking and Data Network (STDN)/ Tracking and Data Relay Satellite System (TDRSS)
Deployed Satellites	Satcom KU-1/PAM-DII
Get-Away Specials	The Environmental Monitoring Package was contributed by Goddard to measure the effects of launch and landing forces on the bridge and, hence, on the internal environment of the GAS containers. Sound levels, vibrations, and temperature were measured by attaching acoustical pickups, accelerometers, strain gauges, and thermocouples to the bridge. These instruments were connected to a GAS container with equipment that controlled the instruments and recorded their data.

Table 3–41 continued

The GAS Bridge Assembly was flown for the first time on this mission. It contained the 12 GAS canisters.

G-310
The objective of this Air Force Academy–sponsored payload was to measure the dynamics of a vibrating beam in a zero-g environment.

G-463, G-464, G-462
Customer: NASA Office of Space Science and Applications
Ultraviolet Experiment was a group of get-away specials designed to measure diffuse ultraviolet background radiation. The two ultraviolet spectrometers were to look into distant space to observe the high-energy spectrum thought to be associated with the origin of the universe. Other observational targets included galaxies, dust areas, Halley's Comet, and selected stars. It was the only set of GAS experiments to fly as a group of three electrically interconnected containers.

G-062
Customer: General Electric Company Space Division
Four student experiments from Pennsylvania State University and sponsored by the General Electric Co. made up this payload. The liquid droplet heat radiator experiment tested an alternative method of heat transfer, which investigated how moving droplets can radiate heat into space. The second experiment studied the effect of microgravity on the surface tension of a fluid. The third experiment studied the effect of convection on heat flow in a liquid by submersing a heat source in a container of liquid.

G-332
Customer: Booker T. Washington High School
This canister contained two contributions from Houston, Texas. The Brine Shrimp Artemia experiment from Booker T. Washington High School determined the behavioral and physiological effects of microgravity on eggs hatched in space. The High School for Engineering provided the Fluid Physics Experiment, which examined the behavior of fluid when heated in microgravity.

G-446
Customer: Alltech Associates, Inc.
This experiment investigated the effect of gravity on particle dispersion of packing material in high-performance liquid chromatography analytical columns. The investigators expected that by reducing gravity, a more efficient column would be produced.

Table 3–41 continued

G-470
Customer: Goddard Space Flight Center
A joint investigation by Goddard and the U.S. Department of Agriculture examined the effects of weightlessness on gypsy moth eggs and engorged female American dog ticks. It was hoped that the data obtained would lead to new means of controlling these insect pests.

G-007
Customer: Alabama Space and Rocket Center
This canister housed four specific payloads that were originally scheduled to fly on STS 41-G. However, it was not turned on during that mission. Postflight investigation determined that the experiments were not at fault, and they were rescheduled for STS 61-C. The experiment included:
1. A study of the solidification of lead-antimony and aluminum-copper alloys
2. A comparative morphological and anatomical study of the primary root system of radish seeds
3. Examination of the growth of metallic-appearing needle crystals in an aqueous solution of potassium tetracyanoplatinate
4. A half-wave dipole antenna installed on the canister's top cover plate that was sponsored by the Marshall Amateur Radio Club

G-449
Customer: St. Mary's Hospital, Milwaukee
The Laser Laboratory at St. Mary's Hospital in Milwaukee sponsored this four-part experiment:
1. The BMJ experiment studied the biological effects of neodymium and helium-neon laser light upon desiccated human tissue undergoing cosmic ray bombardment. Medications also were exposed to laser light and cosmic radiation.
2. LEDAJO was to determine cosmic radiation effects on medications and medical/surgical materials using Lexan detectors.
3. BLOTY analyzed contingencies that develop because of zero-gravity in blood typing. In Earth-bound blood typing, gravity was essential to produce clumping.
4. CROLO evaluated laser optical protective eyewear materials following exposure to cosmic radiation.

G-481
Customer: Vertical Horizons
This payload transported samples of painted linen canvases and other artistic materials into space. The investigators evaluated how unprimed canvas, prepared linen canvas, and portions of oil painted canvas reacted to space travel.

Table 3–41 continued

Experiments	G-494 Customer: National Research Council of Canada This payload was co-sponsored by the Canada Centre for Space Science and the National Research Council of Canada. The experiment consisted of seven filtered photometers that measured oxygen, oxide, and continuum emissions in the terrestrial night glow and in the Shuttle night glow. Materials Science Laboratory-2 (MSL-2). Primary mission objectives were the engineering verification of the MSL payload carrier and of the three materials processing facilities. Secondary objectives were the acquisition of flight specimens and experimental data for scientific evaluation. The MSL-2 held the following experiments: 1. Electromagnetic Levitator. This experiment studied the effects of material flow during solidification of a melted material in the microgravity environment. 2. Automated Directional Solidification Furnace. This experiment investigated the melting and solidification process of four materials. 3. Three-Axis Acoustic Levitator. Twelve liquid samples were suspended in sound pressure waves, and rotated and oscillated in a low-gravity, nitrogen atmosphere. Investigators studied the degree of sphericity attainable and small bubble migration similar to that found in the refining of glass. Comet Halley Active Monitoring Program. This was supposed to investigate the dynamical/morphological behavior as well as the chemical structure of Comet Halley. The 35mm camera that was to photograph Comet Halley did not function properly because of battery problems. Infrared Imaging Experiment. This acquired radiometric pictures/information of selected terrestrial and celestial targets. Initial Blood Storage Experiment. This experiment investigated the factors that limit the storage of human blood. The experiment attempted to isolate factors such as sedimentation that occurred under standard blood bank conditions. A comparison was made of changes in whole blood and blood components that had experienced weightless conditions in orbit with similar samples stored in otherwise comparable conditions on the ground.

Table 3–41 continued

	Hitchhiker G-1. This was the first of a generic class of small payloads under the Small Payload Accommodation program. These payloads were located in the orbiter bay on the starboard side and used specially designed carriers, which attached to the existing GAS attach fittings. This supported three instruments: 1. Particle Analysis Cameras for the Shuttle provided film images of particle contamination around the Shuttle in support of future DOD infrared telescope operations. 2. Capillary Pump Loop provided a zero-g test of a new heat transport system. 3. Shuttle Environment Effects on Coated Mirrors was a passive witness mirror-type experiment that determined the effects of contamination and atomic oxygen on ultraviolet optics material. Shuttle Student Involvement Program: 1. Argon Injection as an Alternative to Honeycombing was a material processing experiment that examined the ability to produce a lightweight, honeycomb structure superior to the Earth-produced structures. 2. Formation of Paper in Microgravity studied the formation of cellulose fibers in a fiber mat under weightless conditions. 3. Measurement of Auxin Levels and Starch Grains in Plant Roots investigated the geotropism of a corn root growth in microgravity and determined whether starch grains in the root cap were actually involved with auxin production and transport.
Mission Success	Successful

Table 3–42. STS 51-L Mission Characteristics

Vehicle	*Challenger* (OV-099)
Crew	Cmdr: Francis R. Scobee
	Pilot: Michael J. Smith
	MS: Judith A. Resnik, Ellison S. Onizuka, Ronald E. McNair
	PS: Gregory B. Jarvis
	Teacher in Space Project: Sharon Christa McAuliffe
Launch	January 28, 1986, 11:38:00, EST, Kennedy Space Center
	The first Shuttle liftoff scheduled for January 22 was slipped to January 23, then January 24, because of delays in STS 61-C. The launch was reset for January 25 because of bad weather at the transoceanic abort landing site in Dakar, Senegal. To use Casablanca (not equipped for night landings) as an alternate transoceanic abort landing site, T-zero was moved to a morning liftoff time. The launch was postponed 1 day when launch processing was unable to meet the new morning liftoff time. A prediction of unacceptable weather at Kennedy led to the launch being rescheduled for 9:37 a.m., EST, January 27. The launch was delayed 24 hours when the ground-servicing equipment hatch-closing fixture could not be removed from the orbiter hatch. The fixture was sawed off and the attaching bolt drilled out before closeout was completed. During the delay, cross winds exceeded return-to-launch-site limits at Kennedy's Shuttle Landing Facility. The January 28 launch was delayed 2 hours when the hardware interface module in the launch processing system, which monitors the fire detection system, failed during liquid hydrogen tanking procedures. An explosion 73 seconds after liftoff claimed the crew and vehicle.
Orbital Altitude & Inclination	2,778.8 km (planned)/28.5 degrees (planned)
Launch Weight	121,778.4 kg
Landing & Post-landing Operations	No landing
Rollout Distance	N/A
Rollout Time	N/A
Mission Duration	73 seconds
Landed Revolution No.	N/A
Mission Support	Spaceflight Tracking and Data Network (STDN)/ Tracking and Data Relay Satellite System (TDRSS)
Deployed Satellites	None
Get-Away Specials	None
Experiments	None
Mission Success	Unsuccessful

Table 3–43. STS-26 Mission Characteristics

Vehicle	*Discovery* (OV-103)
Crew	Cmdr: Frederick H. Hauck
	Pilot: Richard O. Covey
	MS: John M. Lounge, David C. Hilmers, George D. Nelson
Launch	September 29, 1988, 11:37:00 a.m., EDT, Kennedy Space Center The launch was delayed 1 hour, 38 minutes to replace the fuses in the cooling systems of two of the crew's flight pressure suits and because of lighter than expected upper atmospheric winds. Suit repairs were successful, and the countdown continued after a waiver of a wind condition constraint.
Orbital Altitude & Inclination	376 km/28.5 degrees
Launch Weight	115,489.3 kg
Landing & Post-landing Operations	October 3, 1988, 9:37:11 a.m., PDT, Runway 17, Edwards AFB Orbiter was returned to Kennedy October 8, 1988.
Rollout Distance	2,271.1 m
Rollout Time	46 seconds
Mission Duration	4 days, 1 hour, 0 minutes, 11 seconds
Landed Revolution No.	64
Mission Support	Spaceflight Tracking and Data Network (STDN)/ Tracking and Data Relay Satellite System (TDRSS)
Deployed Satellites	TDRS-3/IUS
Get-Away Specials	None
Experiments	Physical Vapor Transport of Organic Solids. This experiment by 3M scientists produced organic thin films with ordered crystalline structures to study their optical, electrical, and chemical properties. The results could eventually be applied to the production of specialized thin films on Earth or in space.
	Protein Crystal Growth (PCG) experiments. A team of industry, university, and government research investigators explored the potential advantages of using protein crystals grown in space to determine the complex, three-dimensional structure of specific protein molecules. Knowing the precise structure of these complex molecules would aid in understanding their biological function and could lead to methods of altering or controlling the function in ways that may result in new drugs.
	Infrared Communications Flight Experiment. Using the same kind of invisible light that remotely controls home TV sets and VCRs, mission specialist Nelson conducted experimental voice communications with his crewmates via infrared, rather than standard radio-frequency, waves. One major objective of the experiment was to demonstrate the feasibility of the secure transmission of information via infrared light. Unlike radio-frequency signals, infrared waves will not pass through the orbiter's windows; thus, a secure voice environment would be created if infrared waves were used as the sole means of communications within the orbiter. Infrared waves can also carry data as well as voice (such as biomedical information).

Table 3–43 continued

Automated Directional Solidification Furnace. This special space furnace developed and managed by Marshall Space Flight Center demonstrated the possibility of producing lighter, stronger, and better performing magnetic composite materials in a microgravity environment.

Aggregation of Red Blood Cells. Blood samples from donors with such medial conditions as heart disease, hypertension, diabetes, and cancer flew in this experiment developed by Australia and managed by Marshall. The experiment provided information on the formation rate, structure, and organization of red cell clumps and on the thickness of whole blood cell aggregates at high and low flow rates. It helped determine whether microgravity could play a beneficial role in new and existing clinical research and medical diagnostic tests. Results obtained in the Shuttle microgravity environment were compared with results from a ground-based experiment to determine what effects gravity had on the kinetics and morphology of the sampled blood.

Isoelectric Focusing. This was a type of electrophoresis experiment that separated proteins in an electric field according to their surface electrical charge.

Mesoscale Lightning Experiment. This obtained nighttime images of lightning to better understand the effects of lightning discharges on each other, on nearby storm systems, and on storm microbursts and wind patterns and to determine interrelationships over an extremely large area.

Phase Partitioning Experiment. This investigated the role gravity and other physical forces played in separating—that is, partitioning—biological substances between two unmixable liquid phases.

OASIS Instrumentation. This collected and recorded a variety of environmental measurements during various in-flight phases of the orbiter. The information was used to study the effects on the orbiter of temperature, pressure, vibration, sound, acceleration, stress, and strain. It also was used to assist in the design of future payloads and upper stages.

Earth-Limb Radiance Experiment. Developed by the Barnes Engineering Co., this photographed Earth's "horizon twilight glow" near sunrise and sunset. The experiment provided photographs of Earth's horizon that allowed scientists to measure the radiance of the twilight sky as a function of the Sun's position below the horizon. This information allowed designers to develop better, more accurate horizon sensors for geosynchronous communications satellites.

Table 3–43 continued

	Shuttle Student Involvement Program: 1. Utilizing a Semi-Permeable Membrane to Direct Crystal Growth attempted to control crystal growth through the use of a semi-permeable membrane. Lead iodide crystals were formed as a result of a double replacement reaction. Lead acetate and potassium iodide reacted to form insoluble lead iodide crystals, potassium ions, and acetate ions. As the ions traveled across a semi-permeable membrane, the lead and iodide ions collided, forming the lead iodide crystal. 2. Effects of Weightlessness on Grain Formation and Strengthening Metals heated a titanium alloy metal filament to near the melting point to observe the effect of weightlessness on crystal reorganization within the metal. It was expected that heating in microgravity would produce larger crystal grains and thereby increase the inherent strength of the metal filament.
Mission Success	Successful

Table 3–44. STS-27 Mission Characteristics

Vehicle	*Atlantis* (OV-104)
Crew	Cmdr: Robert L. Gibson
	Pilot: Guy S. Gardner
	MS: Richard M. Mullane, Jerry L. Ross, William M. Shepherd
Launch	December 2, 1988, 9:30:34 a.m., EST, Kennedy Space Center
	The launch, set for December 1, 1988, during a classified window lying within a launch period between 6:32 a.m. and 9:32 a.m., was postponed because of unacceptable cloud cover and wind conditions and reset for the same launch period on December 2.
Orbital Altitude & Inclination	Altitude classified/57.0 degrees
Launch Weight	Classified
Landing & Post-landing Operations	December 6, 1988, 3:36:11 p.m., PST, Runway 17, Edwards AFB Orbiter was returned to Kennedy December 13, 1988
Rollout Distance	2,171.1 m
Rollout Time	43 seconds
Mission Duration	4 days, 9 hours, 5 minutes, 37 seconds
Landed Revolution No.	Not available
Mission Support	Spaceflight Tracking and Data Network (STDN)/ Tracking and Data Relay Satellite System (TDRSS)
Deployed Satellites	Not available
Get-Away Specials	None
Experiments	Not available
Mission Success	Successful

Table 3–45. Return to Flight Chronology

Date	Event
January 28, 1986	Moments after the *Challenger* (STS 51-L) explosion, all mission data, flight records, and launch facilities are impounded. Within an hour, Associate Administrator for Space Flight Jesse Moore names an expert panel to investigate.
January 29	Interim Mishap Investigation Board is named and approved by NASA Acting Administrator William R. Graham.
February 3	President Ronald Reagan announces the formation of a presidential commission to investigate the *Challenger* accident. Commission is to be headed by former Secretary of State William Rogers.
February 5	Acting Administrator Graham establishes the 51-L Data and Design Analysis Task Force to assist the Rogers Commission, designating the Associate Administrator for Space Flight as chairperson.
February 18	U.S. Senate holds first of a series of hearings on the *Challenger* accident.
February 20	Rear Admiral Richard H. Truly is appointed Associate Administrator for Space Flight.
February 25	Administrator James M. Beggs, on leave since December 4, 1985, pending disposition of indictment, resigns. Indictment is later dismissed, and Beggs receives apology from Attorney General Edwin Meese.
March 5	First Program Management Review for Space Shuttle program is held at Marshall. Reviews are planned for every 6 weeks.
March 13	NASA begins review of Failure Modes Effects Analysis and Critical Items Lists.
March 24	Admiral Truly, in NASA Headquarters Office of Space Flight memorandum "Strategy for Safely Returning Space Shuttle to Flight Status," outlines actions required prior to next flight, first flight/first year operations, and development of sustainable flight rate. Truly directs Marshall to form solid rocket motor joint redesign team with National Research Council (NRC) oversight. Truly initiates review of National Space Transportation System (NSTS) management structure. First system design review is conducted to identify changes to improve flight safety.
March 28	Arnold D. Aldrich, manager of NSTS, initiates review of all items on Critical Items List.
March	NASA Flight Rate Capability Working Group is established.
April 7	NASA initiates Shuttle crew egress and escape review.
May 12	James C. Fletcher is sworn in as NASA Administrator.
June 6	*Report to the President by the Presidential Commission on the Space Shuttle Challenger Accident* (Rogers Commission) is released. It recommends: • Redesign faulty joint seal (either eliminate joint or redesign seal to more stringent standards) • Provide independent redesign oversight by NRC • Review Shuttle management to redefine the program manager's responsibility, place astronauts in management positions, and establish an STS Safety Advisory Panel • Improve criticality review and hazard analysis (An audit panel from NRC should verify the adequacy of the effort.)

Table 3-45 continued

Date	Event
June 6 cont.	• Establish a safety office headed by a NASA associate administrator to oversee safety, reliability, maintainability, and quality assurance functions with viable problem reporting, documentation, and resolution • Improve communication, especially from Marshall, develop launch constraints policy, and record Flight Readiness Review (FRR) and Mission Management Team meetings (Flight crew commander should attend FRR.) • Improve landing safety, including tire brake and nosewheel steering, and conditions for Kennedy landing, with landing area weather forecasts more than an hour in advance, and create crew escape system for controlled gliding flight and launch abort possibilities in case of main engine failures early in ascent • Establish flight rate to be consistent with NASA resources, and create firm payload assignment policy • Implement maintenance safeguards, especially for Criticality I items
June 11	Admiral Truly testifies before the House Committee on Science and Technology on status of work in response to Rogers Commission recommendations and announces small group to examine overall Space Shuttle program management, to be headed by astronaut Robert L. Crippen.
June 13	President Reagan writes to NASA requesting the implementation of Rogers Commission recommendations.
June 19	Centaur upper stage is terminated because of safety concerns.
June 25	Astronaut Robert L. Crippen is directed to form a fact-finding group to assess Shuttle management structure and implement effective management and communications.
June 30	Andrew J. Stofan is appointed Associate Administrator of the Office of Space Station at NASA Headquarters.
July 8	NASA establishes an Office of Safety, Reliability, Maintainability, and Quality Assurance and appoints George A. Rodney Associate Administrator.
July 11	NASA Report to the President, *Actions to Implement Recommendations of the Presidential Commission on the Space Shuttle Challenger*, announces return to flight for first quarter of 1988. Fletcher states NASA has responded favorably to the Rogers Commission recommendations in every area and promises another report in 1 year.
July 24	NASA announces abandonment of lead center concept for space station.
August 15	President Reagan issues statement announcing intent to build a fourth Shuttle orbiter as a replacement and that NASA will no longer launch private satellites.
September 10	Astronaut Brian D. O'Connor is appointed chair of the Space Flight Safety Panel.
September 29	James R. Thompson is appointed Director of Marshall Space Flight Center.
October 1	Lt. Gen. Forrest S. McCartney is appointed Director of Kennedy Space Center.

Table 3–45 continued

Date	Event
October 3	Revised NASA manifest is published incorporating president's new policy on commercialization of space and changes in priorities for flying on the Shuttle.
October 6	Dale D. Myers is appointed Deputy Administrator of NASA.
October 12	Aaron Cohen is appointed Director of Johnson Space Center.
October 14	Astronaut Frederick D. Gregory is appointed Chief, Operational Safety Branch, Safety Division, Office of Safety, Reliability, Maintainability, and Quality Assurance at NASA Headquarters.
October 29	U.S. House of Representatives Committee on Science and Technology releases its report: *Investigation of the Challenger Accident*.
November 11	Shuttle management is reorganized. NSTS manager Aldrich is appointed Director of NSTS in the Office of Space Flight at NASA Headquarters. Two NSTS deputy director positions are established: Richard H. Kohrs as Deputy Director for NSTS program and Robert L. Crippen as Deputy Director of NSTS operations. Shuttle project office manager at Marshall is to report directly to the deputy director for NSTS program.
December 30	Former Apollo program manager Brig. Gen. Samuel C. Phillip's study of NASA management is presented to the NASA administrator.
January 7, 1987	Administrator Fletcher issues "State of NASA" memorandum and reestablishes Project Approval Document as a management tool.
January 9	Flight crew is selected for first Space Shuttle mission (STS-26, *Discovery*) after accident: commander—Frederick H. Hauck; pilot—Richard O. Covey; and mission specialists—John M. Lounge, George D. Nelson, and David C. Himmers.
January 25	Public Opinion Laboratory publishes *The Impact of the Challenger Accident on Public Attitudes Toward the Space Program: A Report to the National Science Foundation*. Findings include: • Accident increased an already strong national pride in the Shuttle program. Public responded to the deaths of the *Challenger* astronauts with a sense of personal loss. • Public viewed accident as a minor setback, with universal expectation of a return to flight. • Cost-benefit assessment increased significantly as a result of the accident. • There was a willingness to support increased federal funds for space. • Rogers Commission discussion and criticism did not erode positive views of NASA held by public. • Net effect of accident was a more positive attitude toward the space program.
February 25	NASA publishes *Responses to the Recommendations of the House of Representatives Committee on Science and Technology Report of the Investigation of the Challenger Accident*, which includes a summary of activities undertaken in response to the Rogers Commission investigation.
February	Crew begins training for STS-26 mission.
March 9	Former NASA Deputy Associate Administrator (1965–1975) Willis H. Shapley is appointed Associate Deputy Administrator (Policy).

Table 3–45 continued

Date	Event
May 29	John M. Klineberg is appointed Director of Lewis Research Center.
June 22	Noel W. Hinners is appointed Associate Deputy Administrator (Institution).
June 22	John W. Townsend is appointed Director of Goddard Space Flight Center.
June 30	Administrator Fletcher submits report to the president on status of NASA's work to implement Rogers Commission recommendations. Report details changes to solid rocket motor design, management structure and communications, criticality review and hazards analysis, safety organization, landing safety, launch abort and crew escape, flight rate maintenance safeguards, and related return to flight safeguards.
July 22	Second interim progress report of NRC's Committee on Shuttle Criticality Review and Hazard Analysis Audit is issued.
July 31	Replacement orbiter contract is awarded to Rockwell International, and production of OV-105 is initiated.
August 4	STS-26 begins power-up.
August 17	*Leadership and America's Future in Space* (Ride Report) is released.
August 30	First major test occurs on redesigned solid rocket motor.
August	Advanced solid rocket motor design and definition study contracts are awarded to five aerospace firms by Marshall.
October 22	NASA issues first mixed fleet manifest for Space Shuttle missions and expendable launch vehicles.
October 29	Vice President Bush, in speech at Marshall, pledges to reestablish the National Space Council if elected president.
November 19	Testing begins on escape system that could be activated during controlled gliding flight.
February 11, 1988	White House issues the *President's Space Policy Directive and Commercial Space Initiative,* declaring it is the president's policy to establish long-range goals to expand the human presence and activity beyond Earth orbit into the solar system, to create opportunities for U.S. commerce in space, and to continue the national commitment to a permanently manned space station.
September 29	Successful launch of STS-26, *Discovery*, signals NASA's "return to flight." Mission launches TDRS-C, lasts 4 days, 1 hour, 57 seconds, and orbits Earth 64 times.

Table 3–46. Sequence of Major Events of the Challenger Accident

Mission Time (GMT, in hr:min:sec)	Event	Elapsed Time (sec.)	Source
16:37:53.444	ME 3 Ignition Command	6.566	GPC
37:53.564	ME 2 Ignition Command	6.446	GPC
37:53.684	ME 1 Ignition Command	6.326	GPC
38:00.010	SRM Ignition Command (T=0)	0.000	GPC
38:00.018	Holddown Post 2 PIC firing	0.008	E8 Camera
38:00.260	First continuous vertical motion	0.250	E9 Camera
38:00.688	Confirmed smoke above field joint on RH solid rocket motor	0.678	E60 Camera
38:00.846	Eight puffs of smoke (from 0.836 through 2.500 sec MET)	0.836	E63 Camera
38:02.743	Last positive evidence of smoke above right aft solid rocket booster/ external tank attach ring	2.733	CZR-1 Camera
38:03.385	Last positive visual indication of smoke	3.375	E60 Camera
38:04.349	SSME 104% Command	4.339	E41M2076D
38:05.684	RH solid rocket motor pressure 11.8 psi above nominal	5.674	B47P2302C
38:07.734	Roll maneuver initiated	7.724	V90R5301C
38:19.869	SSME 94% Command	19.859	E41M2076D
38:21.134	Roll maneuver completed	21.124	VP0R5301C
38:35.389	SSME 65% Command	35.379	E41M2076D
38:37.000	Roll and yaw attitude response to wind (36.990 to 62.990 sec)	36.990	V95H352nC
38:51.870	SSME 104% Command	51.860	E41M2076D
38:58.798	First evidence of flame on RH solid rocket motor	58.788	E207 Camera
38:59.010	Reconstructed Max Q (720 psf)	59.000	BET
38:59.272	Continuous well-defined plume on RH solid rocket motor	59.262	E207 Camera
38:59.763	Flame from RH solid rocket motor in +Z direction (seen from south side of vehicle)	59.753	E204 Camera
39:00.014	SRM pressure divergence (RH vs. LH)	60.004	B47P2302
39:00.248	First evidence of plume deflection, intermittent	60.238	E207 Camera
39:00.258	First evidence of solid rocket booster plume attaching to external tank ring frame	60.248	E203 Camera
39:00.998	First evidence of plume deflection, continuous	60.988	E207 Camera
39:01.734	Peak roll rate response to wind	61.724	V90R5301C
39:02.094	Peak TVC response to wind	62.084	B58H1150C
39:02.414	Peak yaw response to wind	62.404	V90R5341C
39:02.494	RH outboard elevon actuator hinge moment spike	62.484	V58P0966C
39:03.934	RH outboard elevon actuator delta pressure change	63.924	V58P0966C
39:03.974	Start of planned pitch rate maneuver	63.964	V90R5321C

SPACE TRANSPORTATION/HUMAN SPACEFLIGHT

Table 3–46 continued

Mission Time (GMT, in hr:min:sec)	Event	Elapsed Time (sec.)	Source
39:04.670	Change in anomalous plume shape (LH₂ tank leak near 2058 ring frame)	64.660	E204 Camera
39:04.715	Bright sustained glow on sides of external tank	64.705	E204 Camera
39:04.947	Start of SSME gimbal angle large pitch variations	64.937	V58H1100A
39:05.174	Beginning of transient motion from changes in aero forces due to plume	65.164	V90R5321C
39:06.774	Start of external tank LH₂ ullage pressure deviations	66.764	T41P1700C
39:12.214	Start of divergent yaw rates (RH vs. LH solid rocket booster)	72.204	V90R2528C
39:12.294	Start of divergent pitch rates (RH vs. LH solid rocket booster)	72.284	V90R2525C
39:12.488	SRB major high rate actuator command	72.478	V79H2111A
39:12.507	SSME roll gimbal rate 5 deg/sec	72.497	V58H1100A
39:12.535	Vehicle max +Y lateral acceleration (+.227 g)	72.525	V98A1581C
39:12.574	SRB major high rate actuator motion	72.564	B58H1151C
39:12.574	Start of H₂ tank pressure decrease with two flow control valves open	72.564	T41P1700C
39:12.634	Last state vector downlinked	72.624	Data reduction
39:12.974	Start of sharp MPS LOX inlet pressure drop	72.964	V41P1330C
39:13.020	Last full computer frame of TDRS data	73.010	Data reduction
39:13.054	Start of sharp MPS LH₂ inlet pressure drop	73.044	V41P1100C
39:13.055	Vehicle max; Y lateral acceleration (.254 g)	73.045	V98A1581C
39:13.134	Circumferential white pattern on external tank aft dome (LH₂ tank failure)	73.124	E204 Camera
39:13.134	RH solid rocket motor pressure 19 psi lower than LH solid rocket motor	73.124	B47P2302C
39:13.147	First hint of vapor at intertank	73.137	E207 Camera
39:13.153	All engine systems start responding to loss of fuel and LOX inlet pressure	73.143	SSME team
39:13.172	Sudden cloud along external tank between intertank and aft dome	73.162	E207 Camera
39:13.201	Flash between orbiter and LH₂ tank	73.191	E204 Camera
39:13.221	SSME telemetry data interference from 73.211 to 73.303	73.211	
39:13.223	Flash near solid rocket booster forward attach and brightening of flash between orbiter and external tank	73.213	E204 Camera
39:13.292	First indication of intense white flash at solid rocket booster forward attach point	73.282	E204 Camera

Table 3–46 continued

Mission Time (GMT, in hr:min:sec)	Event	Elapsed Time (sec.)	Source
39:13.337	Greatly increased intensity of white flash	73.327	E204 Camera
39:13.387	Start of RCS jet chamber pressure fluctuations	73.377	V42P1552A
39:13.393	All engines approaching HPFT discharge temp redline limits	73.383	E41Tn010D
39:13.492	ME 2 HPFT discharge temp Chan. A vote for shutdown; two strikes on Chan.	B73.482	MEC data
39:13.492	ME 2 controller last time word update	73.482	MEC data
39:13.513	ME 3 in shutdown from HPFT discharge temperature redline exceedance	73.503	MEC data
39:13.513	ME 3 controller last time word update	73.503	MEC data
39:13.533	ME 1 in shutdown from HPFT discharge temperature redline exceedance	73.523	Calculation
39:13.553	ME 1 last telemetered data point	73.543	Calculation
39:13.628	Last validated orbiter telemetry measurement	73.618	V46P0120A
39:13.641	End of last reconstructed data frame with valid synchronization and frame count	73.631	Data reduction
39:14.140	Last radio-frequency signal from orbiter	74.130	Data reduction
39:14.597	Bright flash in vicinity of orbiter nose	74.587	E204 Camera
39:16.447	RH solid rocket booster nose cap separation/chute deployment	76.437	E207 Camera
39:50.260	RH solid rocket booster RSS destruct	110.250	E202 Camera
39:50.262	LH solid rocket booster RSS destruct	110.252	E230 Camera

Table 3–47. Chronology of Events Prior to Launch of Challenger (STS 51-L) Related to Temperature Concerns

Date and Time (EST)	Key Participants	Event
Jan. 27, 1986 12:36 p.m.	NASA project managers and contractor support personnel (including Morton Thiokol)	*Launch Scrub.* Decision made to scrub because of high crosswinds at launch site.
Jan. 27 1:00 p.m.	Same as above	*Postscrub Discussion.* All appropriate personnel are polled as to feasibility to launch again with 24-hour cycle. Result in no solid rocket booster constraints for launch at 9:38 a.m., January 28: • Request is made for all participants to report any constraints.
Jan. 27 1:00 p.m.	At Kennedy Space Center: Boyd C. Brinton, manager, space booster project, Thiokol; Lawrence O. Wear, manager, solid rocket motor project office, Marshall Space Flight Center At Morton Thiokol, Utah: Arnold R. Thompson, supervisor, rocket motor cases; Robert Ebeling, manager, ignition system and final assembly, solid rocket motor project	*Conversation.* Wear asks Brinton if Thiokol had any concerns about predicted low temperatures and above what Thiokol had said about cold temperature effects following January 1985 flight 51-C: • Brinton telephones Thompson and other Thiokol personnel to ask them to determine whether there were concerns based on predicted weather conditions. Ebeling and other engineers are notified and asked for evaluation.
Jan. 27 2:00 p.m.	NASA Level I and Level II management. At Kennedy: Jesse W. Moore, associate administrator for space flight, NASA Headquarters; Arnold D. Aldrich, manager, space transportation programs, Johnson Space Center; Larry Mulloy, manager, solid rocket booster projects office, Marshall; William Lucas, director, Marshall	*Mission Management Team Meeting.* Discussion includes temperature at the launch facility and weather conditions predicted for launch at 9:38 a.m. on Jan. 28, 1986.

Table 3–47 continued

Date and Time (EST)	Key Participants	Event
Jan. 27 2:30 p.m.	*At Thiokol, Utah:* R. Boisjoly, seal task force, Morton Thiokol, Utah; Robert Ebeling, manager, ignition system and final assembly, solid rocket motor project	Boisjoly learns of cold temperatures at Cape at meeting convened by Ebeling.
Jan. 27 4:00 p.m.	*At Kennedy:* A.J. McDonald, manager, solid rocket motor project, Morton Thiokol; Carver Kennedy, vice president, space services, at Kennedy for Morton Thiokol *At Thiokol, Utah:* Robert Ebeling, manager solid rocket motor project office, igniter and final assembly, Thiokol, Utah	*Telephone Conversation.* McDonald receives call from Ebeling expressing concern about performance of solid rocket booster field joints at low temperature: • McDonald indicates he will call back latest temperature predictions up to launch time. • Carver Kennedy calls Launch Operations Center and receives latest temperature information. • McDonald transmits data to Utah and indicates he will set up telecon and asks engineering to prepare.
Jan. 27 5:15 p.m.	*At Kennedy:* Al McDonald, manager, solid rocket motor project, Morton Thiokol; Cecil Houston, manager, Marshall resident office at Kennedy	Telephone Conversion. McDonald calls Houston informing him that Morton Thiokol engineering had concerns regarding O-ring temperatures: • Houston indicates he will set up teleconference with Marshall and Morton Thiokol.

Table 3–47 continued

Date and Time (EST)	Key Participants	Event
Jan. 27 5:25 p.m.	*At Kennedy:* Cecil Houston, manager, Marshall resident office at Kennedy *At Marshall:* Judson A. Lovingood, deputy manager, Shuttle projects office, at Marshall	*Telephone Conversation:* Houston calls Lovingood informing him of the concerns about temperature effects on the O-rings and asks him to establish a telecon with: Stanley R. Reinartz, manager, Shuttle projects office, Marshall at Kennedy; Lawrence B. Mulloy, manager, solid rocket booster project, Marshall at Kennedy; George Hardy, deputy director, science and engineering, at Marshall; and Thiokol personnel.
Jan. 27 5:30 p.m.	*At Kennedy:* Stanley R. Reinartz, manager, Shuttle projects office, Marshall *At Marshall:* Jud Lovingood, deputy manager, Shuttle projects office, Marshall	*Telephone Conversation.* Lovingood calls Reinartz to inform him of planned 5:45 p.m. teleconference. Lovingood proposes that Kingsbury (director of science and engineering, Marshall) participate in teleconference.
Jan. 27 5:45 p.m.	*At Kennedy:* Stan Reinartz, manager, Shuttle projects, Marshall *At Marshall:* Jud Lovingood, deputy manager, Shuttle projects office, Marshall *Plus Thiokol and other personnel*	*Teleconference.* The discussion addresses Thiokol concerns regarding the temperature effects on the O-ring seals: • Thiokol is of the opinion launch should be delayed until noon or afternoon. • A decision was made to transmit the relevant data to all of the parties and set up another teleconference for 8:15 p.m. • Lovingood recommends to Reinartz to include Lucas, director, Marshall, and Kingsbury in 8:45 p.m. conference and to plan to go to Level II if Thiokol recommends not launching.

Table 3–47 continued

Date and Time (EST)	Key Participants	Event
Jan. 27 6:30 p.m.	*At Marshall:* Jud Lovingood, deputy manager, Shuttle projects office, Marshall *At Kennedy:* Stan Reinartz, manager, Shuttle projects office, Marshall	*Telephone Conversation.* Lovingood calls Reinartz and tells him that if Thiokol persists, they should not launch: • Lovingood also suggests advising Aldrich, manager, NSTS (Level II), of teleconference to prepare him for Level I meeting to inform of possible recommendation to delay.
Jan. 27 7:00 p.m.	*At Kennedy:* Larry Mulloy, manager, solid rocket booster projects office, Marshall; Stan Reinartz, manager, Shuttle projects office, Marshall; William Lucas, director, Marshall; Vim Kingsbury, director of engineering, Marshall	*Conversion.* Reinartz and Mulloy visit Lucas and Kingsbury in their motel rooms to inform them of Thiokol concern and planned teleconference.
Jan. 27 8:45 p.m.	*Teleconference Participants:* *At Kennedy:* Stan Reinartz, manager, Shuttle projects office, Marshall; Larry Mulloy, manager, solid rocket booster projects office, Marshall; Al McDonald, manager, solid rocket motor project, Morton Thiokol *At Marshall:* Jud Lovingood, deputy manager, Shuttle project office, Marshall; George Hardy, deputy director, science and engineering, Marshall *At Thiokol, Utah:* Jerry Mason, senior vice president, Thiokol, Wasatch; Joe Kilminster, vice president/manager, Shuttle projects, Thiokol, Wasatch; Robert Lund, vice president, engineering, Thiokol; Roger Boisjoly, seal task force—structures, Thiokol; Arnie Thompson, supervisor, structures, Thiokol *Plus other personnel*	*Teleconference.* Telefaxes of charts presenting history of O-ring erosion and blow-by for the primary seal in the solid rocket booster field joints from previous flights, as well as results of subscale tests and static tests of solid rocket motors: • The data show that the timing function of the O-rings would be slower from lower temperatures and that the worst blow-by occurred on solid rocket motor 15 (STS 51-C) in January 1985 with O-ring temperatures of 53 degrees F. • Recommendation by Thiokol was not to launch *Challenger* (STS 51-L) until the temperature of the O-ring reached 53 degrees F, which was the lowest O-ring temperature of any previous flight.

Table 3–47 continued

Date and Time (EST)	Key Participants	Event
Jan. 27 8:45 p.m. cont.		• Mulloy asks for recommendation from Kilminster. • Kilminster states that based upon the recommendation, he can not recommend launch. • Hardy is reported by both McDonald and Boisjoly to have said he is "appalled" by Thiokol's recommendation. • Reinartz comments that he is under the impression that solid rocket motor is qualified from 40 degrees F to 90 degrees F. • NASA personnel challenge conclusions and recommendations. • Kilminster asks for 5 minutes off-line to caucus with Thiokol personnel.
Jan. 27 10:30 p.m.	*Thiokol personnel:* Jerry Mason, senior vice president; Joe Kilminster, vice president manager, Shuttle projects; Cal Wiggins, vice president, Space Division; Robert Lund, vice president, engineering; Arnie Thompson, supervisor, structures; Roger Boisjoly, seal task force—structures *Plus other personnel*	*Thiokol Caucus.* Caucus lasts for about 30 minutes at Thiokol, Wasatch, Utah: • Major issues are (1) temperature effects on O-ring and (2) erosion of the O-ring. • Thompson and Boisjoly voice objections to launch, and indication is that Lund also is reluctant to launch. • A final management review is conducted with only Mason, Lund, Kilminster, and Wiggins. • Lund is asked to put on "management hat" by Mason. • Final agreement is: (1) there is a substantial margin to erode the primary O-ring by a factor of three times the previous worst case, and (2) even if the primary O-ring does not seal, the secondary is in position and will.

Table 3–47 continued

Date and Time (EST)	Key Participants	Event
Jan. 27 10:30–11:00 p.m.	*At Kennedy:* Allan J. McDonald, manager, space booster project Morton Thiokol; Lawrence B. Mulloy, manager, solid rocket booster projects, Marshall; Stan Reinartz, manager, shuttle projects, Marshall; Jack Buchanan, manager, Kennedy operations, for Thiokol; Cecil Houston, Marshall resident manager at Kennedy	*Conversation.* McDonald continues to argue for delay: • McDonald challenges Reinartz's rationale that solid rocket motor is qualified at 40 degrees F to 90 degrees F and Mulloy's explanation that propellant mean bulk temperatures are within specifications.
Jan. 27 11:00 p.m.	*Same Kennedy, Marshall, and Thiokol participants as earlier 8:45 p.m. teleconference*	*Teleconference.* Thiokol indicates it had reassessed; temperature effects are a concern, but data are inconclusive: • Kilminster reads the rationale for recommending launch. • Thiokol recommends launch. • Hardy requests that Thiokol puts its recommendation in writing and send it by fax to both Kennedy and Marshall.
Jan. 27 11:15–11:30 p.m.	*At Kennedy:* Allan J. McDonald, manager, space booster project, Lawrence Mulloy, Thiokol; manager, solid rocket booster projects office, Marshall; Stan Reinartz, manager, shuttle projects office, Marshall; Jack Buchanan, manager, Kennedy operations, for Thiokol; Cecil Houston, manager, Marshall resident office at Kennedy	*Conversation:* McDonald argues again for delay, asking how NASA could rationalize launching below qualification temperature: • McDonald indicates if anything happens, he would not want to have to explain it to a board of inquiry. • McDonald indicates he would cancel launch because of the (1) O-ring problem at low temperatures, (2) booster recovery ships heading into wind toward shore because of high seas, and (3) icing conditions on the launch pad. • McDonald is told it is not his concern and that his stated concerns will be passed on in an advisory capacity.

Table 3–47 continued

Date and Time (EST)	Key Participants	Event
Jan. 27 11:30 p.m.	*At Kennedy:* Larry Mulloy, manager, solid rocket booster projects office, Marshall; Stan Reinartz, manager, Shuttle projects, Marshall at Kennedy; Arnold Aldrich, manager, NSTS program office, Johnson Space Center	*Teleconference.* Discussion centers around the recovery hips' activities and brief discussion of the ice condition at the launch complex area: • Discussion does not include concerns about temperature effects on O-rings. • Reinartz and Mulloy place call to Aldrich. • McDonald delivers fax to Jack Buchanan's office at Kennedy and overhears part of conversation. • Aldrich is apparently not informed of the O-ring concerns.
Jan. 27 11:45 p.m.		*Telefax.* Kilminster telefaxes Thiokol's recommendation to launch: • Fax is signed by Kilminster. • McDonald retrieves fax at Kennedy.
Jan. 28 12:01 a.m.		Kennedy meeting breaks up.
Jan. 28 1:30–3:00 a.m.	*At Kennedy:* Charles Stevenson, supervisor of ice crew, Kennedy; B.K. Davis, ice team member, Marshall	*Ice Crew Inspection of Launch Pad B.* Ice crew finds large quantity of ice on fixed service structure, mobile launch platform, and pad apron and reports conditions.
Jan. 28 5:00 a.m.	*At Kennedy:* Larry Mulloy, manager, solid rocket booster projects office, Marshall; William Lucas, director, Marshall; Jim Kingsbury, director of engineering, Marshall	*Conversation.* Mulloy tells Lucas of Thiokol's concerns over temperature effects on O-rings and final resolution: • Lucas is shown copy of Thiokol fax.

Table 3–47 continued

Date and Time (EST)	Key Participants	Event
Jan. 28 7:00–9:00 a.m.	*At Kennedy:* Charles Stevenson, supervisor of ice crew, Kennedy; B.K. David, ice crew member, Kennedy	*Ice Crew Inspection of Launch Pad B.* Ice crew inspects Launch Pad B and *Challenger* for ice formation: • Davis measures temperature on solid rocket boosters, external tank, orbiter, and launch pad with infrared pyrometer. • Left-hand solid rocket booster seems to be about 25 degrees F, and right-hand solid rocket booster seems to be about 8 degrees F near the aft region. • Ice crew is not concerned because there is no Launch Commit Criteria on surface temperatures and does not report. • Crew reports patches of sheet ice on lower segment and skirt of left solid rocket booster.
Jan. 28 8:00 a.m.	*At Marshall:* Jud Lovingood, deputy manager, shuttle projects office, Marshall; Jack Lee, deputy director, Marshall	*Conversation:* Lovingood informs Lee of previous night's discussions: • He indicates that Thiokol had at first recommended not launching and, then after Wasatch conference, recommended launching. • He also informs Lee that Thiokol is providing in writing its recommendation for launch.

Table 3–47 continued

Date and Time (EST)	Key Participants	Event
Jan. 28 9:00 a.m.	*Nominally NASA Level I and Level II Management*	*Mission Management Team Meeting.* Discussion of ice conditions at launch complex. There is no apparent discussion of temperature effects on O-ring seal.
Jan. 28 10:30 a.m.	*At Kennedy:* Charles Stevenson, supervisor of ice crew; B.K. Davis, ice team member	*Ice Crew Inspection of Launch Pad B.* Ice crew inspects Launch Pad B for third time: • Crew removes ice from water troughs, returns to Launch Control Center at T-20 minutes, and reports conditions to Mission Management Team, including fact that ice remains on left solid rocket booster.
Jan. 28 11:38 a.m.		*Launch. Challenger* (STS 51-L) is launched.

Table 3–48. Schedule for Implementation of Recommendations (as of July 14, 1986)

Date of Action/ Target Date	Action	Recommendation
March 1986	Maintenance Safeguards Team is established.	IX - Maintenance Safeguards
March 1986	NASA establishes a Flight Rate Capability Working Group.	VIII - Flight Rate
March 13, 1986	NASA initiates review of all Shuttle program Failure Modes and Effects Analyses and associated Critical Items Lists.	III - Critical Item Review and Hazard Analysis
March 24, 1986	Marshall is directed to form a solid rocket motor joint redesign team.	I - Solid Rocket Motor
April 7, 1986	NASA initiates a Shuttle crew egress and escape review.	VII - Launch Abort and Crew Escape
May 5–6, 1986	Formal Program Management Review for Space Shuttle program with managers of all Shuttle program activities is held at Marshall.	Related investigation
June 19, 1986	Termination of Centaur upper stage development is announced.	Related investigation
June 25, 1986	Second formal Program Management Review for Space Shuttle program with managers of all Shuttle program activities is held at Kennedy.	Related investigation
June 25, 1986	Astronaut Robert Crippen is directed to form a fact-finding group to assess the Space Shuttle management structure and communications procedures.	II & IV - Shuttle Management Structure and Communications
July 8, 1986	George Rodney is appointed Associate Administrator for Safety, Reliability, Maintainability, and Quality Assurance.	IV - Safety Organization
Aug. 15, 1986	Flight Rate Capability Working Group recommendations are due to the Office of Space Flight.	VIII - Flight Rate
Aug. 15, 1986	Management and communications fact-finding group is to report to the associate administrator for space flight.	II & IV - Shuttle Management Structure and Communications
Sept. 1986	Solid Rocket Motor Preliminary Design Review is conducted.	I - Solid Rocket Motor
Sept. 1, 1986	Deadline arrives for establishment of a Shuttle Safety Panel.	II - Shuttle Management Structure
Sept. 30, 1986	Maintenance plan is completed.	IX - Maintenance Safeguards

Table 3–48 continued

Date of Action/ Target Date	Action	Recommendation
Oct. 1, 1986	Decision on implementation of recommendations of management fact-finding group is due.	II - Shuttle Management Structure
Oct. 1, 1986	Crew escape and launch abort studies are to be completed.	VII - Launch Abort and Crew Escape
Dec. 1986	Decision on implementation on crew escape and launch aborts is due.	VII - Launch Abort and Crew Escape
March 1987	Final review occurs with NASA Headquarters of Failure Modes and Effects Analyses and Critical Items Lists.	III - Critical Item Review and Hazard Analysis
July 1987	Landing aid implementation is completed.	VI - Landing Safety
Aug. 1987	Interim brake system is delivered.	VI - Landing Safety

Table 3–49. Revised Shuttle Manifest (as of October 3, 1986)

Date	Mission	Purpose	Vehicle
Feb. 18, 1988	STS-26/TDRS-C	NASA tracking and communications satellite	*Discovery*
May 30, 1988	STS-27/DOD	Classified	*Atlantis*
July 15, 1988	STS-28/DOD	Classified	*Columbia*
Sept. 15, 1988	STS-29/TDRS/D	NASA tracking and communications satellite	*Discovery*
Nov. 15, 1988	STS-30/HST	NASA program to observe the universe to gain information about its origin, evolution, and disposition of stars, galaxies, etc., dedicated mission, serviceable on later missions	*Atlantis*
Jan. 15, 1989	STS-31/Astro	Three-mission NASA program designed to obtain ultraviolet data on astronomical objects; igloo plus two pallets	*Columbia*
March 1, 1989	STS-32/DOD	Classified	*Discovery*
May 1, 1989	STS-33/Magellan	NASA mission to acquire radar map of the surface of Venus; planetary probe using IUS	*Atlantis*
June 1, 1989	STS-34/DOD Spacelab	Spacelab mission for Strategic Defense Initiative	*Discovery*
July 1, 1989	STS-35/MSL-3, GPS-1, GPS-2	MSL—NASA mission performs materials processing experiments in low gravity; uses MPESS cross-bay; weighs approximately 3,175 kilograms GPS—DOD navigation and position system; uses PAM-D2 upper stage	*Columbia*
July 15, 1989	STS-36/DOD	Classified	*Atlantis*
Aug. 30, 1989	STS-37/DOD	Classified	*Discovery*
Sept. 15, 1989	STS-38/GPS-3, GPS-4, MSL-4	See STS-35	*Columbia*
Oct. 15, 1989	STS-39/Planetary	Assignments for Galileo and Ulysses to be determined; use IUS	*Atlantis*
Dec. 1, 1989	STS-40/SLS-1	NASA Spacelab module mission to investigate the effects of weightlessness exposure using human and animal specimens	*Discovery*

Table 3–49 continued

Date	Mission	Purpose	Vehicle
Jan. 15, 1990	STS-41/GRO	NASA mission to investigate extraterrestrial gamma-ray sources; free-flyer mounts to Shuttle fittings and provides own propulsion; an ELV candidate	*Columbia*
Feb. 1, 1990	STS-42/DOD	Classified	*Atlantis*
April 1, 1990	STS-43/IML	Commercial maritime communications services; uses PAM-D	*Discovery*
April 15, 1990	STS-44/GPS-5, EOS-1, SHARE	EOS—Commercial mission to produce pharmaceuticals for large-scale tests leading to FDA approval and commercial production; special crossbay structure; weighs approximately 2,722 kilograms SHARE—NASA mission to evaluate on-orbit thermal performance of a heat pipe radiator element designed for Space Station heat rejection system application; 50-foot elements mounts on longeron	*Columbia*
May 30, 1990	STS-45/DOD	Classified	*Atlantis*
July 1, 1990	STS-46/DOD	Classified	*Discovery*
July 15, 1990	STS-47/GPS-6, Skynet-4, MSL-5	Skynet—United Kingdom military communications satellite; uses PAM-D2 upper stage	*Columbia*
Aug. 15, 1990	STS-48/DOD	Classified	*Atlantis*
Sept. 30, 1990	STS-49/Planetary	Assignments for Galileo and Ulysses to be determined; uses IUS	*Discovery*
Oct. 15, 1990	STS-50/GPS-7, INSAT-1D, TSS-1	INSAT—Indian communications and meteorological satellite; uses PAM-D TSS—NASA/Italy cooperative mission to demonstrate system capabilities by deploying and retrieving tethered satellite and measuring engineering data from payload on satellite; pallet	*Columbia*

Table 3–49 continued

Date	Mission	Purpose	Vehicle
Nov. 15, 1990	STS-51/ LDEF RETR, Syncom	LDEF RETR—NASA mission to retrieve and return the LDEF to Earth so results may be analyzed; purpose to avoid uncontrolled reentry; will occupy about half of payload bay; weighs approximately 9,980 kilograms Syncom—Commercial mission to provide communications services under lease to the U.S. Navy (Leasat); weighs 7,711 kilograms with own perigee stage	*Atlantis*
Jan. 15, 1991	STS-52, ATLAS-1, COFS-1	ATLAS—NASA mission to measure long-term variability in the total energy radiated by the Sun and determine the variability in the solar spectrum; igloo plus two pallets COFS—NASA mission to demonstrate structural integrity through deployment, retraction, and restowage and develop techniques for distributed control and adaptive control methods; pallet	*Discovery*
Feb. 1, 1991	STS-53/GPS-8, GPS-9, MSL-6, SSBUV-1	SSBUV—NASA mission to measure ozone characteristics of the atmosphere; mounts on longeron; weighs approximately 453.6 kilograms	*Columbia*
March 1, 1991	STS-54/DOD	Classified	OV-105
March 30, 1991	STS-55/ EURECA, Skynet-4, GPS-10	EURECA—ESA platform placed in orbit for 6 months offering conventional services to experiments; releasable, retrievable cross-bay structure; weighs approximately 3,856 kilograms	*Atlantis*

Table 3–50. Space Station Work Packages

	Work Package 1 Marshall	Work Package 2 Johnson	Work Package 3 Goddard	Work Package 4 Lewis
Lead Center				
Contractors	Boeing Aerospace Co.; Martin Marietta Aerospace	McDonnell Douglas Astronautics Co.; Rockwell International	RCA Astro Electronics; General Electric Co.	Rocketdyne Div., Rockwell International; TRW Federal Systems Div. (contract terminated April 1986)
Responsibilities	Lab Module Hab Module Node Structure Logistics Carriers Environmental Control & Life Support Systems Certain Module Outfitting & Distributed Systems	Truss Structure & Utility Runs Propulsion System Data Management Thermal Control Communication & Tracking Attitude Control Extra Vehicular System	Polar Platform Payload Attach Points & Pointing Systems Flight Telerobotic Servicer	Power Generation Power Management and Distribution

Table 3–51. Japanese Space Station Components

	Pressurized Module	Experiments Logistics Module		Exposed Facility
		Pressurized Section	Exposed Section	
Shape	Cylinder	Cylinder	Box	Box 2 unit
Size	4m diameter	4m diameter	4m x 4m	2.5m (h) x 1.4m (w)
Length	10m	4m	2m	4m
Number of mission payloads	Payload rack 10	Rack 8		Payload 10
Average power		Housekeeping 5 kW, Mission 20 kW		
Data transfer rate		32 Mbps (Max., Optical LAN)		
Type of activity	Materials processing and life sciences experiments	On-orbit storage and transport logistics support		Exchange of experimental equipment and materials and construction of large structures; scientific observations; and communications, scientific/engineering, and materials experiments

CHAPTER FOUR
SPACE SCIENCE

CHAPTER FOUR

SPACE SCIENCE

Introduction

NASA's Space Science and Applications program was responsible for planning, directing, executing, and evaluating NASA projects focused on using the unique characteristics of the space environment for scientific study of the universe, solving practical problems on Earth, and providing the scientific research foundation for expanding human presence into the solar system. The space science part of these responsibilities (the subject of this chapter) aimed to increase scientific understanding through observing the distant universe, exploring the near universe, and understanding Earth's space environment.

The Office of Space Science (OSS) and the Office of Space Science and Applications (OSSA) formed the interface among the scientific community, the president, and Congress. These offices evaluated ideas for new science of sources and pursued those thought most appropriate for conceptual study.[1] They represented the aspirations of the scientific community, proposed and defended programs before the Office of Management and Budget and Congress, and conducted the programs that Congress authorized and funded. NASA's science missions went through definable phases. In the early stages of a scientific mission, the project scientist, study scientist, or principal investigator would take the lead in specifying the science that the proposed mission intended to achieve and determined its feasibility. Once the mission was approved and preparations were under way, the mission element requirements, such as schedule and cost, took priority. However, once the mission was launched and the data began to be transmitted, received, and analyzed, science again became dominant. From 1979 to 1988, NASA had science missions that

---

[1]The ideas for new science came from a variety of sources, among them the various divisions within the science offices, the NASA field installations, the National Academy of Sciences, industry and academia, other U.S. government agencies, international organizations, NASA advisory committees, and the demand caused by shifting national priorities.

were in each of these stages—some in the early conceptual and mission analysis stages, others in the definition, development, and execution stages, and still others in the operational stage, with the data being used by the scientific community.

Thus, although NASA launched only seventeen dedicated space science missions and conducted four science missions aboard the Space Shuttle from 1979 to 1988, compared to the previous decade when the agency flew approximately sixty-five space science missions, the agency also continued to receive and analyze impressive data from earlier launches and prepared for future missions, some delayed following the *Challenger* accident. In addition to the delays caused by the *Challenger* accident, level funding also contributed to the smaller number of missions. NASA chose to invest its resources in more complex and costly missions that investigated a range of phenomena rather than fly a series of missions that investigated similar phenomena.

In addition to those managed by NASA, some NASA-launched missions were for other U.S. government or commercial organizations and some were in partnerships with space agencies or commercial entities from other countries. The following sections identify those scientific missions in which NASA provided only launch-related services or other limited services.

In spite of the small number of missions, NASA's OSS and OSSA were very visible. Almost every Space Shuttle mission had space science experiments aboard in addition to the dedicated Spacelab missions. Furthermore, scientists received spectacular and unprecedented data from the missions that had been launched in the previous decade, particularly the planetary probes.

This chapter describes each space science mission launched during these years as well as those conducted aboard the Space Shuttle. An overview of findings from missions launched during the previous decade is also presented.

The Last Decade Reviewed (1969–1978)

From 1969 to 1978, NASA managed space science missions in the broad areas of physics and astronomy, bioscience, and lunar and planetary science. The majority of NASA's science programs were in the physics and astronomy area, with fifty-three payloads launched. Explorer and Explorer-class satellites comprised forty-two of these investigative missions, which provided scientists with data on gamma rays, x-rays, energetic particles, the solar wind, meteoroids, radio signals from celestial sources, solar ultraviolet radiation, and other phenomena. Many of these missions were conducted jointly with other countries.

NASA launched four observatory-class physics and astronomy spacecraft programs between 1969 and 1978. These provided flexible orbiting platforms for scientific experiments. Participants in the Orbiting Geophysical Observatories gathered information on atmospheric composition. The Orbiting Astronomical Observatory returned volumes of data on the composition, density, and physical state of matter in interstellar space to scientists on Earth. It was the most complex automated spacecraft yet in the space science program. It took the first ultraviolet photographs of the stars and produced the first hard evidence of the existence of black holes in space. The High Energy Astronomy Observatories (HEAO) provided high-quality data on x-ray, gamma ray, and cosmic ray sources. HEAO-1 was the heaviest scientific satellite to date. The Orbiting Solar Observatory missions took measurements of the Sun and were the first satellites to capture on film the beginning of a solar flare and the consequent streamers of hot gases that extended out 10.6 million kilometers. It also discovered "polar ice caps" on the Sun (dark areas thought to be several million degrees cooler than the normal surface temperatures).

NASA launched several other Explorer-class satellites in cooperative projects with other countries or other government agencies. Uhuru, launched from the San Marco launch platform in 1970, scanned 95 percent of the celestial sphere for sources of x-rays and discovered three new pulsars. The bioscience program sponsored only Biosatellite 3, whose objective was to determine the effects of weightlessness on a monkey. In addition, NASA's life scientists designed many of the experiments that were conducted on Skylab.

NASA's Office of Planetary Programs explored the near planets with the Pioneer and Mariner probes. NASA conducted three Mariner projects during the 1970s, which investigated Mars, Mercury, and Venus. Mariner 9 became the first American spacecraft to go into orbit around another planet; it mapped 95 percent of the Martian surface. The two Viking landers became the first spacecraft to soft-land on another planet when they landed on Mars and conducted extended mission operations there while two orbiters circled the planet and mapped the surface.

With the Pioneer program, NASA extended its search for information to the outer planets of the solar system. Pioneer 10 (traveling at the highest velocity ever achieved by a spacecraft) and Pioneer 11 left Earth in the early 1970s, reaching Jupiter in 1973 and Saturn in 1979. Eventually, in 1987, Pioneer 10 would cross the orbit of Pluto and become the first manufactured object to travel outside our solar system. NASA also sent two Voyager spacecraft to the far planets. These excursions produced impressive high-resolution images of Jupiter and Saturn.

Detailed information relating to space science missions from 1969 through 1978 can be found in Chapter 3 of the *NASA Historical Data Book, Volume III*.[2]

---

[2] Linda Neuman Ezell, *NASA Historical Data Book, Volume III: Programs and Projects, 1969–1978* (Washington, DC: NASA SP-4012, 1988).

Space Science (1979–1988)

During the ten-year period from 1979 to 1988, NASA launched seventeen space science missions. These included missions sponsored by OSS or OSSA (after its establishment in 1981), missions launched for other U.S. government agencies, and missions that were part of an international effort. The science missions were primarily in the disciplines of Earth and planetary exploration, astrophysics, and solar terrestrial studies. The Life Sciences Division, while not launching any dedicated missions, participated heavily in the Spacelab missions and other scientific investigations that took place during the decade.

The decade began with the "year of the planets" in space exploration. During 1979, scientists received their first high-resolution pictures of Jupiter and five of its satellites from Voyagers 1 and 2. Pioneer 11 transmitted the first close-up pictures of Saturn and its moon Titan. Both of these encounters revealed previously unknown information about the planets and their moons. Pioneer Venus went into orbit around Venus in December 1978, and it returned new data about that planet throughout 1979. Also, one Viking orbiter on Mars continued to transmit pictures back to Earth, as did one lander on the planet's surface.

Spectacular planetary revelations continued in 1980 with Voyager 1's flyby of Saturn. Dr. Bradford Smith of the University of Arizona, the leader of the Voyager imaging team, stated that investigators "learned more about Saturn in one week than in the entire span of human history."[3] Thousands of high-resolution images revealed that the planet had hundreds, and perhaps thousands, of rings, not the six or so previously observed. The images also showed three previously unknown satellites circling the planet and confirmed the existence of several others.

Scientists also continued receiving excellent data from NASA's two Earth-orbiting HEAOs (launched in 1977 and 1978, respectively). HEAO-2 (also referred to as the Einstein Observatory) returned the first high-resolution images of x-ray sources and detected x-ray sources 1,000 times fainter than any previously observed and 10 million times fainter than the first x-ray stars observed. Scientists studying HEAO data also confirmed the emission of x-rays from Jupiter—the only planet other than Earth known to produce x-rays. Mission operations ceased in 1981, but more than 100 scientific papers per year were still being published using HEAO data in the mid-1990s.

The Solar Maximum Mission, launched in 1980, gathered significant new data on solar flares and detected changes in the Sun's energy output. Scientists stated that a cause-and-effect relationship may exist between sustained changes in the Sun's energy output and changes in Earth's weather and climate. The satellite's observations were part of NASA's

---

[3]"Highlights of 1980 Activities," *NASA News*, Release 80-199, December 24, 1980.

solar monitoring program, which focused on studying the Sun during a nineteen-month period when sunspot activity was at a peak of its eleven-year cycle of activity.

During 1981, OSS merged with the Office of Space and Terrestrial Applications to form OSSA. OSSA participated in the Space Shuttle program with its inclusion of the OSTA-1 payload aboard STS-2. This was the first scientific payload to fly on the STS.

Exploration of the solar system continued with Voyager 2's successful encounter with Saturn in August 1981. Building on the knowledge gained by the Voyager 1 encounter, Voyager 2 provided information relating to the ring structure in detail comparable to a street map and enabled scientists to revise their theories of the ring structure. After leaving Saturn's surroundings, Voyager 2 embarked on a trajectory that would bring it to Uranus in 1986.

Pioneer 6 continued to return interplanetary and solar science information while on the lengthiest interplanetary mission to date. Pioneer 10 reached more than 25 thousand million miles from the Sun. Pioneer missions to Venus and Mars also continued transmitting illuminating information about these planets.

Beginning in 1982, an increasing number of space science experiments were flown aboard the Space Shuttle. The Shuttle enabled scientists to conduct a wide variety of experiments without the commitment required of a dedicated mission.[4] Instruments on satellites deployed from the Shuttle also investigated the Sun's ultraviolet energy output, measured the nature of the solar wind, and detected frozen methane on Pluto and Neptune's moon Triton. In addition, the Pioneer and the Viking spacecraft continued to record and transmit data about the planets each was examining.

The Infrared Astronomical Satellite, a 1983 joint venture among NASA, the Netherlands, and the United Kingdom, revealed a number of intriguing discoveries in its ten-month-long life. These included the possibility of a second solar system forming around the star Vega, five undiscovered comets, a possible tenth planet in our solar system, and a solar dust cloud surrounding our solar system.

During 1983, the Space Telescope, then scheduled for launch in 1986, was renamed the Edwin P. Hubble Space Telescope. Hubble was a member of the Carnegie Institute, whose studies of galaxies and discoveries of the expanding universe and Hubble's Constant made him one of America's foremost astronomers.

In 1984, the Smithsonian Institution's National Air and Space Museum became the new owner of the Viking 1 lander, which was parked

---

[4]Tables in Chapter 3 describe many of the experiments conducted aboard the Space Shuttle. Spacelab experiments and OSS and Spacelab missions are described in this chapter. The Office of Space and Terrestrial Applications missions are addressed in Chapter 2, "Space Applications," and OAST-1 is described in Chapter 3, "Aeronautics and Space Research and Technology," both in Volume VI of the *NASA Historical Data Book*.

on the surface of Mars. The transfer marked the first time an object on another planet was owned by a United States museum. Also in 1984, the Hubble Space Telescope's five scientific instruments underwent acceptance testing at the Goddard Space Flight Center in preparation for an anticipated 1986 launch. The acceptance testing represented the completion of the most critical element of the final checkout steps for the instruments before their assembly aboard the observatory. NASA announced the start of the Extreme Ultraviolet Explorer, a new satellite planned for launch from the Space Shuttle in 1988 that eventually was launched in 1992 by a Delta launch vehicle. The mission would make the first all-sky survey in the extreme ultraviolet band of the electromagnetic spectrum.

An encounter with the Comet Giacobini-Zinner by the International Cometary Explorer highlighted NASA's 1985 science achievements. This was the first spacecraft to carry out the on-site investigation of a comet. Also during 1985, Spacelab 3 carried a series of microgravity experiments aboard the Shuttle, and astronauts on Spacelab 2 conducted a series of astronomy and astrophysics experiments. An instrument pointing system on Spacelab 2, developed by the European Space Agency, operated for the first time and provided a stable platform for highly sensitive astronomical instruments.

The *Challenger* accident in January 1986 temporarily halted science that relied on the Shuttle for deploying scientific satellites and for providing a setting for on-board experiments. Four major scientific missions planned for 1986 were postponed, including Astro-1, the Hubble Space Telescope, and two planetary missions—Galileo and Ulysses. The Spartan Halley spacecraft, to be deployed from *Challenger,* was destroyed. However, other science activities still took place. Also, the Space and Earth Science Advisory Committee of the NASA Advisory Council issued a report on the status of space science within NASA. The two-year study, titled "The Crisis in Space and Earth Science, A Time for New Commitment," called for greater attention and higher priority for science programs. The most notable 1986 achievement was Voyager 2's encounter with Uranus in January. This encounter provided data on a planetary body never before examined at such close range. From Uranus, the Voyager continued traveling toward a 1989 rendezvous with Neptune.

In October 1987, NASA issued a revised manifest that reflected the "mixed fleet" concept. This dictated that NASA use the Shuttle only for missions that required human participation or its special capabilities. Some science missions, which had been scheduled for the Shuttle, could be transferred to an expendable launch vehicle with no change in mission objectives. No science missions were launched in 1987.

Only one expendable launch vehicle space science launch took place in 1988, but with the resumption of Space Shuttle flights that spring, NASA prepared for the 1989 launches of several delayed space science missions. This included the Hubble Space Telescope, scheduled for launch in December 1989 (but not deployed until April 1990), which underwent comprehensive ground system tests in June 1988. The

Magellan spacecraft was delivered to the Kennedy Space Center in October 1988. This spacecraft, scheduled for launch in April 1989, would map the surface of Venus. Galileo, scheduled for launch in October 1989, underwent additional minor modifications associated with its most recent Venus-Earth-Earth gravity assist trajectory.

Management of the Space Science Program

NASA managed its space science and applications program from a single office, OSSA, from November 1963 to December 1971. A 1971 reorganization split the office into two organizations: the OSS and the Office of Space and Terrestrial Applications.

Office of Space Science

NASA managed its space science programs from a single office from December 1971 until November 9, 1981 (Figure 4–1). Noel W. Hinners led OSS until his departure from NASA in February 1979. (He returned as director of the Goddard Space Flight Center in 1982.) Thomas A. Mutch led the office from July 1979 through the fall of 1980, when Andrew Stofan became acting associate administrator.

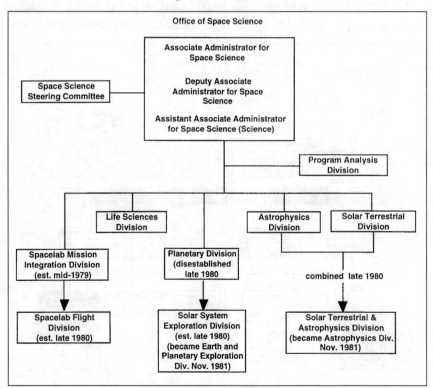

Figure 4–1. *Office of Space Science (Through November 1981)*

In 1979, OSS included divisions for astrophysics, life sciences, planetary science, solar terrestrial science, and program analysis. The Planetary Division was renamed the Solar System Exploration Division in late 1980. This division was disestablished at the time of the reorganization in 1981 and re-formed as the new Earth and Planetary Exploration Division, existing with this title until 1984, when it regained its former title of the Solar System Exploration Division.

The Spacelab Mission Integration Division, which was established in mid-1979, evolved into the Space Flight Division in late 1980. Also in late 1980, the Astrophysics Division and the Solar Terrestrial Division combined into the Solar Terrestrial and Astrophysics Division. This division existed until the reorganization in November 1981, when it re-formed as the Astrophysics Division.

Office of Space Science and Applications

In November 1981, NASA combined OSS and the Office of Space and Terrestrial Applications (OSTA) into the single OSSA (Figure 4–2). NASA Administrator James E. Beggs stated that the consolidation was done because of the program reductions that had occurred in the preceding years and because of the similarity of the technologies that both OSS and OSTA pursued. When the consolidation took place, OSSA consisted of divisions for communications, life sciences, astrophysics, Earth and

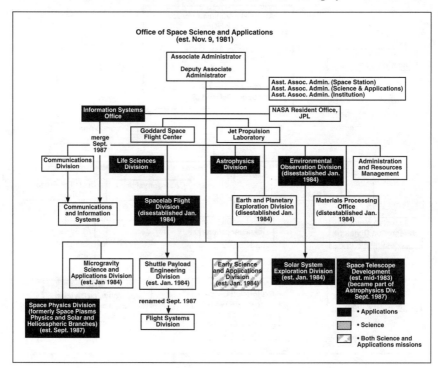

Figure 4–2. Office of Space Science and Applications (Established November 1981)

planetary exploration, Spacelab flight, environmental observation, and administration and resources management; it also had materials processing and information systems offices. The reorganization also placed the Goddard Space Flight Center and the Jet Propulsion Laboratory under the administrative management of OSSA. Andrew Stofan led OSSA as acting associate administrator until the appointment of Burton I. Edelson on February 14, 1982. Lennard A. Fisk succeeded Dr. Edelson in April 1987.

The Earth and Planetary Exploration Division, the Spacelab Flight Division, the Environmental Observation Division, and the Materials Processing Office were disestablished in January 1984. At that time, the Earth and Planetary Exploration Division became the Solar System Exploration Division, and the Spacelab Flight Division became the Shuttle Payload Engineering Division. NASA also established a new Microgravity Sciences and Applications Division and a new Earth Science and Applications Division. In September 1987, the Communications Division and the Information Systems Office merged into the Communications and Information Systems Division. NASA also promoted the Space Plasma Physics Branch and the Solar and Heliospheric Branch to the Space Physics Division. The Space Plasma Physics Branch had been part of the Earth Science and Applications Division, and the Solar and Heliospheric Branch came from the Astrophysics Division. The Space Telescope Development Division, which had been established in mid-1983, became part of the Astrophysics Division. At the same time, the Shuttle Payload Engineering Division was renamed the Flight Systems Division.

Of these divisions, life sciences, astrophysics, Earth and planetary exploration, space physics, solar system exploration, and space telescope development were considered science divisions rather than applications. This chapter covers missions that are managed by these science divisions.

The Life Sciences Division was led by Gerald Soffen through 1983, when he was succeeded by Arnauld Nicogossian. Astrophysics programs were led by Theodrick B. Norris through mid-1979, when Franklin D. Martin assumed the role of director. He remained in place when the division combined with the Solar, Terrestrial Division in 1980 (which had been headed by Harold Glaser) through early 1983. At that time, C.J. Pellerin was named to the post.

Angelo Guastaferro led the Planetary Division until it was disestablished in late 1980. Guastaferro moved to the new Solar System Exploration Division, where he remained through early 1981, when he moved to the Ames Research Center. Daniel Herman served as director of this division until the OSSA reorganization in November 1981, when the division was eliminated. When the Solar Systems Exploration Division was reestablished in 1984, Geoffrey Briggs headed it.

Jesse W. Moore led the Spacelab Mission Integration Division, which became the Spacelab Flight Division, until the November 1981 reorganization. Michael Sander assumed the leadership post at that time and held it until the division was disestablished in 1983. James C. Welch headed the Space Telescope Development Division until it was eliminated in September 1987. The Space Physics Division, which was established in September 1987, was led by Stanley Shawhan.

Office of Chief Scientist

The Office of Chief Scientist was also integral to NASA's science activities. NASA formed this office in 1977 as "a revised role for the [agency's] associate administrator."[5] Its purpose was to "promote across-the-board agency cognizance over scientific affairs and interaction with the scientific community." The chief scientist was responsible for "advising the Administrator on the technical content of the agency's total program from the viewpoint of scientific objectives" and "will serve as a focal point for integrating the agency's programs [and] plans and for the use of scientific advisory committees."[6]

John E. Naugle served as chief scientist through June 1979. The position was vacant until he returned as acting chief scientist in December 1980, remaining until mid-1981. The position was vacant again until the appointment of Frank B. McDonald in September 1982. McDonald served as chief scientist until the appointment of Noel Hinners in 1987, who held that role concurrently with his position as NASA associate deputy administrator–institution.

Office of Exploration

In June 1987, the NASA administrator established the Office of Exploration. Also related to NASA's science activities, this office was to meet the need for specific activities supporting the long-term goal to "expand human presence and activity beyond Earth orbit into the Solar System."[7] The office was responsible for coordinating NASA planning activities, particularly to the Moon and Mars. Major responsibilities were to analyze and define missions proposed to achieve the goal of human expansion of Earth, provide central coordination of technical planning studies that involved the entire agency, focus on studies of potential lunar and Martian initiatives, and identify the prerequisite investments in science and advance technology that must be initiated in the near term to achieve the initiatives. Primary concentrations of the Office of Exploration included mission concepts and scenarios, science opportunities, prerequisite technologies and research, precursor missions, infrastructure support requirements, and exploration programmatic

---

[5]"NASA Reorganization," NASA Special Announcement, October 25, 1977.

[6]Additional responsibilities are listed in NASA Management Instruction 1103.36, "Roles and Responsibilities—Chief Scientist," May 17, 1984.

[7]Office of the Press Secretary, "Presidential Directive on National Space Policy," January 5, 1988.

requirements of resources and schedules. John Aaron served as acting assistant administrator until the appointment of Franklin D. Martin as assistant administrator in December 1988.

Money for Space Science

Although NASA manages its space science missions through divisions that correspond to scientific disciplines, Congress generally allocates funds through broader categories. From 1979 to 1988, NASA submitted its science budget requests and Congress allocated funds through three categories: physics and astronomy, lunar and planetary (called planetary exploration beginning in FY 1980), and life sciences. Each of these broad categories contained several line items that corresponded either to missions such as the space telescope or to activities such as research and analysis.

Some budget category titles exactly match mission names. Other missions that do not appear in the budget under their own names were reimbursable—that is, NASA was reimbursed by another agency for its services and expended minimal funds (relatively speaking) or no funds of its own. These minimal expenses were generally included in other budget categories, such as launch support or ground system support. Still other missions were in-house projects—the work was done primarily by civil servants funded by the Research and Program Management appropriation rather than the Research and Development appropriation. Other science missions could be found in the detailed budget data and the accompanying narratives that NASA's budget office issued. For instance, the FY 1983 Explorer Development budget category under the larger Physics and Astronomy category included the Dynamics Explorer, the Solar Mesosphere Explorer, the Infrared Astronomical Satellite, the Active Magnetospheric Particle Tracer Explorer, the Cosmic Background Explorer, and a category titled "Other Explorers." NASA described the Explorer program as a way of conducting missions with limited, specific objectives that did not require major observatories.

During the period addressed in this chapter, all the launched missions were included under the broad budget category of Physics and Astronomy. The Planetary Exploration budget category funded both the ongoing activities relating to missions launched prior to 1979 and those that would be launched beginning in 1989. The Life Sciences budget category funded many of the experiments that took place on the Space Shuttle and also funded NASA-sponsored experiments on the Spacelab missions. This budget category also paid for efforts directed at maintaining the health of Space Shuttle crews, increasing understanding of the effects of microgravity, and investigating the biosphere of Earth. Funds designated for life sciences programs also contributed heavily to the Space Station program effort.

Over this ten-year period, funding for space science roughly doubled. This almost kept pace with the increase in the total Research and

Development (R&D) and Space Flight, Control and Data Communications (SFC&DC) budgets, which slightly more than doubled. (The R&D appropriation was split into R&D and SFC&DC in 1984.) Thus, even though there were fewer missions over this ten-year period than in the prior ten years, if relative funding is a guide, NASA placed roughly the same importance on space science at the beginning of the decade that it did at its conclusion.

The figures in the tables following this chapter (Tables 4–1 through 4–23) show dollars that have not been inflated. If one considers inflation and real buying ability, then funding for space science remained fairly level over the decade.

Space Science Missions

Prior to the merger of NASA's OSS and OSTA in November 1981, missions could clearly be considered either space science or space applications. However, once the two organizations merged, a clear distinction was not always possible. This chapter includes activities formulated by NASA as space science missions and funded that way by Congress. It also includes science missions managed by other organizations for which NASA provided only launch services or some other nonscientific service.

The first subsection describes physics and astronomy missions, beginning with missions that were launched from 1979 to 1988. The next subsection covers on-board Shuttle missions during the decade. The third subsection contains physics and astronomy missions that were launched during the previous decade but continued to operate in these years and the missions that were under development during this decade but would not be launched until after 1988. The final subsection describes planetary missions—first those that were launched during the previous decade but continued to return data and then those being developed from 1979 to 1988 in preparation for launch after 1988. Table 4–24 lists each science mission that NASA either managed or had some other support role (indicated with an "*") and its corresponding discipline or management area.

Physics and Astronomy Program

The goal of NASA's Physics and Astronomy program was to add to what was already known about the origin and evolution of the universe, the fundamental laws of physics, and the formation of stars and planets. NASA conducted space-based research that investigated the structure and dynamics of the Sun and its long- and short-term variations; cosmic ray, x-ray, ultraviolet, optical, infrared, and radio emissions from stars, interstellar gas and dust, pulsars, neutron stars, quasars, black holes, and other celestial sources; and the laws governing the interactions and processes occurring in the universe. Many of the phenomena being investigated were not detectable from ground-based observatories because of the obscuring or distorting effects of Earth's atmosphere. NASA accomplished the objectives of the program with a mix of large, complex, free-flying space missions, less complex Explorer spacecraft, Shuttle and Spacelab flights, and suborbital activities.

Spacecraft Charging at High Altitudes

The Spacecraft Charging at High Altitudes mission was part of a U.S. Air Force program seeking to prevent anomalous behavior associated with satellites orbiting Earth at or near geosynchronous altitudes of 37,000 kilometers. NASA provided the launch vehicle and launch vehicle support as part of a 1975 agreement between OSS (representing NASA) and the Space and Missile Systems Organization (representing the Air Force). OSS also provided three scientific experiments. Each experiment investigated electrical static discharges that affected satellites in geostationary orbit. The experiments measured electrons, protons, and alpha particles, the surface charging and discharging of the satellite, and anomalous currents flowing through the spacecraft's wires at any given time. This mission's characteristics are listed in Table 4–25.

UK-6

The launch of UK-6 (also called Ariel) marked the one hundredth Scout launch. This was a fully reimbursable mission under the terms of a March 16, 1976, contract between NASA and the United Kingdom Science Research Council. NASA provided the launching and tracking services required for the mission. The project provided scientists with a large body of information about heavy nuclei. These invisible cosmic bullets supplied clues to the nature and origin of the universe. The experiments aboard the satellite examined cosmic rays and x-rays emitted by quasars, supernovas, and pulsars in deep space. UK-6's characteristics are in Table 4–26.

High Energy Astronomy Observatory-3

HEAO-3 was the third in a series of three Atlas-Centaur-launched satellites to survey the entire sky for x-ray sources and background of about one millionth of the intensity of the brightest known source, SCO X1. It also measured the gamma ray flux, determined source locations and line spectra, and examined the composition and synthesis of cosmic ray nuclei.

HEAO-3 carried three instruments that performed an all-sky survey of cosmic rays and gamma rays, similar to the earlier HEAO missions except at a higher orbital inclination. This higher orbital inclination allowed instruments to take advantage of the greater cosmic ray flux near Earth's magnetic poles. One objective was to measure the spectrum and intensity of both diffuse and discrete sources of x-ray and gamma ray radiation. In addition, HEAO-3 carried an instrument that observed high atomic number relativistic nuclei in the cosmic rays and measured the elemental composition and energy spectra of these nuclei to determine the abundance of the individual elements.

HEAO-3 operated until May 30, 1981, when it expended the last of its supply of thruster gases used for attitude control and was powered down. With twenty months of operating time in orbit, HEAO-3 became the third HEAO spacecraft to perform for more than twice its intended design life. Its characteristics are in Table 4–27; Figures 4–3 through 4–5 show diagrams of three HEAO instruments.

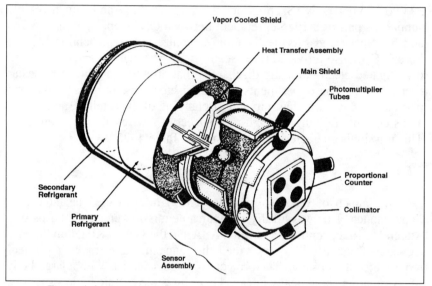

Figure 4–3. HEAO High-Spectral Resolution Gamma Ray Spectrometer

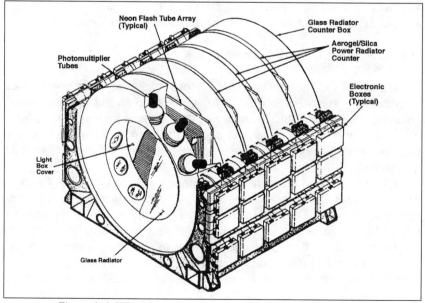

Figure 4–4. HEAO Isotopic Composition of Primary Cosmic Rays

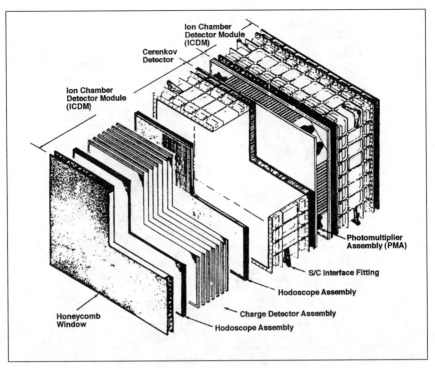

Figure 4–5. HEAO Heavy Nuclei Experiment

Solar Maximum Mission

The Solar Maximum Mission (also known as Solar Max) observatory was an Earth-orbiting satellite that continued NASA's solar observatory research program, which had begun in 1962. The satellite was a three-axis inertially stabilized platform that provided precise stable pointing to any region on the Sun to within five seconds of arc. The mission studied a specific set of solar phenomena: the impulsive, energetic events known as solar flares and the active regions that were the sites of flares, sunspots, and other manifestations of solar activity. Solar Max allowed detailed observation of active regions of the Sun simultaneously by instruments that covered gamma ray, hard and soft x-ray, ultraviolet, and visible spectral ranges. Table 4–28 lists the mission's characteristics, and Figure 4–6 contains a diagram of Solar Max's instruments.

Solar Max was part of an international program involving a worldwide network of observatories. More than 400 scientists from approximately sixty institutions in seventeen foreign nations and the United States participated in collaborative observational and theoretical studies of solar flares. In the solar science community, 1980 was designated the "Solar Maximum Year" because it marked the peak of sunspot activity in the Sun's eleven-year cycle of activity.

The first months of the mission were extremely successful. Careful

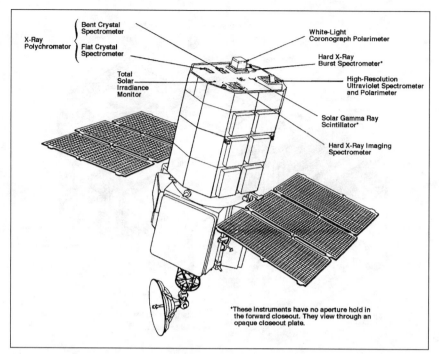

Figure 4–6. Solar Maximum Instruments

orchestration of the instruments resulted in the most detailed look at solar flares ever achieved. The instruments recorded hundreds of flares, and the cumulative new data base was unsurpassed. Solar Max instruments set new standards of accuracy and precision and led scientists to a number of firsts and new answers to old questions. However, nine months into the mission, fuses in the attitude control system failed, and the satellite lost its ability to point with fine precision at the Sun. Although a few instruments continued to send valuable data despite the loss of fine pointing, most of the instruments were useless, and those still operating lost the benefits of operating in a coordinated program. The mission was declared a success, however, because its operation, although abbreviated, fulfilled the success criteria established before launch. Nevertheless, its reduction from the expected two years to nine months meant a significant loss to solar science.

NASA designed Solar Max to be serviced in space by a Space Shuttle crew. Thus, in April 1984, the crew of STS 41-C successfully repaired Solar Max. Following its repair, Solar Max operated successfully until November 1989. A description of the STS 41-C repair mission is in Chapter 3.

Dynamics Explorer 1 and 2

The Dynamics Explorer 1 and 2 satellites provided data about the coupling of energy, electric currents, electric fields, and plasmas (ionized atomic particles) among the magnetosphere, the ionosphere, and the atmosphere. The two spacecraft worked together to examine the processes by which energy from the Sun flows through interplanetary space and entered the region around Earth, controlled by the magnetic forces from Earth's magnetic field, to produce the auroras (northern lights) that affect radio transmissions and possibly influence basic weather patterns.

The two satellites were stacked on a Delta launch vehicle and placed into coplanar (in the same plane but at different altitudes) orbits. Dynamics Explorer 1 was placed in a higher elliptical orbit than Dynamics Explorer 2. The higher orbit allowed for global auroral imaging, wave measurements in the center of the magnetosphere, and crossing of auroral field lines at several Earth radii. Dynamics Explorer 2's lower orbit allowed for neutral composition and temperature and wind measurements, as well as an initial apogee to allow measurements above the interaction regions for suprathermal ions and plasma flow measurements at the base of the magnetosphere field lines. The two satellites carried a total of fifteen instruments, which took measurements in five general categories:

- Electric field-induced convection
- Magnetosphere-ionosphere electric currents
- Direct energy coupling between the magnetosphere and the ionosphere
- Mass coupling between the ionosphere and the magnetosphere
- Wave, particle, and plasma interactions

The Dynamics Explorer mission complemented the work of two previous sets of satellites, the Atmosphere Explorers and the International Sun-Earth Explorers. The three Atmosphere Explorer satellites studied the effects of the absorption of ultraviolet light waves by the upper atmosphere at altitudes as low as a satellite can orbit (about 130 kilometers). The three International Sun-Earth Explorer satellites studied how the solar wind interacted with Earth's magnetic field to transfer energy and ionized charged particles into the magnetosphere. The Dynamics Explorer mission also was to set the stage for a fourth program planned for later in the 1980s that would provide a comprehensive assessment of the energy balance in near-Earth space. The mission's characteristics are in Table 4–29.

Solar Mesospheric Explorer

The Solar Mesospheric Explorer, launched in 1981, was part of the NASA Upper Atmospheric Research program. NASA developed this program under the congressional mandates in the FY 1976 NASA Authorization Act and the Clean Air Act Amendments of 1977. It focused on developing a solid body of knowledge of the physics, chemistry, and dynamics of the upper atmosphere. From an initial emphasis on assessments of the impacts of chlorofluoromethane releases, Shuttle exhausts, and aircraft effluents on stratospheric ozone, the program evolved into extensive field measurements, laboratory studies, theoretical developments, data analysis, and flight missions.

The Solar Mesospheric Explorer was designed to supply data on the nature and magnitude of changes in the mesospheric ozone densities that resulted from changes in the solar ultraviolet flux. It examined the interrelationship between ozone and water vapor and its photo dissociation products in the mesosphere and among ozone, water vapor, and nitrogen dioxide in the upper stratosphere.

The University of Colorado's Laboratory for Atmospheric and Space Physics provided the science instruments for this mission. The laboratory, under contract to the Jet Propulsion Laboratory, was also responsible for the observatory module, mission operations, the Project Operations Control Center, and science data evaluation and dissemination. Ball Aerospace's Systems Division provided the spacecraft bus and satellite integration and testing. The science team was composed of seventeen members from four institutions. A science data processing system, located at the Laboratory for Atmospheric and Space Physics, featured an on-line central processing and analysis system to perform the majority of data reduction and analysis for the science investigations.

The spacecraft consisted of two sections (Figure 4–7). The spacecraft bus carried communication, electrical, and command equipment. A notable feature was the 1.25-meter diameter disc used for mounting the 2,156 solar cells directed toward the Sun to feed power into the two nickel cadmium batteries. A passive system that used insulating material and a network of stripes on the outer surface kept internal temperatures within limits. The satellite body was spin-stabilized.

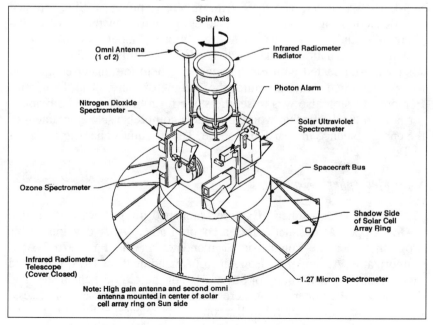

Figure 4–7. Solar Mesospheric Explorer Satellite Configuration

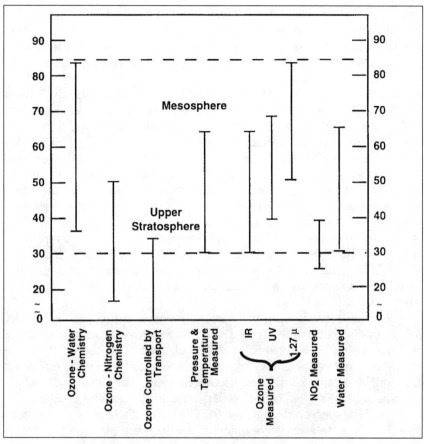

Figure 4–8. Altitude Regions to Be Measured by Solar Mesospheric Explorer Instruments

The observatory module carried the instruments. Four limb scanning instruments measured ozone, water vapor, nitrogen dioxide, temperature, and pressure in the upper stratosphere and mesosphere at particular altitudes (Figure 4–8). Two additional instruments monitored the Sun. The Solar Mesospheric Explorer spun about its long axis at ninety degrees to its orbital plane so that on every turn, the instruments scanned the atmosphere on the horizon between twenty and eighty kilometers. Data from the rotating science instruments are gated (cycled "on") once each revolution. Table 4–30 lists the characteristics of each instrument, and Table 4–31 lists the mission's characteristics.

Infrared Astronomy Satellite

The Infrared Astronomy Satellite (IRAS) was the second Netherlands-United States cooperative satellite project, the first being the Astronomical Netherlands Satellite launched in 1974. A memorandum of understanding between the Netherlands Agency for Aerospace Programs and NASA established the project on October 4, 1977. The United Kingdom also participated in the program under a separate memorandum of understanding between the United Kingdom's Science and Engineering Research Council and the Netherlands Agency for Aerospace Programs.

Under the terms of the memorandum of understanding, the United States provided the infrared telescope system, the tape recorders, the Delta launch vehicle, the scientific data processing, and the U.S. co-chair and members of the Joint IRAS Science Working Group. The Netherlands Agency for Aerospace Programs provided the other co-chair and European members of the Joint IRAS Science Working Group, the spacecraft, the Dutch additional experiment (DAX), and the integration, testing, and launch preparations for the flight satellite. The Netherlands Agency for Aerospace Programs and the Science and Engineering Research Council provided spacecraft command and control and primary data acquisition with a ground station and control center located at Chilton, England. The United States provided limited tracking, command, and data acquisition by stations in the NASA Ground Spacecraft Tracking and Data Network.

IRAS was the first infrared satellite mission. It produced an all-sky survey of discrete sources in the form of sky and source catalogues using four broad photometry channels between eight and 120 micrometers. The mission performed the all-sky survey, provided additional observations on the more interesting known and discovered sources, and analyzed the data.

The satellite system consisted of two major systems: the infrared telescope and the spacecraft (Figure 4–9). The infrared telescope system consisted of the telescope, cryogenics equipment, electronics, and a focal-plane detector array. The detector array consisted of a primary set

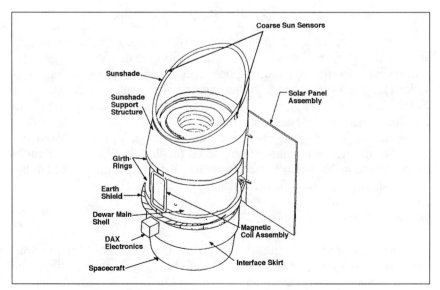

Figure 4–9. *Infrared Astronomy Satellite Configuration*

of infrared detectors, a set of photodiodes for use as aspect sensors, and a DAX. The DAX comprised a low-resolution spectrometer, a chopped photometric channel, and a short wavelength channel. The spacecraft provided the support functions of electrical power, attitude control, computing, and telecommunications.

During its all-sky survey, IRAS observed several important phenomena. It detected a new comet, named Comet IRAS-Araki-Alcock (1983d), which was distinguished by its very close approach to Earth, 5 million kilometers on May 11, 1983, the closest approach to Earth of a comet in 200 years. IRAS discovered a second, extremely faint comet (1983f) on May 12. This comet was a million times fainter than the first and was leaving the solar system. IRAS also discovered very young stars (protostars) no more than a million years old. It also observed two closely interacting galaxies that were being disrupted by each other's gravitational forces. IRAS made approximately 200,000 observations and transmitted more than 200 billion bits of data, which scientists have continued to examine and analyze.

IRAS revolutionized our understanding of star formation, with observations of protostars and of interstellar gas in star-forming regions. It discovered the "interstellar cirrus" of wispy cool far-infrared emitting dust throughout our galaxy. It discovered infrared emissions in spiral galaxies, including a previously unknown class of "ultraluminous infrared galaxies" in which new stars were forming at a very great rate. IRAS also showed that quasars emit large amounts of far-infrared radiation, suggesting the presence of interstellar dust in the host galaxies of those objects.

IRAS operated successfully until November 21, 1983, when it used the last of the super-fluid helium refrigerant that cooled the telescope. IRAS represented as great an improvement over ground-based telescopes as the Palomar 200-inch telescope was over Galileo's telescope. The unprecedented sensitivity of IRAS provided a survey of a large, unexplored gap in the electromagnetic spectrum. The international IRAS science team compiled a catalogue of nearly 250,000 sources measured at four infrared wavelengths—including approximately 20,000 new galaxies and 16,000 small extended sources—and the Jet Propulsion Laboratory's Infrared Processing and Analysis Center produced IRAS Sky Maps. IRAS successfully surveyed more than 96 percent of the sky. Its mission characteristics are in Table 4–32.

The Plasma Interaction Experiment (PIX-II) also rode on the Delta launch vehicle that deployed IRAS. A Lewis Research Center investigation, PIX-II evaluated the effects of solar panel area on the interactions between the space charged-particle environment and surfaces at high potentials (+/–one keV). PIX-II was the second experiment to investigate the effects of space plasma on solar arrays, power system conductors, insulators, and other exposed spacecraft components. The experiment remained with the second stage of the Delta launch vehicle in orbit at an altitude of 640 kilometers. Data from PIX-II were transmitted to two tracking stations.

European X-Ray Observatory Satellite

NASA launched the European X-Ray Observatory Satellite (EXOSAT) for the European Space Agency (ESA), which reimbursed NASA for the cost of providing standard launch support in accordance with the terms of a launch services agreement signed March 25, 1983. A Delta 3914 placed the satellite in a highly elliptical orbit that required approximately four days to complete. This orbit provided maximum observation periods, up to eighty hours at a time, while keeping the spacecraft in full sunlight for most of the year, thereby keeping thermal conditions relatively stable and simplifying alignment procedures. The orbit also allowed practically continuous coverage by a single ground station.

EXOSAT supplied detailed data on cosmic x-ray sources in the soft x-ray band four one-hundredths keV to eighty keV. The principal scientific objectives involved locating x-ray sources and studying their spectroscopic and temporal characteristics. The location of x-ray sources was determined by the use of x-ray imaging telescopes. The observatory also mapped diffuse extended sources such as supernova remnants and resolve sources within nearby galaxies and galaxies within clusters. The spacecraft performed broad-band spectroscopy, or "color" cataloguing of x-ray sources, and studied the time variability of sources over time scales ranging from milliseconds to days.

The EXOSAT observatory was a three-axis stabilized platform with an inherent orbit correction capability. It consisted of a central body covered with super-insulating thermal blankets and a one-degree-of-freedom rotatable solar array. The platform held the four experiments, which were co-aligned with the optical axis defined by two star trackers, each mounted on an imaging telescope (Figure 4–10). Table 4–33 contains the mission's characteristics.

Shuttle Pallet Satellite

The Shuttle Pallet Satellite (SPAS)-01 was a reusable platform built by the German aerospace firm Messerschmitt-Bolkow-Blohm (MBB) and carried on STS-7 as part of an agreement with MBB. The agreement provided that, in return for MBB's equipping SPAS-01 for use in testing the deployment and retrieval capabilities of the remote manipulator arm, NASA would substantially reduce the launching charge for SPAS-01. The platform contained six scientific experiments from the West German Federal Ministry of Research and Technology, two from ESA, and three from NASA along with several cameras.

The first satellite designed to be recaptured by the Shuttle's robot arm, SPAS-01 operated both inside and outside the orbiter's cargo bay. In the cargo bay, the satellite demonstrated its system performance and served as a mounted platform for operating scientific experiments. Seven scientific experiments were turned on during the third day of the flight and ran continuously for about twenty-four hours.

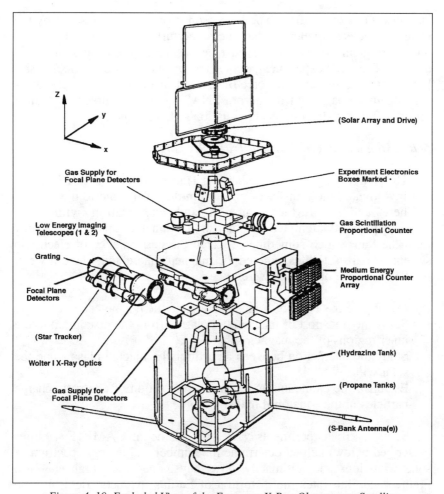

Figure 4–10. Exploded View of the European X-Ray Observatory Satellite

In the free-flyer mode, SPAS-01 was used as a test article to demonstrate the orbiter's capability to deploy and retrieve satellites in low-Earth orbit. During this phase of the mission, crew members operated two German and three NASA experiments. MBB built the platform to demonstrate how spaceflights could be used for private enterprise purposes. The West German Federal Ministry of Research and Technology supported the SPAS-01 pilot project and contributed to mission funding. Mission characteristics are in Table 4–34.

Hilat

The Air Force developed Hilat to gather data on ionospheric irregularities and auroras (northern lights) in an effort to improve the effectiveness of Department of Defense communications systems. The interaction of charged particles, ionized atmospheric gases, and magnetic fields can degrade radio communications and radar system performance at high latitudes. Four of the five experiments on board were sponsored by the Defense Nuclear Agencies. They measured turbulence caused by ionospheric irregularities and observed electron, ion, proton, and magnetic activity. The fifth experiment, sponsored by the Air Force Geophysics Laboratory at Hanscom Air Force Base, used an auroral ionospheric mapper to gather imagery of the auroras. NASA was reimbursed for launch services. Table 4–35 contains the mission's characteristics.

Active Magnetospheric Particle Tracer Explorers

The Active Magnetospheric Particle Tracer Explorers (AMPTE) project investigated the transfer of mass from the solar wind to the magnetosphere and its further transport and energization within the magnetosphere. It attempted to establish how much of this immense flow from the Sun, which sometimes affected the performance of electronic systems aboard satellites, entered the magnetosphere and where it went. AMPTE mission objectives were to:

- Investigate the entry of solar wind ions to the magnetosphere
- Study the transport of magnetotail plasma from the distant tail to the inner regions of the magnetosphere
- Study the interaction between an artificially injected plasma and the solar wind
- Establish the elemental and charge composition of energetic charge particles in the equatorial magnetosphere

The scientific experiments carried aboard the three AMPTE satellites (described below) helped determine the number and energy spectrum of solar wind ions and, ultimately, how they gained their high energies. Figure 4–11 illustrates the distortion of Earth's magnetic field into the magnetosphere.

AMPTE also investigated the interaction of two different flowing plasmas in space, another common astronomical phenomenon. AMPTE studied in detail the local disturbances that resulted when a cold dense plasma was injected and interacted with the hot, rapidly flowing natural plasmas of the solar wind and magnetosphere. The AMPTE spacecraft injected tracer elements into near-Earth space and then observed the motion and acceleration of those ions. One expected result was the formation of artificial comets, which were observed from aircraft and from the ground. In this respect, AMPTE's active interaction with the environment made it different from previous space probes, which had passively measured their surrounding environment.

This international cooperative mission consisted of three spacecraft: (1) a German-provided Ion Release Module (IRM), which injected artificial tracer ions (lithium and barium) inside and outside Earth's magnetosphere; (2) a U.S.-provided Charge Composition Explorer (CCE), which detected and monitored these ions as they convected and diffused

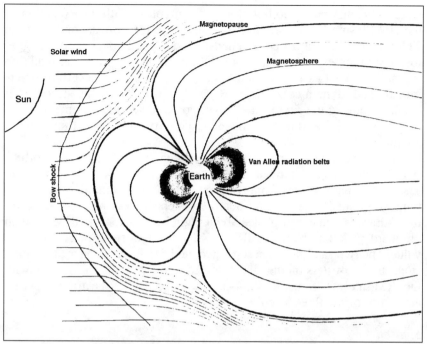

Figure 4–11. Distortion of Earth's Magnetic Field
(The solar wind distorts Earth's magnetic field, in some cases pushing field lines from the day side of Earth back to the night side.)

through the inner magnetosphere; and (3) a United Kingdom-provided subsatellite (UKS), which detected and monitored these ions within a few hundred kilometers of the release point. Each of the spacecraft contributed to the achievement of the mission objectives. The IRM released tracer ions in the solar wind and attempted to detect them with the CCE inside the magnetosphere. This was done four times under different solar wind conditions and with different tracer ions.

The IRM also released barium and lithium ions into the plasma sheet and observed their energy spectrum at the CCE. Four such releases took place. In addition to the spacecraft observations, ground stations and aircraft in the Northern and Southern Hemispheres observed the artificial comet and tail releases. No tracer ions were detected in the CCE data, a surprising result, because, according to accepted theories, significant fluxes of tracer ions should have been observed at the CCE. However, in the case of the last two tail releases, the loss of the Hot Plasma Composition Experiment instrument on April 4, 1985, severely restricted the capability of the CCE to detect low-energy ions. The spacecraft also formed two barium artificial comets. In both instances, a variety of ground observation sites in the Northern and Southern Hemispheres obtained good images of these comets.

Observations relating to the composition, charge, and energy spectra of energetic particles in the near equatorial orbit plane of the CCE were to occur for a period of at least six months. With the exception of the Hot Plasma Composition Experiment, the instruments on board the CCE acquired the most comprehensive and unique data set on magnetospheric ions ever collected. For the first time, the ions that made up the bulk of Earth's ring current were identified, their spectrum determined, and dynamics studied. Several major magnetic storms that occurred during the first year of operation allowed measurements to be taken over a wide range of magnetic activity indices and solar wind conditions.

The three AMPTE spacecraft were launched into two different orbits. A Delta launch vehicle released the three satellites in a stacked fashion. The CCE separated first from the group of three, and the IRM and UKS remained joined. The CCE on-board thrusters fired to position the satellite in Earth's equatorial plane. About eight hours later, the IRM fired an on-board rocket to raise the IRM/UKS orbit apogee to twice its initial value. The two satellites then separated, and for the remainder of the mission, small thrusters on the UKS allowed it to fly in close formation with the IRM satellite. Tables 4–36, 4–37, and 4–38 list the specific orbit characteristics of the three satellites.

Spartan 1

Spartan 1 was the first of a continuing series of low-cost free-flyers designed to extend the observing time of sounding-rocket-class experiments from a few minutes to several hours. The Astrophysics Division of NASA's OSSA sponsored the satellite. The Naval Research Laboratory provided the scientific instrument through a NASA grant. The instrument, a medium-energy x-ray scanner, had been successfully flown several times on NASA sounding rockets. It scanned the Perseus Cluster, Galactic Center, and Scorpius X-2 to provide x-ray data over the energy range of a half keV to fifteen keV (Figure 4–12).

The June 1985 launch was NASA's second attempt to launch Spartan 1. It had previously been manifested on STS 41-F for an August 1984 flight, but was demanifested because of problems with the launch of *Discovery*.

Researchers could use the Spartan family of reusable satellites for a large variety of astrophysics experiments. The satellites were designed to be deployed and retrieved by the Shuttle orbiter using the remote manipulator system. Once deployed, the Spartan satellite could perform scientific observations for up to forty hours. All pointing sequences and satellite control commands were stored aboard the Spartan in a microcomputer controller. A 10^{10}-bit tape recorder recorded all data, and no command or telemetry link was provided. Once the Spartan satellite completed its observations, it "safed" all systems and placed itself in a stable attitude to allow for retrieval by the orbiter and a return to Earth for data analysis and preparation for a new mission. Table 4–39 lists Spartan 1's mission characteristics.

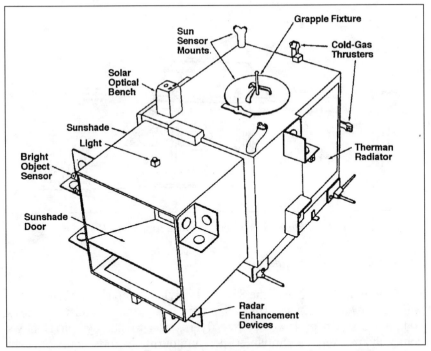

Figure 4–12. Spartan 1

Plasma Diagnostic Package

The Plasma Diagnostics Package (PDP) flew on two Shuttle missions—STS-3 as part of the OSS-1 payload and STS 51-F as part of the Spacelab 2 mission. On its first flight, it made measurements while mounted in the Shuttle payload bay and while suspended from the remote manipulator arm. It successfully measured electromagnetic noise created by the Shuttle and detected other electrical reactions taking place between the Shuttle and the ionospheric plasma.

On STS 51-F, the PDP made additional measurements near the Shuttle and was also released as a free-flyer on the third day of the mission to measure electric and magnetic fields at various distances from the orbiter. During the maneuvers away from the Shuttle, called a "fly-around," a momentum wheel spun the satellite to fix it in a stable enough position for accurate measurements. As the orbiter moved away to a distance of approximately a half kilometer, an assembly of instruments mounted on the PDP measured various plasma characteristics, such as low-energy electron and proton distribution, plasma waves, electric field strength, electron density and temperature, ion energy and direction, and pressure of unchanged atoms. This was the first time that ambient plasma was sampled so far from the Shuttle. The survey helped investigators determine how far the orbiter's effects extended. Figure 4–13 illustrates PDP experiment hardware, and Table 4–40 describes characteristics of the PDP on STS 51-F. PDP characteristics on STS-3 were very similar.

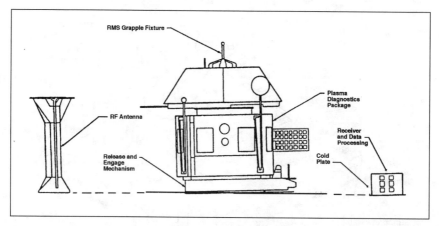

Figure 4–13. Plasma Diagnostics Package Experiment Hardware

Spartan 203 (Spartan Halley)

Spartan 203 was one of the STS 51-L payloads aboard *Challenger* that was destroyed in January 1986. Spartan Halley, the second in NASA's continuing series of low-cost free-flyers, was to photograph Halley's comet and measure its ultraviolet spectrum during its forty hours of flight in formation with the Shuttle. The spacecraft was to be deployed during the second day of the flight and retrieved on the fifth day. Both operations would use the remote manipulator system. The instruments being used had flown on sounding rockets as well as on the Mariner spacecraft. The mission was to take advantage of Comet Halley's location of less than 107.8 million kilometers from the Sun during the later part of January 1986. This period was scientifically important because of the increased rate of sublimation as the comet neared perihelion, which would occur on February 9. As Halley neared the Sun, temperatures would rise, releasing ices and clathrates, compounds trapped in ice crystals.

NASA's Goddard Space Flight Center and the University of Colorado's Laboratory for Atmospheric and Space Physics recycled several instruments and designs to produce a low-cost, high-yield spacecraft. Two spectrometers, derived from backups for a Mariner 9 instrument that studied the Martian atmosphere in 1971, were rebuilt to survey the comet in ultraviolet light from 128- to 340-nanometer wavelength. The spectrometers were not to produce images but would reveal the comet's chemistry through the ultraviolet spectral lines they recorded. From these data, scientists would have gained a better understanding of how (1) chemical structure of the comet evolved from the coma and proceeded down the tail, (2) species changed with relation to sunlight and dynamic processes within the comet, and (3) dominant atmospheric activities at perihelion related to the comet's long-term evolution. Figure 4–14 shows the Spartan Halley configuration, and Table 4–41 lists the mission's characteristics.

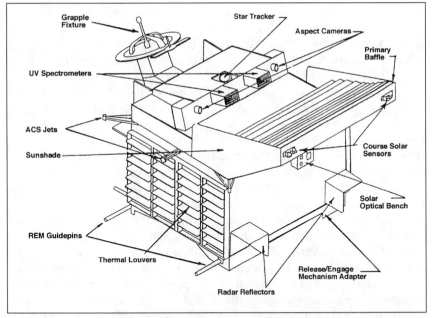

Figure 4–14. Spartan Halley Configuration

Polar BEAR

The Polar Beacon Experiments and Auroral Research satellite (Polar BEAR) mission, a follow-on to the 1983 Hilat mission, conducted a series of experiments for the Department of Defense that studied radio interference caused by the Aurora Borealis. Launched by NASA on a Scout launch vehicle, the satellite had hung in the Smithsonian for more than fifteen years. The retooled Oscar 17 satellite was built in the mid-1960s by the Navy as a spare but never launched. Polar BEAR's characteristics are in Table 4–42.

San Marco D/L

The San Marco D/L spacecraft, one element of a cooperative satellite project between Italy and the United States, explored the relationship between solar activity and meteorological phenomena, with emphasis on lower atmospheric winds of the equatorial thermosphere and ionosphere. This information augmented and was used with data obtained from ground-based facilities and other satellites. The San Marco D/L project was the fifth mission in a series of joint research missions conducted under an agreement between NASA and the Italian Space Commission. The first memorandum of understanding (MOU) between Italy's Italian Commissione per le Ricerche Spaziali and NASA initiated the program in May 1962. The first flight under this agreement took place in March 1964 with the successful launch by the Centro Ricerche Aerospaziali of a two-stage Nike sounding rocket from the Santa Rita launch platform off Kenya's coast. This vehicle carried the basic elements of the San Marco science instrumentation, flight-qualified the components, and provided a means of checking out range instrumentation and equipment.

This launch was followed by the December 1964 launch of the fully instrumented San Marco-I spacecraft from Wallops Island, Virginia. This marked the first time in NASA's international cooperative program that a satellite launch operation had been conducted by a non-U.S. team and the first use of a satellite fully designed and built in Western Europe. This launch also qualified the basic spacecraft design and confirmed the usefulness and reliability of the drag balance device for accurate determinations of air density values and satellite attitude.

Implementation of the agreement continued with the launch of San Marco-II into an equatorial orbit from the San Marco platform off the coast of Kenya in April 1967. This was the first satellite to be placed into equatorial orbit. The San Marco-II carried the same instrumentation as the San Marco-I, but the equatorial orbit permitted a more detailed study of density variations versus altitude in the equatorial region. The successful launch also qualified the San Marco range as a reliable facility for future satellite launches.

A second MOU between Centro Ricerche Aerospaziali and NASA signed in November 1967 provided for continued cooperation in satellite measurements of atmospheric characteristics and the establishment of the San Marco C program. The effort enhanced and continued the drag balance studies of the previous projects and initiated complementary mass spectrometer investigations of the equatorial neutral particle atmosphere. This phase enabled simultaneous measurements of atmospheric density from one satellite by three different techniques: direct particle detection, direct drag, and integrated drag. The San Marco C1 was launched on April 24, 1971, and the San Marco C2 was launched on February 18, 1974, both from the San Marco platform. The platform had also been used earlier in 1970 to launch Uhuru, an Explorer satellite that scanned 95 percent of the celestial sphere for sources of x-rays. It discovered three new pulsars that had not previously been identified.

NASA and Centro Ricerche Aerospaziali signed a third MOU in August 1974, continuing and extending their cooperation in satellite measurements of atmospheric characteristics and establishing the San Marco/Atmosphere Explorer Cooperative Project. This effort measured diurnal variations of the equatorial neutral atmosphere density, composition, and temperature for correlation with the Explorer 51 data for studies of the physics and dynamics of the thermosphere.

The San Marco D MOU was signed by Centro Ricerche Aerospaziali in July 1976 and by NASA in September 1976. This MOU assigned project management responsibility for the Italian portion of the project to Centro Ricerche Aerospaziali, while the Goddard Space Flight Center assumed project responsibility for the U.S. portion. There was also an

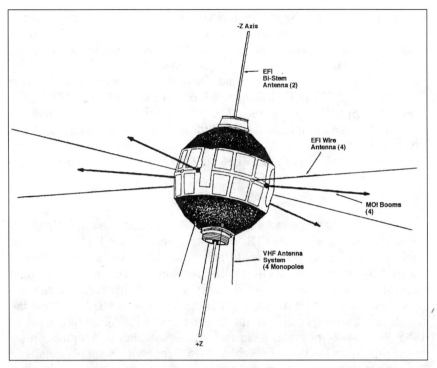

Figure 4–15. San Marco D/L Spacecraft

auxiliary cooperative agreement between the University of Rome and the Deutsche Forschungs Versuchsanstat für Luft und Raumfahrt (DFVLR) of the Federal Republic of Germany. This activity would explore the possible relationship between solar activity and meteorological phenomena to further define the structure, dynamics, and aeronomy of the equatorial thermosphere. Although initially both a low-orbit and an upper orbit spacecraft were planned, the program was reduced to a single spacecraft program—the low-orbit San Marco D/L (Figure 4–15).

In accordance with the MOU, the Centro Ricerche Aerospaziali provided the spacecraft, its subsystems, and an air drag balance system. The Deutsche Forschungs Versuchsanstat fur Luft und Raumfahrt provided an airglow solar spectrometer. NASA provided an ion velocity instrument, a wind/temperature spectrometer, and an electric field instrument. NASA also provided the Scout launch vehicle and technical and consultation support to the Italian project team. Mission characteristics of the San Marco D/L are in Table 4–43.

Attached Shuttle Payload Bay Science Missions

Beginning with the launch of STS-1 in April 1981, NASA had an additional platform available for performing scientific experiments. No longer did it have to deploy a satellite to obtain the benefits of a microgravity environment. Now, the payload bay on the Space Shuttle could provide this type of environment. NASA used these surroundings for a variety of smaller experiments, small self-contained payloads, and large experimental missions. These larger missions were sponsored by NASA's OSS, OSTA, OSSA, and Office of Aeronautics and Space Technology (OAST). This chapter addresses the OSS and OSSA missions (the Spacelab missions). The OSTA missions are included in Chapter 2, "Space Applications," and the mission sponsored by OAST is discussed in Chapter 3, "Aeronautics and Space Research and Technology," both in Volume VI of the *NASA Historical Data Book*.

Spacelab Missions

NASA conducted three joint U.S./ESA Spacelab missions. Spacelab 1 (STS-9) and Spacelab 2 (STS 51-G) were verification flights. Spacelab 3 (STS 51-B) was an operational flight. Spacelab 1 was the largest international cooperative space effort yet undertaken and concluded more than ten years of intensive work by some fifty industrial firms and ten nations. Spacelab 1 cost the ESA approximately $1 billion. NASA also flew the first Spacelab reimbursable flight, Deutschland-1 (D-1), on STS 61-A in 1985. Table 4–44 provides a chronology of Spacelab development prior to the first Spacelab mission. Tables 4–45 through 4-48 supply details of the experiments flown on each mission.

Spacelab 1. The Spacelab 1 mission, which flew on STS-9, exemplified the versatility of the Space Shuttle. Payload specialist Ulf Merbold of ESA summed up the mission: "That was science around the clock and round the earth."[8] Payload specialists conducted science and applications investigations in stratospheric and upper atmospheric physics, materials processing, space plasma physics, biology, medicine, astronomy, solar physics, Earth observations, and lubrication technology. The broad discipline areas included atmospheric physics and Earth observations, space plasma physics, astronomy and solar physics, material sciences and technology, and life sciences (Table 4–45).

Atmospheric physics and Earth observations, space plasma physics, and solar physics investigators used the Spacelab 1 orbiting laboratory to study the origin and influence of turbulent forces that sweep by Earth causing visible auroral displays and disturbing radio broadcasts, civilian and military electronics, power distribution, and satellite systems. The astronomy investigations studied astronomical sources in the ultraviolet and x-ray wavelengths. These wavelengths were not observable on Earth because of absorption by the ionosphere or ozone layer. The materials science and technology investigations demonstrated the capability of Spacelab as a technological development and test facility. The experi-

---

[8]"Spacelab Utilization Future Tasks," *MBB/ERNO Report*, Vol. 9, No. 1, April 1984, p. 8, NASA Historical Reference Collection, NASA Headquarters, Washington, DC.

ments in this group took advantage of the microgravity conditions to perform studies on materials and mechanisms that are adversely affected on Earth by gravity. The life sciences investigations studied the effects of the space environment (microgravity and high-energy radiation) on human physiology and on the growth, development, and organization of living systems. Figures 4–16, 4–17, and 4–18 show the locations of the Spacelab 1 experiments.

Spacelab 3. Spacelab 3, conducted on STS 51-B, was the first operational Spacelab mission. It used several new mini-laboratories that would be used again on future flights. Investigators evaluated two crystal growth furnaces, a life support and housing facility for small animals, and two types of apparatus for the study of fluids on this flight. Most of the experiment equipment was contained inside the laboratory, but instruments that required direct exposure to space were mounted outside in the open payload bay of the Shuttle. Figure 4–19 shows the experiment module layout, and Table 4–46 lists Spacelab 3's experiments.

Materials science was a major thrust of Spacelab 3. Spacelab served as a microgravity facility in which processes could be studied and materials produced without the interference of gravity. A payload specialist with special expertise in crystal growth succeeded in producing the first crystal grown in space. Studies in fluid mechanics also took advantage of the microgravity environment. Investigations proved the concept of "containerless" processing for materials science experiments with the successful operation of the Drop Dynamics Module.

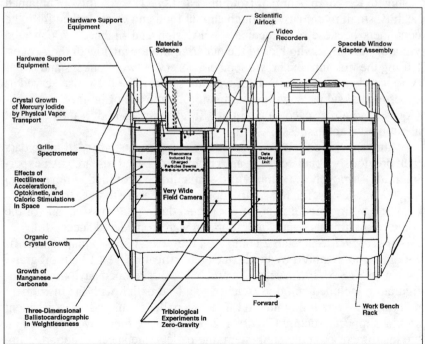

Figure 4–16. Spacelab 1 Module Experiment Locations (Port Side)

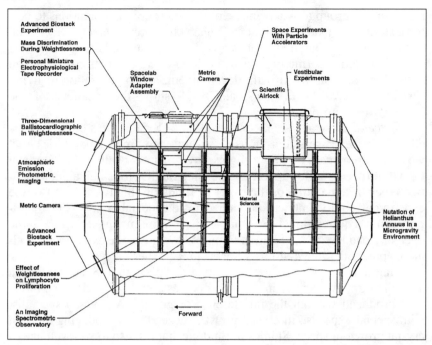

Figure 4–17. Spacelab 1 Module Experiment Locations (Starboard Side)

Spacelab 3 carried a contingent of animals living in the newly designed Research Animal Holding Facility. This facility maintained healthy, small mammals, although animal food and waste leaked from the containers because of inadequate seal design and higher than expected vigor of monkeys, who kicked the material into the airflow of their cages. During the mission, the crew members observed two monkeys and twenty-four rodents for the effects of weightlessness. The crew also served as experimental subjects, with investigations in the use of biofeedback techniques to control space sickness and in changes in body fluids brought about by weightlessness.

Atmospheric physics and chemistry experiments provided more data than previously obtained in decades of balloon-based research. An experimental atmospheric modeling machine provided more than 46,000 images useful for solar, Jupiter, and Earth studies. In all, more than 250 billion bits of data were returned during the mission, and of the fifteen experiments conducted, fourteen were considered successful.

Spacelab 2. Spacelab 2 completed the second of two planned verification flights required by the Spacelab Verification Test Flight program. Flown on STS 51-F, Spacelab 2 was a NASA-developed payload. Its configuration included an igloo attached to a lead pallet, with the instrument pointing subsystem mounted on it, a two-pallet train, and an experiment special support structure (Figure 4–20). The experiments were located on the instrument pointing subsystem, the pallets, the special support structure, and the middeck of the orbiter, and one was based on the ground.

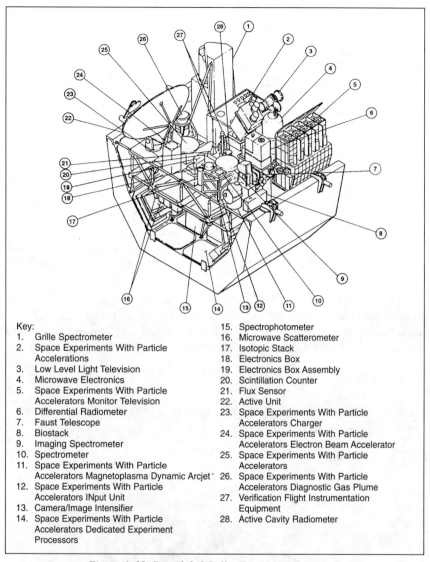

Key:
1. Grille Spectrometer
2. Space Experiments With Particle Accelerations
3. Low Level Light Television
4. Microwave Electronics
5. Space Experiments With Particle Accelerators Monitor Television
6. Differential Radiometer
7. Faust Telescope
8. Biostack
9. Imaging Spectrometer
10. Spectrometer
11. Space Experiments With Particle Accelerators Magnetoplasma Dynamic Arcjet*
12. Space Experiments With Particle Accelerators INput Unit
13. Camera/Image Intensifier
14. Space Experiments With Particle Accelerators Dedicated Experiment Processors
15. Spectrophotometer
16. Microwave Scatterometer
17. Isotopic Stack
18. Electronics Box
19. Electronics Box Assembly
20. Scintillation Counter
21. Flux Sensor
22. Active Unit
23. Space Experiments With Particle Accelerators Charger
24. Space Experiments With Particle Accelerators Electron Beam Accelerator
25. Space Experiments With Particle Accelerators
26. Space Experiments With Particle Accelerators Diagnostic Gas Plume
27. Verification Flight Instrumentation Equipment
28. Active Cavity Radiometer

Figure 4–18. Spacelab 1 Pallet Experiment Locations

The pallets provided mounting and support for experiments that required an atmosphere-free environment. The special support structure was specially designed to support the Elemental Composition and Energy Spectral of Cosmic Ray Nuclei experiment.

Fourteen experiments supported by seventeen principal investigators were conducted (Table 4–47). The experiments were in the fields of life sciences, plasma physics, infrared astronomy, high-energy physics, solar physics, atmospheric physics, and technology.

Spacelab D-1. Spacelab D-1, the "German Spacelab," concentrated on scientific experiments on materials in a microgravity environment.

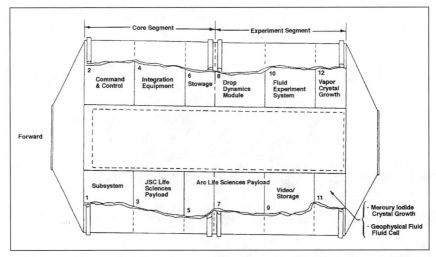

Figure 4–19. Spacelab 3 Experiment Module Layout (Looking Down From the Top)

This mission, flown on STS 61-A, was the second flight of the Materials Experiment Assembly (the first was on STS-7). Experiments included investigations of semiconductor materials, miscibility gap materials, and containerless processing of glass melts (Table 4–48).

OSS-1 (STS-3)

The OSS-1 mission objectives were to conduct scientific observations that demonstrated the Space Shuttle's research capabilities and that

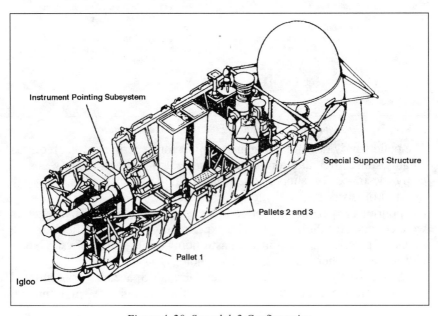

Figure 4–20. Spacelab 2 Configuration

were appropriate for flight on an early mission; to conduct supplementary observations of the orbiter's environment that had specific applicability to plasma physics and astronomical payloads; and to evaluate technology that may have application in future experiments in space. The experiments obtained data on the near-Earth space environment, including the degree of contamination (gases, dust, and outgassing particles) introduced by the orbiter itself.

The OSS-1 payload, also designated the "Pathfinder Mission," was a precursor to the Spacelab missions. It was developed to characterize the environment around the orbiter associated with the operation of the Shuttle and to demonstrate the Shuttle's research capability for science applications and technology in space. It verified that research measurements could be carried out successfully on future Shuttle missions and performed scientific measurements using the Shuttle's unique capabilities.

The mission included scientific investigations in space plasma physics, solar physics, astronomy, life sciences, and space technology. Six of the nine experiments were designed by scientists at five American universities and one British university and were operated under their supervision during the mission. One experiment was developed by the Naval Research Laboratory, and two were developed by the Goddard Space Flight Center (Table 4–49). The OSS-1 experiments being flown in the orbiter's payload bay were carried on a special U-shaped structure called an orbital flight test pallet. The three-meter-by-four-meter aluminum frame and panel structure weighing 527 kilograms was a Spacelab element that would be used later in the STS program (Figure 4–21).

Other Physics and Astronomy Missions

The following sections describe physics and astronomy missions that were launched prior to 1979 and continued operating into the 1980s, followed by a discussion of missions that underwent development from 1979 to 1988 but did not launch until later. Readers can find details of the early stages of the ongoing science missions in Volume III of the *NASA Historical Data Book*.[9]

Ongoing Physics and Astronomy Missions

International Ultraviolet Explorer. The International Ultraviolet Explorer (IUE) mission was a joint enterprise of NASA, ESA, and the British Science Research Council. IUE 1, launched into geosynchronous orbit on January 26, 1978, on a Delta launch vehicle, allowed hundreds of users at two locations to conduct spectral studies of celestial ultraviolet sources. It was the first satellite totally dedicated to ultraviolet astron-

---

[9]Ezell, *NASA Historical Data Book, Volume III.*

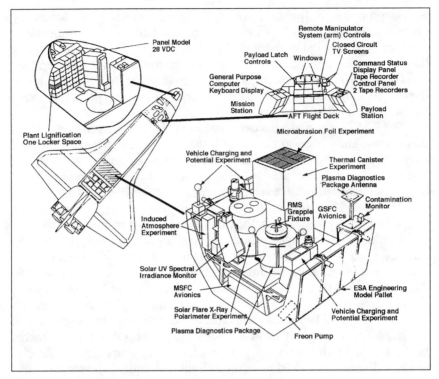

Figure 4–21. OSS-1 Payload Configuration

omy. The IUE mission objective was to conduct spectral distribution studies of celestial ultraviolet sources. The scientific goals were to:

- Obtain high-resolution spectra of stars
- Study gas streams
- Observe faint stars, galaxies, and quasars
- Observe the spectra of planets and comets
- Make repeated observations that showed variable spectra
- Define more precisely the modifications of starlight caused by interstellar dust and gas

NASA provided the IUE spacecraft, the optical and mechanical components of the scientific instruments, the U.S. ground observatory, and the spacecraft control software. ESA contributed the solar arrays needed as a power source and the European ground observatory in Spain. The British Science Research Council oversaw the development of the spectrograph television cameras and, with the United States, the image processing software.

Targets of IUE's investigations included faint stars, hot stars, quasars, comets, gas streams, extragalactic objects, and the interstellar medium. A forty-five-centimeter Ritchey Chretien telescope aided in the investigations. Geosynchronous orbit permitted continuous observations and real-time data by the investigators at the two ground observatories. Objects observed by IUE included planets, stars, and galaxies. IUE specialized in targets of opportunity, such as comets, novae, and supernovae.

Often, IUE allowed simultaneous data acquisition and was used in conjunction with other telescopes from around the world. In its later years of operation, these collaborations involved such spacecraft as the Hubble Space Telescope, the German Roentgen Satellite, the Compton Gamma Ray Observatory, the Voyager probes, the Space Shuttle's Astro-1 mission, the Extreme Ultraviolet Explorer, and Japan's ASCA satellite, as well as numerous ground-based observatories.

In 1979, IUE produced the first evidence confirming the existence of a galactic halo, consisting of high-temperature, rarefied gas extending far above and below the Milky Way. In 1980, it verified expectations that space between isolated galaxies was highly transparent and contributed very little to the total mass of the universe. Extensive observation of active binary stars demonstrated that stellar magnetic fields and rotation probably combined to cause the tremendous levels of solar-like activity in many classes of such stellar systems. Studies using IUE data also indicated a consistent and continuous evolution of coronas, wind characteristics, and mass-loss rates, varying from the hot, fast winds and low mass-loss rate of the Sun to the slow, cool winds and high mass-loss rate of the coolest giant and supergiant stars. In addition, IUE provided the first detailed studies of comets throughout their active cycle in the inner solar system, providing new clues to their internal composition. Observations also confirmed the discovery of a hot halo of gas surrounding the Milky Way.

In 1986, IUE provided space-based observations of Halley's Comet and its tail during the Japanese, European, and Soviet missions to its nucleus and later initiated periodic observations of Supernova 1987a. The observations provided the key data required to identify the true progenitor of the supernova. As it continued to observe Supernova 1987a, IUE discovered the remnant shell from the red supergiant stage of the supernova as well as determined the changing properties of the ejecta from continuing observations. The spacecraft made the best determination of the light curve and its implications concerning the nature of the energy source.

When launched in 1978, the IUE spacecraft had a stated lifetime expectancy of three to five years. It was shut down on September 30, 1996, after more than eighteen years of mission elapsed time.

International Sun-Earth Explorer/International Cometary Explorer. The International Sun-Earth Explorer (ISEE) program was a collaborative three-spacecraft program with ESA. ISEE 3 was injected into a "halo" orbit in November 1978 about the Earth-Sun libration point, from which it observed the solar wind an hour before it reached Earth's magnetosphere. This capability could provide advance warning of impending magnetospheric and ionosphere disturbances near Earth, which the ISEE 1 and 2 spacecraft monitored. ISEE 3 also observed electrons that carried energy from Earth's bow shock toward the Sun.

Although Earth's magnetic field diverted most of the solar wind, some interacted, producing plasma waves; some transferred energy inside the magnetosphere; and some was hurled back toward the Sun.

ISEE 3 completed its original mission of monitoring the solar wind in 1983 and was maneuvered into an orbit swinging through Earth's magnetic tail and behind the Moon, using the Moon's gravity to boost the spacecraft toward rendezvous with a comet. ISEE 3 obtained the first *in situ* field and particle measurements in Earth's magnetotail. Also in 1983, NASA renamed ISEE 3 the International Cometary Explorer (ICE). It left its Earth orbit on December 22, 1983, to encounter the Comet Giacobini-Zinner on September 11, 1985. ICE passed within 8,000 kilometers of the comet's nucleus and through the comet's tail. It provided the first spacecraft data on a comet's magnetic field, plasma environment, and dust content.

Orbiting Astronomical Observatories. The Orbiting Astronomical Observatory-3, named Copernicus, continued to furnish information on an apparent black hole detected in the constellation Scorpius until its operations were shut down on December 31 1980, because of degradation in the experiment's detection system. Its work also included discoveries of clumpy structures and shocked million-degree gas in the interstellar medium and measurements of the ultraviolet spectra of the chromospheres and coronas of stars other than the Sun.

Physics and Astronomy Missions Under Development From 1979 to 1988

Hubble Space Telescope. The history of the Hubble Space Telescope can be traced back as far as 1962, when the National Academy of Sciences published a report recommending the construction of a large space telescope. In 1973, NASA established a small scientific and engineering steering committee to determine which scientific objectives would be feasible for a proposed space telescope. C. Robert O'Dell of the University of Chicago headed the team. He viewed the project as an opportunity to establish a permanent orbiting observatory. In 1978, responsibility for the design, development, and construction of the space telescope went to the Marshall Space Flight Center. The Goddard Space Flight Center was chosen to lead the development of the scientific instruments and the ground control center. Marshall selected Perkin-Elmer of Danbury, Connecticut, over Itek and Kodak to develop the optical system and guidance sensors. Lockheed Missiles and Space Company of Sunnyvale, California, was selected over Martin Marietta and Boeing to produce the protective outer shroud and the support systems module for the telescope, as well as to assemble and integrate the finished product.

ESA agreed to furnish the spacecraft solar arrays, one of the scientific instruments (Faint Object Camera), and personnel to support the Space Telescope Science Institute in exchange for 15 percent of the observing time and access to the data from the other instruments. Goddard scientists were selected to develop one instrument, and scientists at the California Institute of Technology, the University of California at San Diego, and the

University of Wisconsin were selected to develop three other instruments. The telescope's construction was completed in 1985.

Because of Hubble's complexity, NASA established two new facilities under the direction of Goddard that were dedicated exclusively to the scientific and engineering operation of the telescope. The Space Telescope Operations Control Center at Goddard would serve as the ground control facility for the telescope. The Space Telescope Science Institute, located on the campus of Johns Hopkins University, would perform the science planning for the telescope.

Hubble was originally scheduled for a 1986 launch. The destruction of *Challenger* in 1986, however, delayed the launch for several years. Engineers used the interim period to subject the telescope to intensive testing and evaluation. A series of end-to-end tests involving the Space Telescope Science Institute, Goddard, the Tracking and Data Relay Satellite System, and the spacecraft were performed during that time, resulting in overall improvements in system reliability. The launch would finally occur on April 25, 1990.

After launch, it was discovered that the telescope's primary mirror had a "spherical aberration" that caused out-of-focus images. A mirror defect only one-twenty-fifth the width of a human hair prevented Hubble from focusing all light to a single point. In addition, problems with the solar panels caused degradation in the spacecraft's pointing stability. At first many believed that that the spherical aberration, which was undetected during manufacturing because of a flawed measuring device, would cripple the telescope, but scientists quickly found a way to use computer enhancement to work around the abnormality. A repair mission aboard STS-61 in December 1993 replaced the solar panels and installed corrective lenses, which greatly improved the quality of the images. Table 4–50 outlines the development of the Hubble mission.

The scientific objectives of the Hubble mission were to investigate the composition, physical characteristics, and dynamics of celestial bodies, to examine the formation, structure, and evolution of stars and galaxies, to study the history and evolution of the universe, and to provide a long-term space-based research facility for optical astronomy. In addition, the Space Telescope Advisory Committee identified three key Hubble projects: (1) determine distances to galaxies and the Hubble Constant, (2) conduct a medium-deep survey of the sky, and (3) study quasar absorption lines.

The Hubble Space Telescope is a large Earth-orbiting astronomical telescope designed to observe the heavens from above the interference and turbulence of Earth's atmosphere. It is composed of a 2.4-meter Ritchey-Chretien reflector with a cluster of five scientific instruments at the focal plane of the telescope and the fine guidance sensors. Its scientific instruments can make observations in the ultraviolet, visible, and near-infrared parts of the spectrum (roughly 120-nanometer to one-millimeter wavelengths), and it can detect objects as faint as magnitude 31, with an angular resolution of about one-tenth arcsecond in the visible part

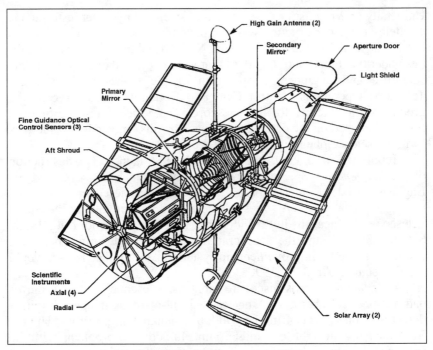

Figure 4–22. Hubble Space Telescope

of the spectrum. The spacecraft is to provide the first images of the surfaces of Pluto and its moon Charon and, by looking back in time and space, to determine how galaxies evolved in the initial period after the Big Bang. The telescope relays data to Earth via the high-gain antennae.

The Hubble Space Telescope is distinguished from ground-based observatories by its capability to observe light in the ultraviolet and near infrared. It also has an order of magnitude better resolution than is capable from within Earth's atmosphere. The telescope has a modular design, allowing on-orbit servicing via the Space Shuttle (Figure 4–22). Over the course of its anticipated fifteen-year operational lifetime, NASA plans several visits by Space Shuttle crews for the installation of new instruments, repairs, and maintenance. Hubble is about the size of a bus—it has a weight of approximately 11,000 kilograms and length of more than thirteen meters. It travels in a 611-kilometer circular orbit with an inclination of twenty-eight and a half degrees.

Compton Gamma Ray Observatory. NASA initiated the Compton Gamma Ray Observatory (CGRO) mission in 1981. It would be the second of NASA's orbiting Great Observatories, following the Hubble Space Telescope. During 1984, NASA completed the critical design reviews on all the instruments, and flight instrument hardware fabrication and assembly began. Also in 1984, NASA completed the spacecraft preliminary design review. In 1985, the design was completed, and NASA conducted the observatory critical design review. Manufacturing began on the structure and mechanisms and nearly completed fabrication of all hardware for the four scientific instruments. Manufacturing of the mechanical components and electronic systems approached completion during 1987, and the primary structure for the observatory was fabricated and assembled.

CGRO was a NASA cooperative program. The Federal Republic of Germany (the former West Germany), with co-investigator support from The Netherlands, ESA, the United Kingdom, and the United States, had principal investigator responsibility for one of the four instruments. Germany also furnished hardware elements and co-investigator support for a second instrument. NASA provided the remaining instruments and named the observatory in honor of Dr. Arthur Holly Compton, who won the Nobel Prize in physics for work on scattering of high-energy photons by electrons. This process was central to the gamma ray detection techniques of all four instruments.

CGRO was launched on April 5, 1991, aboard the Space Shuttle *Atlantis.* Dedicated to observing the high-energy universe, it would be the heaviest astrophysical payload flown to that time, weighing 15,422 kilograms, or more than fifteen metric tons (Figure 4–23). While Hubble's instruments would operate at visible and ultraviolet wavelengths, CGRO would carry a collection of four instruments that together could detect an unprecedented broad range of gamma rays. These instruments were the Burst and Transient Source Experiment, the Oriented Scintillation Spectrometer Experiment, the imaging Compton Telescope (known as COMPTEL), and the Energetic Gamma Ray Experiment Telescope.

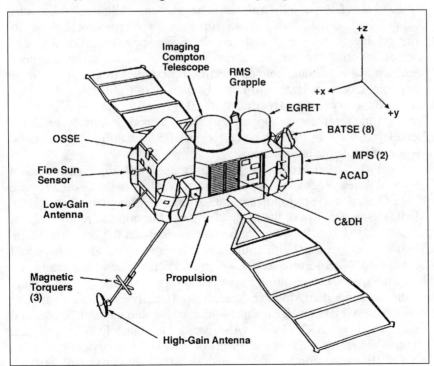

Figure 4–23. Compton Gamma Ray Observatory Configuration

These four instruments would be much larger and more sensitive than any gamma ray telescopes previously flown in space. The large size was necessary because the number of gamma ray interactions that could be recorded was directly related to the mass of the detector. Because the number of gamma ray photons from celestial sources was very small when compared to the number of optical photons, large instruments were needed to detect a significant number of gamma rays in a reasonable amount of time. The combination of these instruments would detect photon energies from 20,000 electron volts to more than 30 billion electron volts. For each of the instruments, an improvement in sensitivity of better than a factor of ten was realized over previous missions.

CGRO mission objectives were to measure gamma radiation from the universe and to explore the fundamental physical processes powering it. The observational objectives of CGRO were to search for direct evidence of the synthesis of the chemical elements, to observe high-energy astrophysical processes occurring in supernovae, neutron stars, and black holes, to locate gamma ray burst sources, to measure the diffuse gamma ray radiation for cosmological evidence of its origin, and to search for unique gamma ray emitting objects. The observatory had a diverse scientific agenda, including studies of very energetic celestial phenomena: solar flares, cosmic gamma ray bursts, pulsars, nova and supernova explosions, accreting black holes of stellar dimensions, quasar emission, and interactions of cosmic rays with the interstellar medium.

Extreme Ultraviolet Explorer. The Extreme Ultraviolet Explorer (EUVE) was an Earth-orbiting sky survey and spectroscopy mission. Its primary objectives were to produce a definitive sky map and catalogue of sources covering the extreme ultraviolet portion of the electromagnetic spectrum and to conduct pointed spectroscopy studies of selected extreme ultraviolet targets. Scientists from the University of California at Berkeley proposed the sky survey experiment for EUVE in 1975 in response to two NASA Announcements of Opportunity. NASA conditionally accepted the Berkeley concept in 1977, pending receipt of adequate funding and completion of implementation studies.

In 1981, the Jet Propulsion Laboratory assumed project management responsibilities. NASA transferred this responsibility to the Goddard Space Flight Center in 1986, following a decision to retrieve the Multimission Modular Spacecraft from the Solar Maximum Mission and refurbish it for use with EUVE. In 1986, when it became evident that the Solar Maximum Mission would reenter Earth's atmosphere before a retrieval mission could be mounted, NASA exercised its option to procure a new spacecraft from Fairchild Space. The resulting Explorer Platform was an upgraded version of the Multimission Modular Spacecraft. Initially, this spacecraft bus would have a dual-launch capability—that is, it could use both Shuttle and Delta launch vehicles. In 1988, NASA decided to launch EUVE on a Delta. Figure 4–24 shows the major elements of the EUVE observatory.

EUVE would conduct the first detailed all-sky survey of extreme ultraviolet radiation between 100 and 900 angstroms, a previously unex-

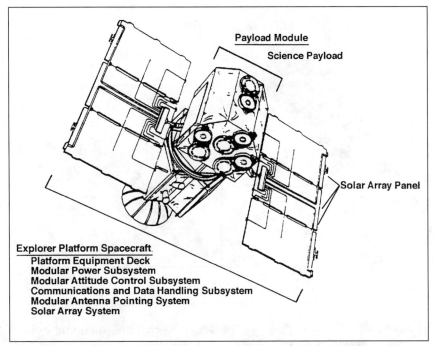

Figure 4–24. Extreme Ultraviolet Explorer Observatory

plored portion of the electromagnetic spectrum. EUVE would be a two-phase mission, with the first six months devoted to scanning the sky to locate and map sources emitting radiation in the extreme ultraviolet range and the remainder of the mission (about twenty-four months) devoted to detailed spectroscopy of sources located during the first phase (Figure 4–25). NASA launched EUVE on a Delta launch vehicle in June 1992. Upon completion of the EUVE mission, plans were to have the Shuttle rendezvous with the Explorer Platform and replace the EUVE payload with the X-ray Timing Explorer (XTE), which would monitor changes in the x-ray luminosity of black holes, quasars, and x-ray pulsars and would investigate physical processes under extreme conditions.[10]

Roentgen Satellite. The Roentgen Satellite (ROSAT) was a cooperative project of the West Germany, the United Kingdom, and the United States to perform high-resolution imaging studies of the x-ray sky. The mission's objectives were to study coronal x-ray emissions from stars of all spectral types, to detect and map x-ray emissions from galactic supernova remnants, to evaluate the overall spatial and source count distributions for various x-ray sources, to perform a detailed study of various populations of active galaxy sources, to perform a morphological study of the x-ray emitting clusters of galaxies, and to

---

[10]The Shuttle was not used to launch the X-ray Timing Explorer, which was launched on a Delta rocket in December 1995.

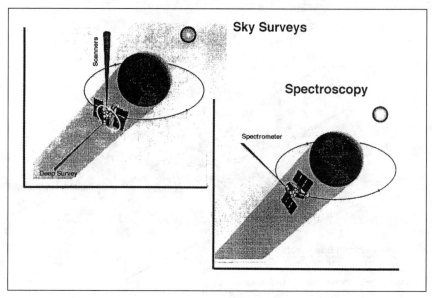

Figure 4–25. Two Phases of the Extreme Ultraviolet Explorer Mission

perform detailed mapping of the local interstellar medium by the extreme ultraviolet survey.

The United States would provide a high-resolution imaging instrument and launch services. West Germany would contribute the spacecraft and the main telescope, and the United Kingdom would provide the widefield camera. The ROSAT project originated from a 1975 proposal to the Bundeministerium für Forschungs und Technologie (BMFT) from scientists at the Max Planck Institut fuer Extraterrestrische Physik (MPE). The original objective was to conduct an all-sky survey with an imaging x-ray telescope of moderate angular resolution. Between 1977 and 1982, German space companies carried out extensive advance studies and preliminary analyses. Simultaneously, the Carl Zeiss Company in Germany initiated the development of a large x-ray mirror system, and MPE began to develop the focal plane instrumentation.

In 1979, following the regulations of ESA convention, BMFT announced the opportunity for ESA member states to participate by offering the possibility of flying a small, autonomous experiment together with the large x-ray telescope. In response to this announcement, a consortium of United Kingdom institutes led by Leicester University proposed an extreme ultraviolet wide-field camera to extend the spectral band measured by the x-ray telescope to longer wavelengths. The British Science and Engineering Research Council approved this experiment, and in 1983, BMFT and the council signed an MOU.

In 1981 and 1982, NASA and BMFT conducted negotiations for U.S. participation in the ROSAT mission, with the resulting MOU signed in 1982. Under this MOU, NASA agreed to provide the ROSAT launch with the Space Shuttle and a focal-point high-resolution imager detector.

BMFT's responsibilities included the design, fabrication, test, and integration of the spacecraft; mission control, tracking, and data acquisition after separation from the Shuttle; and the initial reduction and distribution of data. NASA would provide, at minimal charge, a flight model copy of the high-resolution imager previously flown on the 1978 High Energy Astronomy Observatories mission (HEAO-2). In 1983, NASA Headquarters issued a sole-source contract to the Smithsonian Astrophysical Observatory to build flight and engineering model high-resolution imagers and provide integration and launch support. In May 1985, NASA transferred this contract to the Goddard Space Flight Center for administration and implementation.

The *Challenger* accident led to a reconsideration of schedules and the launch vehicle. In 1987, NASA and BMFT decided to launch with a Delta launch vehicle. Germany redesigned the spacecraft appropriately, and the United States developed a new three-meter fairing for the Delta II nose section to accommodate ROSAT's maximum cross-sectional dimension. ROSAT was launched on a Delta rocket in June 1990. Figure 4–26 shows the ROSAT flight configuration.

Cosmic Background Explorer. The development of the Cosmic Background Explorer (COBE) began during fiscal year 1982. Developed by NASA's Goddard Space Flight Center, COBE would measure the diffuse infrared and microwave radiation from the early universe, to the limits set by our astrophysical environment. The spacecraft would carry out a definitive, all-sky exploration of the infrared background radiation of the universe between the wavelengths of one micrometer and 9.6 millimeters. The detailed information that COBE was to provide on the spectral and

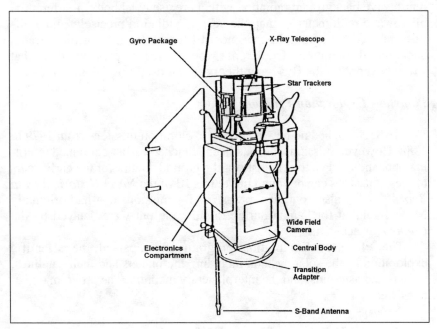

Figure 4–26. ROSAT Flight Configuration

spatial distribution of low-energy background radiation was expected to yield significant insight into the basic cosmological questions of the origin and evolution of the universe. COBE would measure the residual three-Kelvin background radiation believed to be a remnant of the "Big Bang" origin of the universe.

COBE, as initially proposed, was to have been launched by a Delta rocket. However, once the design was under way, the Shuttle was adopted as the NASA standard launch vehicle. After the *Challenger* accident occurred in 1986, ending plans for Shuttle launches from the west coast, NASA redesigned the spacecraft to fit within the weight and size constraints of the Delta. Three of the subsystems that on the Shuttle would have been launched as fixed components—the solar arrays, radio-frequency/thermal shield, and antenna—had to be replaced by deployable systems. The final COBE satellite had a total mass of 2,270 kilograms, a length of 5.49 meters, and a diameter of 2.44 meters with Sun-Earth shield and solar panels folded (8.53 meters with the solar panels deployed) rather than the 4,990 kilograms in weight and 4.3 meters in diameter allowed with a Shuttle launch. (Figure 4–27 shows the COBE observatory.) In 1988, instrument development was completed, the flight hardware delivered, and the observatory integration completed.

COBE was launched aboard a Delta rocket on November 18, 1989, from the Western Space and Missile Center at Vandenberg Air Force Base, California, into a Sun-synchronous orbit. Its orbital alignments are shown in Figure 4–28. COBE carried three instruments: a far-infrared absolute spectrophotometer to compare the spectrum of the cosmic microwave background radiation with a precise blackbody, a differential microwave radiometer to map the cosmic radiation precisely, and a diffuse infrared background experiment to search for the cosmic infrared background radiation. COBE has transmitted impressive data that strongly supports the Big Bang theory of the origin of the universe.

Planetary Exploration Program

NASA launched no new planetary exploration missions from 1979 to 1988. However, missions that had been launched earlier continued returning outstanding data to scientists on the ground. Details of the early years of these missions can be found in Volume III of the *NASA Historical Data Book*.[11] NASA also continued preparing for missions that had originally been scheduled for launch during this decade but were delayed by the *Challenger* accident.

The Planetary Exploration program encompassed the scientific exploration of the solar system, including the planets and their satellites, comets and asteroids, and the interplanetary medium. The program objectives were to:

---

[11]Ezell, *NASA Historical Data Book, Volume III.*

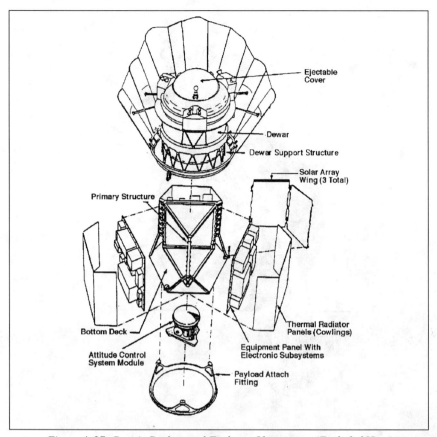

Figure 4–27. *Cosmic Background Explorer Observatory (Exploded View)*

- Determine the nature of planets, comets, and asteroids as a means for understanding the origin and evolution of the solar system
- Understand Earth better through comparative studies with the other planets
- Understand how the appearance of life in the solar system was related to the chemical history of the solar system
- Provide a scientific basis for the future use of resources available in near-Earth space

NASA's strategy emphasized equally the Earth-like inner planets, the giant gaseous outer planets, and the small bodies (comets and asteroids). Missions to these planetary bodies began with reconnaissance and exploration to achieve the most fundamental characterization of the bodies and proceeded to detailed study. In general, the reconnaissance phase of inner planet exploration began in the 1960s and was completed by the late 1970s. Most activities that occurred in the 1980s involved more detailed study of the inner planetary bodies or the early stages of study about the outer planets and small bodies.

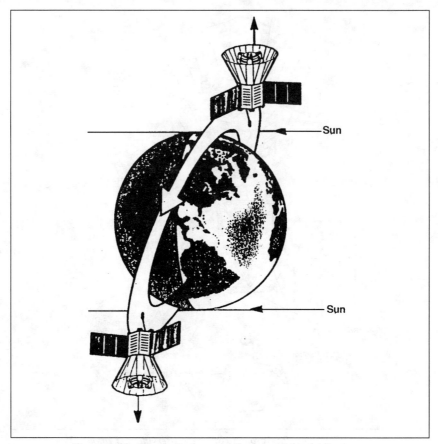

Figure 4–28. Cosmic Background Explorer Orbital Alignments

Voyager Program

The objectives of the Voyager missions were to conduct comparative studies of the Jupiter and Saturn planetary systems, including the satellites and Saturn's rings, and to study the interplanetary medium between Earth and Saturn. Voyager 1 encountered both planets, using Jupiter's gravity to go on to Saturn in 1980, scanned Saturn's primary moon Titan, and was flung by Saturn's gravity up out of the ecliptic plane. Voyager 2 followed Voyager 1 to Jupiter and Saturn, and it then proceeded to Uranus and Neptune, using the gravity of each previous planet to go on to the next one. This outer planet "grand tour" required a planetary alignment that repeats only once every 176 years.[12]

NASA launched Voyager 1 on September 5, 1977. It began its measurements of the Jovian system on January 6, 1979, with its closest

---

[12]"Handy Facts," *The Voyager Neptune Travel Guide,* NASA Jet Propulsion Laboratory, JPL Publication 89-24, June 1, 1989.

approach occurring on March 5, 1979, when it reached within 277,400 kilometers of the surface. During that year, the spacecraft returned more than 18,000 images of Jupiter and its four Galilean planets and mapped the accessible portion of Jupiter's complex magnetosphere.

Voyager discovered the presence of active volcanoes on the Galilean moon Io. Volcanic eruptions had never before been observed on a world other than Earth. The Voyager cameras identified at least nine active volcanoes on Io, with plumes of ejected material extending as far as 280 kilometers above the moon's surface. Io's orange and yellow terrain probably resulted from the sulfur-rich materials brought to the surface by volcanic activity that resulted from tidal flexing caused by the gravitational pull among Io, Jupiter, and the other three Galilean moons.

The spacecraft encountered Saturn in November 1980, approaching within 123,910 kilometers of the surface. Voyager 1 found hundreds, and perhaps thousands, of elliptical rings and one that appeared to be seven twisted or braided ringlets. It passed close to its ring system and to Titan, and it also provided a first close-up view of several of its other moons. Voyager 1 determined that Titan had a nitrogen-based atmosphere with methane and argon—one more similar to Earth's in composition than the carbon dioxide atmosphere of Mars and Venus. Titan's surface temperature of −179 degrees Celsius implied that there might be water-ice islands rising above oceans of ethane-methane liquid or sludge. However, Voyager 1's cameras could not penetrate the moon's dense clouds. Following this encounter, the satellite began to travel out of the solar system as its instruments studied the interplanetary environment.

A Titan-Centaur launched Voyager 2 on August 20, 1977. Its closest approach to Jupiter occurred on July 9, 1979, when it reached 277,400 kilometers from Jupiter's surface. The spacecraft provided patterns of Jupiter's atmosphere and high-resolution views of volcanoes erupting on Io and views of other Galilean satellites and clear pictures of Jupiter's ring.

Voyager 2 came closest to Saturn on August 25, 1981, approaching 100,830 kilometers, and returned thousands of high-resolution images and extensive data. It obtained new data on the planets, satellites, and rings, which revolutionized concepts about the formation and evolution of the solar system. Additional scientific detail on the planet returned by the spacecraft suggested that the rings around Saturn were alternating bands of material at increased and decreased densities. Saturn's eighteenth moon was discovered in 1990 from images taken by Voyager 2 in 1981.

Leaving Saturn's neighborhood, the spacecraft continued on its trip and approached Uranus on January 24, 1986, at a distance of 81,440 kilometers. It was the first spacecraft to look at this giant outer planet. From Uranus, Voyager 2 transmitted planetary data and more than 7,000 images of the planet, its rings, and moons. Voyager 2 discovered ten new moons, twenty new rings, and an unusual magnetic field around the planet. Voyager 2 discovered that Uranus's magnetic field did not follow the usual north-south axis found on the other planets. Instead, the field was tilted sixty degrees and offset from the planet's center. Uranus's atmosphere consisted mainly of hydrogen, with approximately 12 percent helium and small amounts of ammonia, methane, and water vapor. The planet's blue color occurred because the methane in its atmosphere absorbed all other colors.

On its way from Uranus to Neptune, Voyager 2 continued providing data on the interplanetary medium. In 1987, Voyager 2 observed Supernova 1987A and continued intensive stellar ultraviolet astronomy in 1988. Toward the end of 1988, Voyager 2 returned its first color images of Neptune. Its closest approach to Neptune occurred on August 25, 1989, approaching within 4,850 kilometers. The spacecraft then flew to the moon Triton. During the Neptune encounter, it became clear that the planet's atmosphere was more active than that of Uranus. Voyager 2 also provided data on Neptune's rings. Observations from Earth indicated that there were arcs of material in orbit around the planet. It was not clear from Earth how Neptune could have arcs and how these could be kept from spreading out into even, unclumped rings. Voyager 2 detected these arcs, but discovered that they were, in fact, part of thin, complete rings. Leaving Neptune's environment, Voyager 2 continued its journey away from the Sun.

Viking Program

The objective of Vikings 1 and 2 were to observe Mars from orbit and direct measurements in the atmosphere and on the surface, with emphasis on biological, chemical, and environmental data relevant to the existence of life on the planet. NASA had originally scheduled Viking 1 for an equatorial region and Viking 2 for the middle latitudes. NASA launched Viking 1 on August 20, 1975, and followed with the launch of Viking 2 on September 9. Their landings on Mars in the summer of 1976 set the stage for the next step of detailed study of the planet, the Mars Observer mission, which NASA approved in 1984.

The Viking orbiters and landers exceeded their design lifetime of 120 and ninety days, respectively. Viking Orbiter 2 was the first to fail on July 24, 1978, when a leak depleted its attitude-control gas. Viking Lander 2 operated until April 12, 1980, when it was shut down because of battery degeneration. Viking Orbiter 1 quit on August 7, 1980, when the last of its attitude-control gas was used up. Viking Lander 1 ceased functioning on November 13, 1983.

Pioneer Program

Pioneers 10 and 11. NASA launched Pioneers 10 and 11 in the 1972 and 1983, respectively, and the spacecraft continued to return data throughout the 1980s. Their objectives were to study interplanetary characteristics (asteroid/meteoroid flux and velocities, solar plasma, magnetic fields, and cosmic rays) beyond two astronomical units and to determine characteristics of Jupiter (magnetic fields, atmosphere, radiation balance, temperature distribution, and photopolarization). Pioneer 11 had the additional objective of traveling to Saturn and making detailed observations of the planet and its rings.

The flybys of Jupiter by Pioneers 10 and 11 returned excellent data, which contributed significantly to the success of the 1979 flybys of two Voyager spacecraft through the Jovian system. The spacecraft made numerous discoveries as a result of these encounters, and they demonstrated that a safe, close passage by Saturn's rings was possible. The first close-up examination of Saturn occurred in September 1979, when Pioneer 11 reached within 21,400 kilometers of that planet after receiving a gravity-assist at Jupiter five years earlier.

During 1979, Pioneer 10 traveled 410 million kilometers on its way out of the solar system and continued to return basic information about charged particles and electromagnetic fields of interplanetary space where the Sun's influence was fading. It crossed Uranus's orbit in July 1979 on its trip out of the solar system. The spacecraft crossed Neptune's orbit in May 1983, and on June 13, 1983, it became the first artificial object to leave the solar system, heading for the star Aldebaran of the constellation Taurus. During 1985, it returned data on the interstellar medium at a distance of nearly thirty-five astronomical units from the Sun. This was well beyond the orbit of Neptune and in the direction opposite to the solar apex, which is the direction of the Sun's motion with respect to nearby stars. Through 1985 and 1986, it continued to return data, aiming to detect the heliopause, the boundary between the Sun's magnetic influence and interstellar space, and to measure the properties of the interplanetary medium well outside the outer boundary of the solar system.

Pioneer 11, launched in 1973, headed in the opposite direction and completed the first spacecraft journey to Saturn in September 1979. It discovered that the planet radiates more heat than it received from the Sun and also discovered Saturn's eleventh moon, a magnetic field, and two new rings. The spacecraft continued to operate and return data as it moved outward from the Sun during the next several years. By 1987, Pioneer 11 was approaching the orbit of Neptune.

Pioneer Venus. In 1978, NASA launched two Pioneer probes to Venus. Their objectives were to jointly conduct a comprehensive investigation of the atmosphere of Venus. Pioneer Venus 1 would determine the composition of the upper atmosphere and ionosphere, observe the interaction of the solar wind with the ionosphere, and measure the planet's gravitational field. Pioneer Venus 2 would conduct its investigations with hard-impact probes—one large probe, three small probes, and the spacecraft bus would take in situ measurements of the atmosphere on their way to the surface to determine the nature and composition of clouds, the composition and structure of the atmosphere, and the general circulation patterns of the atmosphere.

Pioneer Venus 1 went into orbit around Venus in late 1978 and completed its primary mission in August 1979. A radio altimeter provided the first means of seeing through the planet's dense cloud cover and determining surface features over almost the entire planet. It also observed the comets and obtained unique images of Halley's Comet in 1986, when the comet was behind the Sun and unobservable from Earth. The spacecraft also measured the solar wind interaction, which was found to be comet-like.

Pioneer Venus 2 released its payload of hard-landers in November 1978. These probes were designated for separate landing zones so that investigators could take on-site readings from several areas of the planet during a single mission.

The Pioneer Venus mission carried the study of the planet beyond the reconnaissance stage to the point where scientists were able to make a basic characterization of the massive cloud-covered atmosphere of Venus, which contained large concentrations of sulfur compounds in the lower atmosphere. This characterization also provided some fundamental data about the formation of the planet. However, because of the opacity of the atmosphere, information about the Venus surface character remained sparse. Therefore, in 1981, NASA proposed the Venus Orbiting Imaging Radar mission, which would use a synthetic aperture radar instrument on a spacecraft in low circular orbit to map at least 70 percent of the surface of Venus at a resolution better than about 400 meters. The radar sensor was also to collect radio emission and altimetry data over the imaged portions of Venus's surface. However, the Venus Orbiting Imaging Radar mission was canceled in 1982.

Magellan

In 1983, NASA replaced the Venus Orbiting Imaging Radar mission with a more focused, simpler mission, provisionally named the Venus Radar Mapper. Nonradar experiments were removed from the projected payload, but the basic science objectives of the Venus Orbiting Imaging Radar mission—investigation of the geological history of the surface and the geophysical state of the interior of Venus—were retained. NASA selected Hughes Aircraft Company as the prime contractor for the radar system, Martin Marietta Astronautics Group had responsibility for the spacecraft, and the Jet Propulsion Laboratory managed the mission. In 1986, NASA renamed the mission Magellan in honor of Ferdinand Magellan.

The objective of the Magellan mission was to address fundamental questions regarding the origin and evolution of Venus through global radar imagery of the planet. Magellan was also to obtain altimetry and gravity data to accurately determine Venus's topography and gravity field, as well as internal stresses and density variations. The detailed surface morphology of Venus was to be analyzed to compare the evolutionary history of Venus with that of Earth. The spacecraft configuration is shown in Figure 4–29.

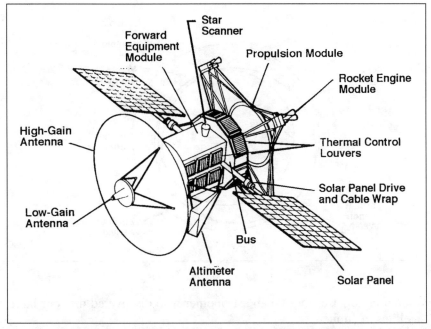

Figure 4–29. Magellan Spacecraft Configuration

Originally scheduled for a 1988 launch, NASA remanifested Magellan after the *Challenger* accident and the elimination of the Centaur upper stage. The launch took place on May 4, 1989, on STS-30, with an inertial upper stage boosting the spacecraft into a Venus transfer orbit (Figure 4–30). Magellan would reveal a landscape dominated by volcanic features, faults, and impact craters. Huge areas of the surface would show evidence of multiple periods of lava flooding with flows lying on top of previous ones. The Magellan mission would end on October 12, 1994, when the spacecraft was commanded to drop lower into the fringes of the Venusian atmosphere during an aerodynamic experiment, and it burned up, as expected. Magellan would map 98 percent of the planet's surface with radar and compile a high-resolution gravity map of 95 percent of the planet.

Project Galileo

Project Galileo had its genesis during the mid-1970s. Space scientists and NASA mission planners at that time were considering the next steps in outer planet exploration. Choosing Jupiter, which was the most readily accessible of the giant planets, as the next target, they realized that an advanced mission should incorporate a probe to descend into the atmosphere and a relatively long-lived orbiter to study the planet, its satellites, and the Jovian magnetosphere. NASA released the Announcement of Opportunity in 1976. The science payload was tentatively selected in August 1977 and confirmed in January 1979. Congress approved the Jupiter orbiter-probe mission in 1977. The program was renamed Project

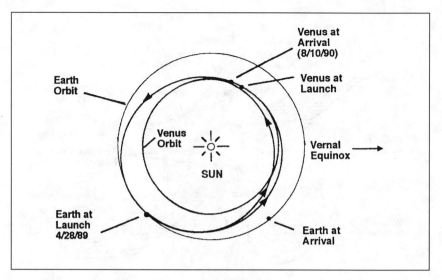

Figure 4–30. Magellan Orbit

Galileo in honor of the Italian astronomer who discovered the four large satellites of Jupiter.

Project Galileo was a cooperative effort between the United States and the Federal Republic of Germany (West Germany). A wide range of science experiments, chosen to make maximum progress beyond the Voyager finds, was selected. The mission was originally planned for an early 1985 launch on a Shuttle/Centaur upper stage combination but was delayed first to 1986 and then to 1989 because of the *Challenger* accident and the cancellation of the Centaur upper stage. Planned to operate for approximately twenty months, the Galileo spacecraft was launched October 18, 1989, on STS-34, assisted by an inertial upper stage on a trajectory using gravity assists at Venus and Earth. The orbiter would be able to make as many as ten close encounters with the Galilean satellites.

Project Galileo would send a sophisticated, two-part spacecraft to Jupiter to observe the planet, its satellites, and its space environment. The objective of the mission was to conduct a comprehensive exploration of Jupiter and its atmosphere, magnetosphere, and satellites through the use of both remote sensing by an orbiter and in situ measurements by an atmospheric probe. The scientific objectives of the mission were based on recommendations by the National Academy of Sciences to provide continuity, balance, and orderly progression of the exploration of the solar system.

Galileo would make three planetary gravity-assist swingbys (one at Venus and two at Earth) needed to carry it out to Jupiter in December 1995. (Figure 4–31 shows the Galileo trajectories.) There, the spacecraft would be the first to make direct measurements from a heavily instrumented probe within Jupiter's atmosphere and the first to conduct long-term observations of the planet, its magnetosphere, and its satellites from orbit.

The Galileo spacecraft would have three segments to investigate the planet's atmosphere, the satellites, and the magnetosphere. The probe

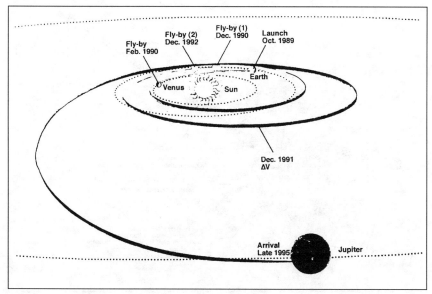

Figure 4–31. Galileo Mission

would descend into the Jovian atmosphere; a nonspinning section of the orbiter carrying cameras and other aimed sensors would image the planet and its satellites; and the spinning main orbiter spacecraft that carried fixed instruments would sense and measure the environment directly as the spacecraft flew through it (Figure 4–32). Unfortunately, after launch, the high-gain antenna on the probe would fail, reducing the amount of data that could be transmitted. Even so, the Galileo orbiter continued to transmit data from the probe throughout 1996.

Ulysses

The International Solar Polar Mission (renamed Ulysses in 1984) was a joint mission of NASA and ESA, which provided the spacecraft and some scientific instrumentation. NASA provided the remaining scientific instrumentation, the launch vehicle and support, tracking support, and the radioisotope thermoelectric generator. The mission was designed to obtain the first view of the Sun above and below the plane in which the planets orbit the Sun. The mission would study the relationship between the Sun and its magnetic field and particle emissions (solar wind and cosmic rays) as a function of solar latitude to provide a better understanding of solar activity on Earth's weather and climate. Figure 4–33 shows the spacecraft configuration.

The basis for the Ulysses project was conceived in the late 1950s by J.A. Simpson, a professor at the University of Chicago. Initially planned as a two-spacecraft mission between NASA and ESA, this mission, called "Out of Ecliptic," would allow scientists to study regions of the Sun and the surrounding space environment above the plane of the ecliptic that had never before been studied. Later, the project name was changed to the

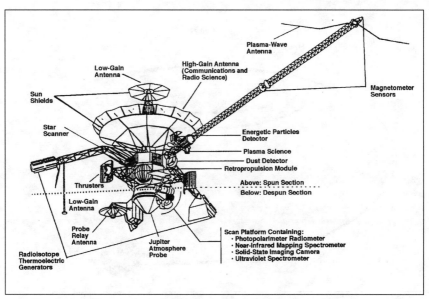

Figure 4–32. Galileo Spacecraft

International Solar Polar Mission. Delays in Shuttle development and concerns over the effectiveness of the inertial upper stage led to a House Appropriations Committee recommendation in the 1980 Supplemental Appropriations Bill that the International Solar Polar Mission be terminated. Later, in 1981, budget cuts led NASA to cancel the U.S. spacecraft contribution to the joint mission, which was restructured to a single ESA spacecraft mission. This was the first time that NASA had reneged on an international commitment. The ESA spacecraft completed its flight acceptance tests in early 1983 and was placed in storage.

In 1984, the International Solar Polar Mission was renamed Ulysses. It was originally scheduled to launch in 1986 but was another victim of the *Challenger* accident and the elimination of the Centaur upper stage. The launch took place in October 1990 using the Shuttle and both an inertial upper stage and payload assist module upper stage. The launch services were contributed by NASA. Table 4–51 presents an overview of the history of the Ulysses project.

Mars Geochemical-Climatology Orbiter/Mars Observer

The Mars Observer mission was the first in a series of planetary observer missions that used a lower cost approach to inner solar system exploration. This approach starts with a well-defined and focused set of science objectives and uses modified production-line Earth-orbital spacecraft and instruments with previous spaceflight heritage. The objectives of the Mars Observer mission were to extend and complement the data

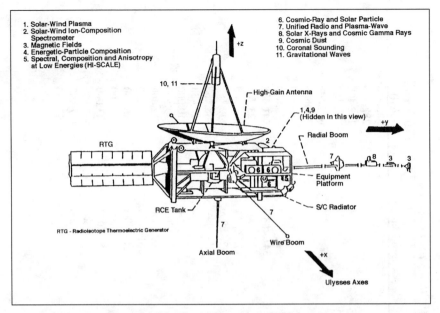

Figure 4–33. Ulysses Spacecraft Configuration

acquired by the Mariner and Viking missions by mapping the global surface composition, atmospheric structure and circulation, topography, figure, gravity, and magnetic fields of Mars to determine the location of volatile reservoirs and observe their interaction with the Martian environment over all four seasons of the Martian year.

The Mars Observer was launched on September 25, 1992. It lost contact with Earth on April 21, 1993, three days before it was to enter orbit around Mars.

Small Planetary Bodies

In 1985, NASA made the first close-up studies of the solar system's comets and asteroids. These objects may represent unaltered original solar system material preserved from the geological and chemical changes that took place in even smaller planetary bodies. By sampling and studying comets and asteroids, scientists could begin to inquire into the origin of the solar system itself. These efforts began with the encounter of Comet Giacobini-Zinner by the International Cometary Explorer spacecraft in September 1985 and continued with the 1986 encounters of Comet Halley by U.S. and foreign spacecraft and by intensive studies of the comet from ground-based observatories coordinated through the International Halley Watch.

Table 4–1. Total Space Science Funding History (in thousands of dollars)

Year and Budget Item	Request	Authorization	Appropriation	Programmed (Actual)
1979 - Space Science	**513,200**	**515,200**	— a	**505,400**
Physics and Astronomy	285,500	285,500	b	282,900
Lunar and Planetary	187,100	187,100	c	182,400
Life Sciences	40,600	42,600		40,100
1980 - Space Science	**601,600**	**601,600**	d	**600,500**
Physics and Astronomy	337,500	337,500	e	336,800
Planetary Exploration	220,200	220,200	f	219,900
Life Sciences	43,900	43,900		43,800
1981 - Space Science	**561,000**	**577,500**	**541,488** g	**541,488**
Physics and Astronomy	346,600 h	352,700	323,700 i	323,700
Planetary Exploration	175,300 j	179,600	175,600 k	175,600
Life Sciences	39,100 l	45,200	42,188 m	42,188
1982 - Space Science	**584,200**	**592,200**	—	**588,133**
Physics and Astronomy	325,400 n	333,400	o	322,433
Planetary Exploration	215,300 p	215,300	q	210,000
Life Sciences	43,500 r	43,500	43,500 s	39,500
1983 - Space Science	**682,000**	**707,000**	**697,800**	**712,400**
Physics and Astronomy	471,700	473,700	461,700 t	470,300
Planetary Exploration	154,600	177,600	180,400 u	186,400
Life Sciences	55,700	55,700	55,700	55,700
1984 - Space Science	**779,000**	**841,500**	**843,000**	**843,000**
Physics and Astronomy	514,600	562,100	578,600	567,600
Planetary Exploration	205,400	220,400	205,400	217,400
Life Sciences	59,000	59,000	59,000	58,000

Table 4–1 continued

Year and Budget Item	Request	Authorization	Appropriation	Programmed (Actual)
1985 - Space Science	**1,027,400**	**1,056,400**	**1,037,400**	**1,030,400**
Physics and Astronomy	677,200	696,200	680,200	677,200
Planetary Exploration	286,900	296,900	293,900	290,900
Life Sciences	63,300	63,300	63,300	62,300
1986 - Space Science	**1,061,400**	**1,042,400**	**1,027,400**	**989,000**
Physics and Astronomy	630,400	620,400	605,400	569,300
Planetary Exploration	359,000	354,000	354,000	353,600
Life Sciences	72,000	68,000 v	68,000	66,100
1987 - Space Science	**973,900**	**978,000**	**972,500**	**985,000**
Physics and Astronomy	529,900 w	529,400	528,500	554,000
Planetary Exploration	374,300 x	374,300	374,300	359,200
Life Sciences	69,700 y	74,300 z	69,700	71,800
1988 - Space Science	**949,000**	**976,700**	**984,000**	**1,014,300**
Physics and Astronomy	567,100	581,800	577,100	614,400
Planetary Exploration	307,300	320,300	332,300	327,700
Life Sciences	74,600	74,600	74,600	72,200

a Undistributed. Total R&D = $3,477,200,000. House Appropriations Committee amount = $284,900,000. Senate Appropriations Committee amount = $265,500,000.
b Undistributed. Total R&D = $3,477,200,000. House Appropriations Committee amount = $181,400,000. Senate Appropriations Committee amount = $177,100,000.
c Undistributed. Total R&D = $3,477,200,000. House Appropriations Committee amount = $40,600,000. Senate Appropriations Committee amount = $40,600.00.
d Undistributed. Total R&D = $4,091,086,000.
e Undistributed. Total R&D = $4,091,086,000.
f Undistributed. Total R&D = $4,091,086,000.
g Reflects recission.
h Amended budget submission. Initial budget submission = $438,700,000.
i Reflects recission.
j Amended budget submission. Initial budget submission = $179,600,000.
k Reflects recission.

Table 4–1 continued

l Amended budget submission. Initial budget submission = $49,700,000.
m Reflects recission.
n Amended budget submission. Initial budget submission = $451,400,000.
o Undistributed. Total FY 1982 R&D basic appropriation = $4,973,100. House Appropriations Committee allocated $325,400,000 for Physics and Astronomy; Senate Appropriations Committee allocated $$340,400,000 for Physics and Astronomy. Effect of General Provision Section 501 reduced R&D appropriation to $4,740,900,000. Report of Conference Committee regarding supplemental appropriation titled "Making Urgent Supplemental Appropriations for the Fiscal Year Ending September 30, 1982, and for Other Purposes," H.R. 6685, dated July 15, 1982, allocates $325,200,000 for Physics and Astronomy (including $40,000,000 for Shuttle-Spacelab payloads).
p Amended budget submission. Initial budget submission = $245,100,000.
q Undistributed. Total FY 1982 R&D basic appropriation = $4,973,100; appropriation reflecting effect of General Provision Section 501 = $4,740,900. Report of Conference Committee regarding supplemental appropriation titled "Making Urgent Supplemental Appropriations for the Fiscal Year Ending September 30, 1982, and for Other Purposes," H.R. 6685, dated July 15, 1982, allocates $205,000,000 for Planetary Exploration.
r Amended budget submission. Initial budget submission = $49,200,000.
s Indicates basic appropriation for Life Sciences. Appropriation that reflects effects of General Provision Section 501 is undistributed. Report of Conference Committee regarding supplemental appropriation titled "Making Urgent Supplemental Appropriations for the Fiscal Year Ending September 30, 1982, and for Other Purposes," H.R. 6685, dated July 15, 1982, allocates $39,500,000 for Life Sciences.
t Senate Appropriations Committee increased amount by $38 million for Physics and Astronomy and for Planetary Exploration, of which not less than $5 million was to be used for Physics and Astronomy. Senate Appropriations Conference Committee retained $5 million for Physics and Astronomy but reduced remaining $33 million to $25 million in final appropriation.
u See footnote "p" above.
v Congressional action reduced authorized amount by $4,000,000 (undistributed).
w Amended budget submission. Original submission = $539,400,000.
x Amended budget submission. Original submission = $323,300,000.
y Amended budget submission. Original submission = $74,700,000.
z Congressional action reduced authorized amount by $400,000 (undistributed) (Authorization Committee acted on original budget submission of $74,700,000).

Table 4–2. Programmed Budget by Budget Category (in thousands of dollars)

Budget Category/Year	1979	1980	1981	1982	1983
Space Science	505,400	600,500	541,488	588,133	712,400
Physics and Astronomy	282,900	336,800	323,700	322,433	470,300
Lunar and Planetary [a]	182,400	219,900	175,600	210,000	186,400
Life Sciences	40,100	43,800	42,188	39,500	55,700

Budget Category/Year	1984	1985	1986	1987	1988
Space Science	843,000	1,030,400	989,000	985,000	1,014,300
Physics and Astronomy	567,600	677,200	569,300	554,000	614,400
Lunar and Planetary	217,400	290,900	353,600	359,200	327,700
Life Sciences	58,000	62,300	66,100	71,800	72,200

[a] Renamed Planetary Exploration in FY 1980.

Table 4–3. High Energy Astronomy Observatories
Development Funding History (in thousands of dollars)

Year (Fiscal)	Submission	Authorization	Appropriation	Programmed (Actual)
1979	11,400	11,400	a	10,647
1980	4,800	4,800	b	2,100

a   Undistributed. House and Senate appropriations committees allocated $11,400,000.
b   Undistributed.

Table 4–4. Solar Maximum Mission Development Funding History
(in thousands of dollars)

Year (Fiscal)	Submission	Authorization	Appropriation	Programmed (Actual)
1979	16,200	16,200	a	16,700
1980	600	600	b	3,100

a   Undistributed. House and Senate appropriations committees allocated $16,200,000.
b   Undistributed.

Table 4–5. Space Telescope Development Funding History
(in thousands of dollars) a

Year (Fiscal)	Submission	Authorization	Appropriation	Programmed (Actual)
1979	79,200	79,200	b	79,200
1980	112,700	112,700	c	112,700
1981	119,300	119,300	119,300	119,300
1982	119,500	119,500	119,500	121,500
1983	137,500	137,500	137,500	182,500
1984	120,600	165,600 d	165,600	195,600
1985	195,000	195,000	195,000	195,000
1986	127,800	127,800	127,800	125,800
1987	95,900 e	95,900	95,900	96,000
1988	98,400	98,400	93,400	93,100

a   Renamed Hubble Space Telescope Development in FY 1986 submission.
b   Undistributed. House Appropriations Committee allocated $64,200,000. Senate Appropriations Committee allocated $79,200,000.
c   Undistributed.
d   House Authorization Committee increased amount for development of space telescope by $47 million; Senate Authorization Committee increased amount for space telescope by $50 million to pay for cost overruns. Conference Committee reduced Senate authorization by $5 million.
e   Amended budget submission. Original submission = $27,900,000.

Table 4–6. Solar Polar Mission Development Funding History (in thousands of dollars) a

Year (Fiscal)	Submission	Authorization	Appropriation	Programmed (Actual)
1979	13,000	13,000	b	12,500
1980	50,000	50,000	c	47,900
1981	39,600 d	39,600	28,000 e	28,000
1982	5,000 f	5,000	g	5,000 h
1983	21,000	21,000	6,000	6,000
1984 i		See Table 4–17		

a   Renamed International Solar Polar Mission in FY 1980.
b   Undistributed. House Appropriations Committee allocated $8,000,000. Senate Appropriations Committee allocated $13,000,000.
c   Undistributed.
d   Amended budget submission. Initial budget submission = $82,600,000. Decrease reflects program descoping that took place in mid-1980 to contain the amount of cost growth because of change in launch date from 1983 to 1985. The change resulted from the FY 1981 budget amendment (*NASA FY 1982 Budget Estimate,* International Solar Polar Mission Development, Objectives and Status, pp. RD 4–12).
e   Reflects recission.
f   Amended budget submission. Initial budget submission = $58,000,000. Decrease reflects NASA's decision to terminate the development of the U.S. spacecraft for the mission.
g   Undistributed. Total FY 1982 R&D appropriation = $4,973,100,000 (basic appropriation).
h   Programmed amount placed under Planetary Exploration funding beginning in FY 1982.
i   Became part of Planetary Exploration program. See Table 4–7.

Table 4–7. Gamma Ray Observatory Development Funding History (in thousands of dollars)

Year (Fiscal)	Submission	Authorization	Appropriation	Programmed (Actual)
1981	19,100	19,100	8,200 a	8,200
1982	8,000 b	8,000	8,000	8,000
1983	34,500	34,500	34,500	34,500
1984	89,800	89,800	89,800	85,950
1985	120,200	120,200	120,200	117,200
1986	87,300	87,300	87,300	85,300
1987	51,500	51,500	51,500	50,500
1988	49,100	49,100	49,100	53,400

a   Reflects recission.
b   Amended budget submission. Initial budget submission = $52,000,000.

*Table 4–8. Shuttle/Spacelab Payload Development Funding History
(in thousands of dollars) a, b*

Year (Fiscal)	Submission	Authorization	Appropriation	Programmed (Actual)
1979	38,300	38,300	c	34,900
1980	41,300	41,300	d	40,600
1981	29,100	29,100	27,400 e	27,400
1982	35,000 f	43,000	g	47,556
1983	81,400	81,400	81,400	81,000
1984	92,900	88,400 h, i	92,900	80,900
1985	105,400	113,400	105,400	105,400
1986	135,500	125,500	110,500	89,400
1987	84,600 j	84,100	84,600	72,800 k
1988	75,400	75,400	80,400	47,800 l

a   Included mission management beginning FY 1981.
b   Incorporated Space Station Payload Development and mission management beginning in FY 1986.
c   Undistributed. Both House and Senate appropriations committees allocated $38,300,000.
d   Undistributed.
e   Reflects recission.
f   Amended budget submission. Initial budget submission = $51,800,000.
g   Undistributed. FY 1982 R&D basic appropriation = $4,973,100. R&D appropriation reflecting effects of General Provision Section 501 = $5,740,900. House Appropriations Committee allocation for Shuttle/Spacelab Payload Development = $35,000,000. Senate Appropriations Committee allocation for Shuttle/Spacelab Payload Development = $40,000,000. Supplemental appropriations bill Conference Committee report indicates allocation of $40,000,000 for Shuttle/Spacelab Payload Development.
h   Senate Authorization Committee reduced amount authorized for solar optical telescope by $1.6 million to offset space telescope increases and added $5 million for space plasma laboratory. Conference Committee added $2.5 million for space plasma laboratory and decreased by $7 million amount authorized for solar optical telescope.
i   Amended budget submission. Original budget submission = $95,400,000.
j   Amended budget submission. Original budget submission = $115,100,000.
k   Included $5 million for astrophysics payloads and $4.6 million for space physics payloads.
l   Additional $8.1 million for astrophysics payloads and $9.9 million for space physics payloads were added to programmed amount.

Table 4–9. Explorer Development Funding History
(in thousands of dollars)

Year (Fiscal)	Submission	Authorization	Appropriation	Programmed (Actual)
1979	29,800	29,800	a	31,288
1980	30,400	30,400	b	32,300
1981	33,000	33,000	33,000	33,300
1982	36,600	36,600	36,600	33,300
1983	34,300	34,300	34,300	34,300
1984	48,700	48,700	48,700	48,700
1985	51,900	51,900	51,900	51,900
1986	55,200	55,200	55,200	48,200
1987	56,700	56,700	56,700	55,700
1988	60,300	70,300	70,300	67,900

a   Undistributed. Both House and Senate appropriations committees allocated $29,800,000 for Explorer Development.
b   Undistributed.

Table 4–10. Physics and Astronomy Mission Operations and Data Analysis Funding History (in thousands of dollars)

Year (Fiscal)	Submission	Authorization	Appropriation	Programmed (Actual)
1979	32,400	32,400	a	25,453
1980	36,500	36,500	b	37,100
1981	38,900	38,900	38,900	38,900
1982	47,000 c	47,000	47,000	45,300
1983	85,600	86,600 d	85,600	61,400
1984	79,500	80,500 e	79,500	68,100
1985	109,100	109,100	109,100	109,100
1986	119,900	119,900	119,900	111,700
1987	125,700 f	125,700	125,700	131,000
1988	128,100	128,100	128,100	140,500

a   Undistributed. Both House and Senate appropriations committees allocated $32,400,000.
b   Undistributed.
d   Amended budget submission. Initial budget submission = $53,500,000.
d   House Authorization Committee reduced amount to be allocated for Space Shuttle/Solar Maximum Mission Spacecraft Retrieval by $9.2 million to $77,400,000 and increased amount by $1 million for data analysis for HEAO and OAO. Senate Authorization Committee increased the amount to $93,600,000 to counter "slow progress in future programs and basic technology areas." (Footnote "d" accompanying *Chronological History of the FY 1983 Budget Submission,* prepared by NASA Comptroller, Budget Operations Division.) Authorization Conference Committee reduced increase to $1 million over submission.
e   House Authorization Committee increased amount for HEAO by $1 million.
f   Amended budget submission. Original budget submission = $172,700,000.

Table 4-11. Physics and Astronomy Research and Analysis Funding History (in thousands of dollars)

Year (Fiscal)	Submission	Authorization	Appropriation	Programmed (Actual)
1979	35,900	35,900	a	44,005
1980	34,300	34,300	b	33,774
1981	36,700 c	42,800	basic: 42,800 reflects Sec. 412: 38,000	37,700
1982	38,000 d	38,000	38,000	22,935
1983	39,200	39,200 e	39,200	28,500
1984	29,800	35,800 f	49,800 g	35,873
1985	36,900	47,900	39,900	111,700
1986	42,300	42,300	42,300	49,000
1987	51,100	51,100	49,700	53,400
1988	60,100	60,100	60,100	82,900 h

a Undistributed. Both House and Senate appropriations committees allocated $35,900,000 for Research and Analysis.
b Undistributed.
c Amended budget submission. Original budget submission = $42,800,000.
d Amended budget submission. Original budget submission = $42,500,000.
e See footnote "c" in Table 4-10.
f House Authorization Committee increased authorization for Universities Basic Research program by $4 million and Universities Research Instrumentation by $2 million. Senate Authorization Committee increased Universities Basic Research by $4 million.
g House and Senate appropriation committees increased appropriation by $20 million for Physics and Astronomy and Planetary Exploration at NASA's discretion.
h Additional $10.3 million for Shuttle Test of Relativity Experiment added to programmed amount.

Table 4-12. Physics and Astronomy Suborbital Programs Funding History (in thousands of dollars)

Year (Fiscal)	Submission	Authorization	Appropriation	Programmed (Actual)
1979	29,300	29,300	a	28,207
1980	26,900	26,900	b	27,226
1981	30,900	30,900	30,900	39,900
1982	35,500c	35,500	35,500	43,842
1983	38,200	39,200d	38,200	48,100
1984	53,300	53,300	52,300	52,477
1985	58,700	58,700	58,700	58,700
1986	62,400	62,400	62,400	59,900
1987	64,400	64,400	64,400	79,100
1988	75,700	80,400	75,700	44,700

a Undistributed. Both House and Senate appropriations committees allocated $29,300,000 for Suborbital Programs.
b Undistributed.
c Amended budget submission. Original budget submission = $37,500,000.
d See footnote "c" in Table 4-10.

*Table 4–13. Space Station Planning Funding History
(in thousands of dollars)*

Year (Fiscal)	Submission	Authorization	Appropriation	Programmed (Actual)
1987 a	—	—	—	18,900
1988	20,000 b	20,000	20,000	15,500

a   Space Station Planning not included in budget estimates or appropriation for FY 1987 as separate budget item. Incorporated in Spacelab/Space Station Payload Development and Mission Management Budget category.
b   Increased budget submission from $0 to $20,000,000.

*Table 4–14. Jupiter Orbiter/Probe and Galileo Programs Funding
History (in thousands of dollars) a*

Year (Fiscal)	Submission	Authorization	Appropriation	Programmed (Actual)
1979	78,700	78,700	b	78,700
1980	116,100	116,100	c	116,100
1981	63,100	63,100	63,100	63,100
1982	108,800	108,000	108,000	115,700
1983	92,600	92,600	91,600	91,600
1984	79,500	79,500	79,500	79,500
1985	56,100	56,100	56,100	58,800
1986	39,700	39,700	39,700	64,200
1987	77,000 d	77,000	77,000	71,200
1988	55,300	55,300	55,300	51,900

a   Renamed Galileo Development in FY 1981.
b   Undistributed. House Appropriations Committee allocated $68,700,000. Senate Appropriations Committee allocated $78,700,000.
c   Undistributed.
d   Reflects budget amendment that increased budget submission from $0 to $77,000,000

*Table 4–15. Venus Radar Mapper/Magellan Funding History
(in thousands of dollars)*

Year (Fiscal)	Submission	Authorization	Appropriation	Programmed (Actual)
1984	29,000	29,000	29,000	29,000
1985	92,500	92,500	92,500	92,500
1986	112,000	112,000	112,000	120,300
1987	69,700 a	69,700	69,700	97,300
1988	59,600	59,600	59,600	73,000

a   Amended budget submission. Original budget submission = $66,700,000.

Table 4–16. Global Geospace Science Funding History
(in thousands of dollars) a

Year (Fiscal)	Submission	Authorization	Appropriation	Programmed (Actual)
1988	—	—	—	18,600

a   Global Geospace Science was previously budgeted under Environmental Observations (Applications). There was no specific budget amount for Global Geospace Science in the FY 1988 budget submission. However, the Senate report, which accompanied the FY 1988 appropriations bill (H.R. 2783, September 25, 1987), indicated that NASA had requested $25,000,000 for the program for FY 1988. NASA's FY 1988 budget submission for Environmental Observations = $393,800,000, the authorization = $393,800,000, and the appropriation = $378,800,00. These figures were compiled prior to the OSSA reorganization. For the FY 1988 budget year that coincided with the OSSA reorganization, Global Geospace Science was moved to Physics and Astronomy.

Table 4–17. International Solar Polar Mission/Ulysses Development
Funding History (in thousands of dollars) a, b

Year (Fiscal)	Submission	Authorization	Appropriation	Programmed (Actual)
1984 c	8,000	8,000	8,000	6,000
1985	9,000	9,000	9,000	9,000
1986	5,600	5,600	5,600	8,800
1987	24,000 d	24,000	24,000	10,300
1988	10,800	10,800	10,800	7,800

a   Renamed International Solar Polar Mission in FY 1980.
b   Renamed Ulysses in FY 1986 submission.
c   Moved from Physics and Astronomy Management (see Table 4–6).
d   Reflects budget amendment that increased budget submission from $0 to 24,000,000.

Table 4–18. Mars Geoscience/Climatology Orbiter Program Funding
History (in thousands of dollars) a

Year (Fiscal)	Submission	Authorization	Appropriation	Programmed (Actual)
1985	16,000	16,000	16,000	13,000
1986	43,800	38,800	38,800	33,800
1987	62,900	62,900	62,900	35,800
1988	29,300	42,300	54,300	53,900

a   Renamed Mars Observer in FY 1986 submission.

Table 4–19. Lunar and Planetary Mission Operations and Data Analysis Funding History (in thousands of dollars)

Year (Fiscal)	Submission	Authorization	Appropriation	Programmed (Actual)
1979	84,400	84,400	a	59,300
1980	59,000	59,000	b	58,800
1981	60,500 c	64,800	basic: 64,800 reflects Sec. 412: 61,800	61,800
1982	45,800 d	45,800	45,800	42,600
1983	26,500	38,500	26,500	38,500
1984	43,400	43,400	43,400	43,400
1985	58,800	58,800	58,800	56,100
1986	95,000	95,000	95,000	67,000
1987	77,200 e	77,200	77,200	75,100
1988	77,000	77,000	77,000	73,792

a   Undistributed. House Appropriations Committee allocated $84,400,000. Senate Appropriations Committee allocated $78,700,000.
b   Undistributed.
c   Amended budget submission. Initial budget submission = $64,800,000.
d   Amended budget submission. Initial budget submission = $50,900,000.
e   Amended budget submission. Initial budget submission = $130,200,000.

Table 4–20. Lunar and Planetary Research and Analysis Funding History (in thousands of dollars)

Year (Fiscal)	Submission	Authorization	Appropriation	Programmed (Actual)
1979	24,000	24,000	a	44,400
1980	45,100	45,100	b	45,000
1981	51,700	51,700	basic: 51,700 reflects Sec. 412: 50,700	50,700
1982	51,500 c	51,500	d	46,700
1983	35,500	46,500	37,300	50,300
1984	45,500	60,500	45,500	59,500
1985	54,500	64,500	61,500	61,500
1986	62,900	62,900	62,900	59,500
1987	63,500	63,500	63,500	69,500
1988	75,300	75,300	75,300	67,308

a   Undistributed. Both House and Senate appropriations committees allocated $24,000,000.
b   Undistributed.
c   Amended budget submission. Original budget submission = $57,200,000.
d   Undistributed. Total R&D (basic appropriation) = $4,973,100.000. R&D appropriation reflecting Sec. 501 = $4,740,900,000.

Table 4–21. Life Sciences Flight Experiments Program Funding History (in thousands of dollars)

Year (Fiscal)	Submission	Authorization	Appropriation	Programmed (Actual)
1979	12,400	14,400	a	15,700
1980	12,900	12,900	b	16,600
1981	12,700 c	14,700	12,700	12,700
1982	14,000 d	14,000	14,000	14,000
1983	24,000	24,000	24,000	24,000
1984	23,000	23,000	23,000	23,000
1985	27,100	27,100	27,100	27,100
1986	33,400	33,400	33,400	32,100
1987	31,700 e	36,700	31,700	30,000
1988	32,900	32,900	32,900	33,800

a   Undistributed. Both House and Senate appropriations committees allocated $12,400,000.
b   Undistributed.
c   Amended budget submission. Initial budget submission = $19,200,000.
d   Amended budget submission. Initial budget submission = $16,500,000.
e   Amended budget submission. Initial budget submission = $36,700,000.

Table 4–22. Life Sciences/Vestibular Function Research Funding History (in thousands of dollars)

Year (Fiscal)	Submission	Authorization	Appropriation	Programmed (Actual) a
1979	3,800	3,800	b	—
1980	3,700	3,700	c	—

a   No amount programmed specifically for Vestibular Function Research. Included in Space Biology Research to be conducted on the orbital flight test or Spacelab 1 mission.
b   Undistributed. Both House and Senate appropriations committees allocated $3,800,000.
c   Undistributed.

Table 4–23. *Life Sciences Research and Analysis Funding History (in thousands of dollars)*

Year (Fiscal)	Submission	Authorization	Appropriation	Programmed (Actual)
1979	24,400	24,400	a	24,400
1980	27,300	27,300	b	27,200
1981	26,400 c	30,500	basic: 30,500 reflects Sect. 412: 29,488	29,488
1982	29,500 d	29,500	29,500	25,500
1983	31,700	31,700	31,700	31,700
1984	36,000	36,000	36,000	35,000
1985	36,200	36,200	36,200	35,200
1986	38,600	38,600	38,600	34,000
1987	63,500	63,500	63,500	41,800
1988	41,700	41,700	41,700	38,400

a   Undistributed. Both House and Senate appropriations committees allocated $24,400,000.
b   Undistributed.
c   Amended budget submission. Initial budget submission = $30,500,000.
d   Amended budget submission. Initial budget submission = $32,700,000.

Table 4–24. Science Missions (1979–1988)

Date	Mission	Discipline/Program Sponsor
Jan. 30, 1979	Spacecraft Charging at High Altitudes	Solar Terrestrial/U.S. Air Force
June 2, 1979	UK-6 (Ariel)*	Astrophysics/U.K. Science Research Council
Aug. 10, 1979	High Energy Astronomy Observatory-3 (HEAO)	Astrophysics
Feb. 14, 1980	Solar Maximum Mission	Solar Terrestrial
Aug. 3, 1981	Dynamics Explorer 1 and 2	Solar Terrestrial and Astrophysics
Oct. 6, 1981	Solar Mesosphere Explorer	Solar Terrestrial and Astrophysics
March 22, 1982	OSS-1 (STS-3)	Spacelab
Jan. 25, 1983	Infrared Astronomy Satellite (IRAS)	Astrophysics
May 26, 1983	European X-Ray Observatory Satellite (EXOSAT)*	Astrophysics/European Space Agency
June 22, 1983	Shuttle Pallet Satellite (SPAS)-01	Platform for science experiments/Germany
June 27, 1983	Hilat*	Astrophysics/U.S. Air Force
Nov. 28, 1983	Spacelab 1 (STS-9)	Spacelab (multidiscipline)
Aug. 16, 1984	Active Magnetospheric Particle Tracer Explorers (AMPTE)	Astrophysics
April 29, 1985	Spacelab 3 (STS 51-B)	Spacelab (multidiscipline)
June 17, 1985	Spartan-1	Astrophysics
July 29, 1985	Spacelab 2 (STS 51-F)	Spacelab (multidiscipline)
July 29, 1985	Plasma Diagnostic Package (PDP)	Earth Sciences and Applications
Oct. 30, 1985	Spacelab D-1 (STS 61-A)	German Spacelab (multidiscipline)
Jan. 23, 1986	Spartan 203 (Spartan-Halley) (failed to reach orbit)	Astrophysics
Nov. 13, 1986	Polar Bear*	Astrophysics/U.S. Air Force
March 25, 1988	San Marco D/L	Astrophysics

* NASA provided launch service or other nonscience role.

Table 4–25. Spacecraft Charging at High Altitudes Characteristics

Launch Date/Range	January 30, 1979/Eastern Test Range
Date of Reentry	Turned off May 28, 1991
Launch Vehicle	Delta 2914
NASA Role	Launch services for U.S. Air Force and three experiments
Responsible (Lead) Center	Goddard Space Flight Center
Mission Objectives	Place the Air Force satellite into a highly elliptical orbit of sufficient accuracy to allow the spacecraft to achieve its final elliptical orbit while retaining sufficient stationkeeping propulsion to meet the mission lifetime requirements
Instruments and Experiments (NASA experiments were the Light Ion Mass Spectrometer, the Electric Field Detector, and the Magnetic Field Monitor)	1. Satellite Surface Potential Monitor measured the potential of a sample surface of various compositions and aspects relative to vehicle ground or to the reference surface by command. 2. Charging Electrical Effect Analyzer measured the electromagnetic background induced in the spacecraft as a result of the charging phenomena. 3. Spacecraft Sheath Electric Fields measured the asymmetric sheath-electric field of the spacecraft, the effects of this electric field on particle trajectories near the spacecraft, and the current to the spherical probe surfaces mounted on booms at distances of 3 meters from the spacecraft surface. 4. Energetic Proton Detector measured the energetic proton environment of the trapped particles at spacecraft altitudes with energies of 20 to 1,000 keV, in six or more differential channels, plus an integral flux in the range from 1 to 2 MeV. 5. High Energy Particle Spectrometer measured the flux, spectra, and pitch angle distribution of the energetic electron plasma in the energy range of 100 keV to >3000 keV, the proton environment at energies between 1 MeV and 100 MeV, and the alpha particle environment between 6 MeV and 60 MeV during the solar particle events. 6. Satellite Electron Beam System consisted of an indirectly heated, oxide-coated cathode and a control grid. It controlled the ejection of electrons from the spacecraft. 7. Satellite Positive Ion Beam System consisted of a Penning discharge chamber ion source and a control grid. It controlled the ejection of ions from the spacecraft. 8. Rapid Scan Particle Detector measured the proton and electron temporal flux variations from 50eV to 60 keV for protons and 50 eV to 10 MeV for electrons, with an ultimate time resolution of milliseconds.

Table 4–25 continued

9. Thermal Plasma Analyzer measured, by retarding potential analysis, the environmental photo and secondary electron densities and temperatures, in the range of 10^{-1} to 10^4 electrons per cubic centimeter, for electrons of energies in the range 0 eV to 100 eV.
10. Light Ion Mass Spectrometer used magnetic mass analysis and retarding potential analysis for temperature determination. It measured the ion density and temperature in the energy range of 0.01 to 100 eV and in the density range of 0.01 to 1,000 ions/cm^3.
11. Energetic Ion Composition Experiment determined momentum and energy per charge and measured ions in the mass range of 1 to 150 AMU per charge with energies of 100 eV to 20,000 eV.
12. San Diego Particles Detectors measured protons and electrons in the energy range 1 eV to 80,000 eV in 64 discrete steps. This experiment measured the particle flux to the spacecraft, overall charge of the spacecraft, differential charge on parts of the spacecraft, and charge accumulated on selected material samples. It also measured the ambient plasma and detected oscillations, enabling better predictions of magnetosphere dynamics.
13. Electric Field Detector measured AC and DC electric fields in the tenuous plasma region of the outer magnetosphere.
14. Magnetic Field Monitor measured the magnetic flux density in the range ±5 milligauss with a resolution of 0.004 milligauss.
15. Thermal Coatings monitored temperatures of insulated material samples to determine the changes that took place in their solar absorptive and emissive characteristics with time exposure in space.
16. Quartz Crystal Microbalance measured the deposition rate of contaminants (mass) as a function of energy in the axial and radial directions, respectively.

Orbit Characteristics:	
Apogee (km)	43,251
Perigee (km)	27,543
Inclination (deg.)	7.81
Period (min.)	1,416.2
Weight (kg)	655
Dimensions	Diameter of 172.7 cm; length of 174.5 cm
Shape	Cylindrical
Power Source	Solar arrays
Prime Contractor	SAMSO, Martin Marietta Aerospace Corp.

Table 4–26. UK-6 (Ariel) Characteristics

Launch Date/Range	June 2, 1979/Wallops Flight Center
Date of Reentry	Switched off March 1982; reentered September 23, 1990
Launch Vehicle	Scout
NASA Role	Launch services for United Kingdom Science Research Council
Responsible (Lead) Center	Langley Research Center
Mission Objectives	Place the UK-6 satellite in an orbit that will enable the successful achievement of the payload scientific objectives: • Measure the charge and energy spectra of galactic cosmic rays, especially the ultraheavy component • Extend the x-ray astronomy to lower levels by examining the spectra, structure, and position of intrinsically low energy sources, extend the spectra of known sources down to low energies, and study the low-energy diffuse component • Study the fast periodic and aperiodic fluctuations in x-ray emissions from a number of low galactic latitude sources and improve the knowledge of the continuum spectra of the sources being observed.
Instruments and Experiments	1. Cosmic Ray Experiment measured the charge and energy spectra of the ultraheavy component of cosmic radiation with particular emphasis on the charge region of atomic weights above 30 (Bristol University). 2. Leicester X-Ray Experiment investigated the periodic and aperiodic fluctuations in emissions from a wide range of x-ray sources, down to submillisecond time scales (Leicester University). 3. MSSL/B X-Ray Experiment studied discrete sources and extended features of the low-energy x-ray sky in the range of 0.1 to 2 keV. It also studied long- and short-term variability of individual x-ray sources (Mullar Space Laboratory of University College, London and Birmingham University). 4. Solar Cell Experiment investigated the performance in orbit of new types of solar cells mounted on a flexible, lightweight support (Royal Aircraft Establishment). 5. CMOS Experiment was a complementary metal oxide semiconductor (CMOS) electronics experiment that investigated the susceptibility of these devices to radiation in a space environment (Royal Aircraft Establishment).

Table 4–26 continued

Orbit Characteristics:	
Apogee (km)	656
Perigee (km)	607
Inclination (deg.)	55.04
Period (min.)	97
Weight (kg)	154.5
Dimensions	n/a
Shape	Cylindrical
Power Source	Solar array and battery power
Prime Contractor	Marconi Space and Defense Systems, Ltd.
Results	The satellite lasted beyond its 2-year design life. However, it lost at least half its data. It suffered from radio interference from Earth, which caused the high-voltage supplies and its tape recorder to switch on and off sporadically and to lose information that should have been stored. The problem was alleviated by using more NASA ground stations, an Italian receiving station in Kenya, and a portable station set up by University College in Australia.

Table 4–27. HEAO-3 Characteristics

Launch Date/Range	September 20, 1979/Eastern Test Range
Date of Reentry	December 7, 1981
Launch Vehicle	Atlas-Centaur
NASA Role	Project management
Responsible (Lead) Center	Marshall Space Flight Center
Mission Objectives	Study gamma ray emission with high sensitivity and resolution over the energy range of about 0.06 MeV to 10 MeV and measure the isotopic composition of cosmic rays from lithium through iron and the composition of cosmic rays heavier than iron
Instruments and Experiments	1. High-Spectral Resolution Gamma Ray Spectrometer (Jet Propulsion Laboratory) explored sources of x-ray and gamma ray line emissions from approximately 0.06 to 10 million electron volts. It also searched for new discrete sources of x-rays and gamma rays and measured the spectrum and intensity of Earth's x-ray and gamma ray albedo (Figure 4–3).
	2. Isotopic Composition of Primary Cosmic Rays (Center for Nuclear Studies, France, and Danish Space Research Institute) measured the isotopic composition of primary cosmic rays with atomic charge Z between Z=4 (beryllium) to Z=26 (iron) and in the momentum range from 2 to 20 giga electron volts per nucleon (Figure 4–4).
	3. Heavy Nuclei Experiment (Washington University, California Institute of Technology, and University of Minnesota) observed rare, high-atomic-number (Z>30), relativistic nuclei in the cosmic rays. It also measured the elemental composition and energy spectra of these nuclei with sufficient resolution to determine the abundance of individual elements from chlorine (Z=17) through at least uranium (Z=92). These data provided information on nucleosynthesis models and on the relative importance of different types of stellar objects as cosmic ray sources (Figure 4–5).
Orbit Characteristics:	
Apogee (km)	504.9
Perigee (km)	486.4
Inclination (deg.)	43.6
Period (min.)	94.5
Weight (kg)	2,904
Dimensions	Diameter of 2.35 m; length of 5.49 m
Shape	Cylindrical with solar panels (two modules: experiment and equipment)
Power Source	Solar arrays and nickel cadmium batteries
Prime Contractor	TRW Systems, Inc.
Results	Mission was highly successful; the satellite returned data for 20 months.

Table 4–28. Solar Maximum Mission

Launch Date/Range	February 14, 1980/Eastern Test Range
Date of Reentry	December 2, 1989
Launch Vehicle	Delta 3910
NASA Role	Project management
Responsible (Lead) Center	Goddard Space Flight Center
Mission Objectives	Observe a sizable number of solar flares or other active-Sun phenomena simultaneously by five or six of the Solar Maximum Mission experiments, with coalignment of the narrow field-of-view instruments, and measure the total radiative output of the Sun over a period of at least 6 months with an absolute accuracy of 0.5 percent and short-term precision of 0.2 percent
Instruments and Experiments (Figure 4–6)	1. Gamma Ray Spectrometer measured the intensity, energy and Doppler shift of narrow gamma ray radiation lines and the intensity of extremely broadened lines. 2. Hard X-Ray Spectrometer helped determine the role that energetic electrons played in the solar flare phenomenon. 3. Hard X-Ray Imaging Spectrometer imaged the Sun in hard x-rays and provided information about the position, extension, and spectrum of the hard x-ray bursts in flares. 4. Soft X-Ray Polychromator investigated solar activity that produced solar plasma temperatures in the 1.5 million to 50 million degree range. It also studied solar plasma density and temperature. 5. Ultraviolet Spectrometer and Polarimeter studied the ultraviolet radiation from the solar atmosphere, particularly from active regions, flares, prominences, and active corona, and studied the quiet Sun. 6. High Altitude Observatory Coronagraph/Polarimeter returned imagery of the Sun's corona in parts of the visible spectrum as part of an investigation of coronal disturbances created by solar flares. 7. Solar Constant Monitoring Package monitored the output of the Sun over most of the spectrum and over the entire solar surface.
Orbit Characteristics:	
Apogee (km)	573.5
Perigee (km)	571.5
Inclination (deg.)	28.5
Period (min.)	96.16
Weight (kg)	2,315.1
Dimensions	Diameter of 2.1 m; length of 4 m
Power Source	Solar arrays
Prime Contractor	Goddard in-house

Table 4–28 continued

Results/Remarks	This mission was judged successful based on the results of the mission with respect to the approved prelaunch objectives. For the first 9 months of operation, the mission continuously gathered data from seven experiments on board. These data represented the most comprehensive information ever collected about solar flares. Project scientists gained valuable insight into the mechanisms that trigger solar flares and significant information about the total energy output from the Sun. The payload of instruments gathered data collectively on nearly 25 flares. After 9 months of normal operation, the satellite's attitude control system lost its capability to point precisely at the Sun. At that point, the spacecraft was placed in a slow spin using a magnetic control mode, which permitted continued operation of three instruments while coarsely pointing at the Sun. This was the first NASA satellite designed to be retrieved and serviced by the Space Shuttle. The Solar Max Repair Mission (STS 41-C) was successful and was completed after 7 hours, 7 minutes of extravehicular activity. Following its repair, Solar Max discovered several comets as well as continuing with its planned solar observations.

Table 4–29. Dynamics Explorer 1 and 2 Characteristics

Launch Date/Range	August 3, 1981/Western Test Range
Date of Reentry	Dynamics Explorer 1 retired February 28, 1991, Dynamics Explorer 2 reentered February 19, 1983
Launch Vehicle	Delta 3913
NASA Role	Project management
Responsible (Lead) Center	Goddard Space Flight Center
Mission Objectives	Investigate the strong interactive processes coupling the hot, tenuous, convecting plasmas of the magnetosphere and the cooler, denser plasmas and gases co-rotating in Earth's ionosphere, upper atmosphere, and plasmasphere
Instruments and Experiments	Dynamics Explorer 1: 1. High Altitude Plasma Instrument (five electrostatic analyzers) measured phase-space distributions of electrons and positive ions from 5 eV to 25 eV as a function of pitch angle. 2. Retarding Ion Mass Spectrometer (magnetic ion mass spectrometer) measured density, temperature, and bulk flow of H+, He+, and O+ in high-altitude mode, and composition in the 1–64 AMU range in low-altitude mode. 3. Spin-Scan Auroral Imager (spin-scan imaging photometers) imaged aurora at visible and ultraviolet and made photometric measurements of the hydrogen corona. 4. Plasma Waves (long dipole antennae and a magnetic loop antenna) measured electric fields from 1 hertz (Hz) to 2 MHz, magnetic fields from 1 Hz to 400 kHz, and the DC potential difference between the electric dipole elements. 5. Hot Plasma Composition (energetic ion mass spectrometer) measured the energy range from 0 keV to 17 keV per unit charge and the mass range from 1 AMU to 138 AMU per unit charge. 6. Magnetic Field Observations (fluxgate magnetometer) measured field-aligned currents in the auroral oval and over the polar cap at two altitudes. Dynamics Explorer 2: 1. Langmuir Probe (cylindrical electrostatic probe) measured electron temperature and electron or ion concentration. 2. Neutral Atmosphere Composition Spectrometer (mass spectrometer) measured the composition of the neutral atmosphere. 3. Retarding Potential Analyzer measured ion temperature, ion composition, ion concentration, and ion bulk velocity. 4. Fabray-Periot Interferometer measured drift and temperature of neutral ionic atomic oxygen. 5. Ion Drift measured bulk motions of ionospheric plasma.

Table 4–29 continued

	6.	Vector Electric Field Instrument (triaxial antennas) measured electric fields at ionospheric altitudes and extra-low-frequency and low-frequency ionosphere irregularities.
	7.	Wind and Temperature Spectrometer (mass spectrometer) measured in-situ, neutral winds, neutral particle temperatures, and the concentration of selected gases.
	8.	Magnetic Field Observations (see Dynamics Explorer 1 above)
	9.	Low Altitude Plasma Instrument (plasma instrument) measured positive ions and electrons from 5 eV to 30 keV.
Orbit Characteristics:	Dynamics Explorer 1	Dynamics Explorer 2
Apogee (km)	23,173	1,012.5
Perigee (km)	569.5	309
Inclination (deg.)	89.91	89.99
Period (min.)	409	97.5
Weight (kg)	424	
Dimensions	Width of 134.6 cm; length of 114.3 cm	
Shape	16-sided polygon	
Power Source	Solar cell arrays	
Prime Contractor	RCA	
Results	The spacecraft achieved a final orbit somewhat lower than planned because of short burn of the second stage in the Delta launch vehicle, but could still carry out the full scientific mission.	

Table 4–30. Solar Mesospheric Explorer Instrument Characteristics

Instrument	Detector	Spectral Range	$\Delta\lambda$	Total Shaft Angle	At Exit Slit	Full Width at Half Maximum	Grating Steps per Scan	$\Delta\lambda$ Per Step
Ultraviolet Ozone Experiment	Channel A, 510 F	1,900–3,100	1,200	14.4°	18Å/mm	15Å	208/11	4.8Å/91Å/ step
	Channel B, 520F	2,300–3,500	1,200		First Order	123Å	512	44Å/step
1.27µ Airglow Experiment	Channel A (Lead Sulfide)	1.1µ–2.5µ	1.4µ	13.4°	384Å/mm	123Å	512	44Å/step
	Channel B (Lead Sulfide)	1.1µ–2.5µ	1.4µ					
Visible NO$_2$ Experiment	Channel A	4,390–4,420	2,800Å	21.6°	28Å/mm	9.8Å	512/438	3.2Å/step
	Channel B, 510 N	2,900–5,700				19.6Å		6.4Å/step
Solar Ultraviolet Monitor	Channel A, 510F	1,800–3,100	1,250Å	9.7°	30Å/mm	14Å	512	2.6Å/step
	Channel B, 510F	1,200–2,540	1,250Å					
Four-Channel Infrared Radiometer	Channel A, 15.1µ	17.2–13.2µ	—	—	—	4.0µ	—	—
	Channel B, 15.5µ	15.7–14.7µ	—			1.0µ		
	Channel C, 9.6µ	10.6–8.6µ	—			2.0µ		
	Channel D, 6.3µ	7.2–6.1µ	—			1.1µ		

Table 4–31. Solar Mesospheric Explorer Characteristics

Launch Date/Range	October 6, 1981/Western Test Range
Date of Reentry	March 5, 1991
Launch Vehicle	Delta 2310
NASA Role	Project management
Responsible (Lead) Center	Jet Propulsion Laboratory
Mission Objectives	Investigate the processes that create and destroy ozone in Earth's mesosphere and upper stratosphere, with the following specific goals:
	• Determine the nature and magnitude of changes in ozone densities that result from changes in the solar ultraviolet flux
	• Determine the interrelationship among the solar flux, ozone, and the temperature of the upper stratosphere and mesosphere
	• Determine the interrelationship between water vapor and ozone
	• Determine the interrelationship between nitrogen dioxide (NO_2) and ozone
	• If a significant number of solar proton events occur, determine the relationship between the magnitude of the decrease in ozone and the flux and energy of the solar protons, the recovery rate of ozone following the event, and the role of water vapor in the solar proton destruction of ozone
	• Incorporate the results of the SME mission in a model of the upper stratosphere and mesosphere that could predict the future behavior of ozone
Instruments and Experiments	1. Ultraviolet Ozone Spectrometer measured ozone between 40 km and 70 km altitude.
	2. 1.27-Micron Spectrometer measured ozone between 50 km and 90 km altitude and hydroxyl between 60 km and 90 km.
	3. Nitrogen Dioxide Spectrometer measured NO_2 between 20 km and 40 km altitude.
	4. Four-Channel Infrared Radiometer measured temperature and pressure between 20 km and 70 km altitudes and water vapor and ozone between 30 km and 65 km altitude.
	5. Ultraviolet Solar Monitor looked 45 degrees from the spacecraft rotation axis to scan through the Sun once each revolution of the spacecraft. The instrument measured the amount of incoming solar radiation from 1,700 Angstroms to 3,100 Angstroms and at 1,216 Angstroms.
	6. Proton Alarm Sensor monitored the amount of integrated solar protons from 30 to 500 million eV.
	7. Spatial Reference Unit controlled the timing for data gating from the instruments.

Table 4–31 continued

Orbit Characteristics:	
Apogee (km)	534
Perigee (km)	533
Inclination (deg.)	98.0
Period (min.)	95.3
Weight (kg)	437
Dimensions	Diameter of 1.25 m; length of 1.7 m
Shape	Cylindrical
Power Source	Solar cell array
Prime Contractor	University of Colorado's Laboratory for Atmospheric and Space Physics, Ball Aerospace Systems Division
Remarks	The mission objective was accomplished by measuring ozone parameters and the processes in the mesosphere and upper stratosphere that determined their values. All mission events occurred as planned and on schedule.

Table 4–32. Infrared Astronomy Satellite Characteristics

Launch Date/Range	January 25, 1983/Western Test Range
Date of Reentry	Ceased operations November 21, 1983
Launch Vehicle	Delta 3910
NASA Role	Provided telescope, tape recorders, launch vehicle, data processing, co-chairman and members of the Joint IRAS Science Working Group
Responsible (Lead) Center	Jet Propulsion Laboratory—overall project management; Ames Research Center—management of the infrared telescope system until integrated with spacecraft
Mission Objectives	Obtain basic scientific data about infrared emissions throughout the total sky, to reduce and analyze these data, and to make these data and results available to the public and the scientific community in a timely and orderly manner
Instruments and Experiments	1. Ritchey-Chretien telescope detected infrared radiation in the region of 9 to 119 microns and observed emissions of infrared energy as faint as one million-trillionth of a watt per square centimeter. 2. Dutch Additional Experiment: • Low-Resolution Spectrometer acquired spectra of strong infrared point sources observed by the main telescope in the wavelength range from 7.4 to 23 microns. • Short-Wavelength Channel Detector obtained information on the distribution of stars in areas of high stellar density. It provided statistical data on the number of infrared sources. 3. Long-Wavelength Photometer mapped infrared sources that radiated in two wavelength bands simultaneously—from 41 to 62.5 microns and from 84 to 114 microns.
Orbit Characteristics:	
Apogee (km)	911
Perigee (km)	894
Inclination (deg.)	99.1
Period (min.)	103
Weight (kg)	1,076
Dimensions	Diameter of 2.1 m; length of 3.7 m
Shape	Cylindrical
Power Source	Two deployable solar panels
Prime Contractor	Ball Aerospace Systems Division in the United States; Fokker Schipol in The Netherlands

Table 4–32 continued

Results	During its 300 days of observations, IRAS carried out the first complete survey of infrared sky. On-board instruments with four broad infrared photometry channels (8 to 120 microns) detected unidentified cold astronomical objects, bands of dust in the solar system, infrared cirrus clouds in interstellar space, infrared radiation from visually inconspicuous galaxies, and possible beginnings of new solar systems around Vega and other stars. IRAS investigated selected galactic and extragalactic sources and mapped extended sources. The mission provided a complete and systematic listing of discrete sources in the form of sky and source catalogs. More than 2x10^11 bits of data were received from IRAS. IRAS also discovered five new comets.

Table 4–33. European X-Ray Observatory Satellite Characteristics

Launch Date/Range	May 26, 1983/Western Space and Missile Center
Date of Reentry	May 6, 1986
Launch Vehicle	Delta 3914
NASA Role	Launch support for European Space Agency
Responsible (Lead) Center	Goddard Space Flight Center
Mission Objectives	Launch the EXOSAT spacecraft into an elliptical polar orbit on a three-stage Delta 3914 launch vehicle with sufficient accuracy to allow the spacecraft to accomplish its scientific mission
Payload Objectives	Make a detailed study of known x-ray sources and identify new x-ray sources
Instruments and Experiments	1. X-Ray Imaging Telescopes (2) 2. Large Area Proportional Counter Array 3. Gas Scintillation Proportional Counter Spectrometer
Orbit Characteristics:	
Apogee (km)	194,643
Perigee (km)	6,726
Inclination (deg.)	72.5
Period (min.)	58,104 (4.035 days)
Weight (kg)	510
Dimensions	Diameter of 2.1 m; height of 1.35 m
Shape	Box
Power Source	Solar array
Prime Contractor	European Cosmos Consortium headed by Messerschmitt-Bolkow-Blohm (MBB)

Table 4–34. Shuttle Pallet Satellite-01 Characteristics

Launch Date/Range	Released from cargo bay June 22, 1983
Date of Reentry	Retrieved June 24, 1983
NASA Role	Provided Shuttle launch for Messerschmitt-Bolkow-Blohm (MBB), BMFT, and European Space Agency, for reduced fee
Launch Vehicle	STS-7 (*Challenger*)
Responsible (Lead) Center	n/a
Mission Objectives	Launch and retrieve the reusable SPAS
Instruments and Experiments	1. Microgravity experiments with metal alloys 2. Microgravity experiments with heat pipes 3. Microgravity experiments with pneumatic conveyors 4. An instrument that can control a spacecraft's position by observing Earth below 5. Remote sensing "push-broom" scanner that can detect different kinds of terrain and land/water boundaries 6. Mass spectrometer for monitoring gases in the cargo bay and around the orbiter's thrusters 7. Experiment for calibrating solar cells 8. A series of tests in which the Remote Manipulator System arm released the pallet to fly in space and then retrieved it and restowed it in the cargo bay
Orbit Characteristics:	
Apogee (km)	300
Perigee (km)	295
Inclination (deg.)	28.5
Period (min.)	90.5
Weight (kg)	2,278
Dimensions	Length of 4.8 m; height of 3.4 m; width of 1.5 m
Shape	Rectangular
Power Source	Battery power while outside orbiter; orbiter power while in cargo bay
Prime Contractor	MBB
Remarks	All experiment activities, planned detailed test objectives, and detailed secondary objectives were accomplished on schedule. The mission carried out successful detached and attached operations. It performed scientific experiments, tested the remote manipulator arm, and photographed *Challenger*.

Table 4–35. Hilat Characteristics

Launch Date/Range	June 27, 1983/Western Test Range
Date of Reentry	n/a
Launch Vehicle	Scout
NASA Role	Launch services for U.S. Air Force
Responsible (Lead) Center	n/a
Mission Objectives	Place the satellite in orbit to permit the achievement of Air Force objectives and satellite evaluation of certain propagation effects of disturbed plasmas on radar and communications systems
Instruments and Experiments	1. Beacon measured signal scintillation. 2. Magnetometer measured field-aligned currents. 3. Particle detector measured precipitating electrons in the 10,000–20,000 eV range. 4. Auroral/ionospheric mapper measured the visible and ultraviolet auroras. 5. Drift meter determined the electronic field from ion drift measurements.
Orbit Characteristics:	
Apogee (km)	819
Perigee (km)	754
Inclination (deg)	82
Period (min)	100.6
Weight (kg)	101.6
Dimensions	n/a
Shape	n/a
Power Source	Solar arrays
Prime Contractor	Applied Physics Laboratory, Johns Hopkins University

Table 4–36. Charge Composition Explorer Characteristics

Launch Date/Range	August 16, 1984/Cape Canaveral
Date of Reentry	Stopped transmitting data January 1989; was officially terminated July 14, 1989; has not reentered the atmosphere
Launch Vehicle	Delta 3924
NASA Role	Provided instrument for cooperative international mission; project management; launch services
Responsible (Lead) Center	Goddard Space Flight Center
Mission Objectives	Place the satellite in near-equatorial elliptical orbit to detect "tracer" ions released by the Ion Release Module within Earth's magnetosphere
Instruments and Experiments	1. Hot Plasma Composition Experiment monitored the natural low-energy magnetospheric tracer elements and detected artificially injected tracer ions at the Charge Composition Explorer over the low-energy range. 2. Charge-Energy-Mass Spectrometer measured the composition, charge state, and energy spectrum of the natural particle population of the ionosphere. 3. Medium Energy Particle Analyzer measured very small fluxes of lithium tracer ions over a wide energy range in the presence of the intense background of protons, alpha particles, and electrons while maintaining as large a geometry factor and as low an energy threshold as possible. 4. Magnetometer measured high-frequency magnetic fluctuations. 5. Plasma Wave Spectrometer provided first-order correlative information for studies of strong wave-particle interactions that develop close to the magnetic equator or have maximum effectiveness there. 6. Additional magnetic field and plasma ray experiments were conducted.
Orbit Characteristics:	
Apogee (km)	49,618
Perigee (km)	1,174
Inclination (deg.)	2.9
Period (min.)	939.5
Weight (kg)	242
Dimensions	122 cm across the flat sides and 40.6 cm high
Shape	Closed right octagonal prism
Power Source	Solar cell array, redundant nickel cadmium batteries, redundant battery charge controllers, and power switching and conditioning elements
Prime Contractor	Applied Physics Laboratory, Johns Hopkins University

Table 4–37. Ion Release Module Characteristics

Launch Date/Range	August 16, 1984/Cape Canaveral
Date of Reentry	November 1987
Launch Vehicle	Delta 3924
NASA Role	See Table 4–36
Responsible (Lead) Center	Goddard Space Flight Center; satellite provided by Federal Republic of Germany
Mission Objectives	Place the satellite in a highly elliptical orbit for the study of Earth's magnetosphere and release barium and lithium atoms into the solar wind and distant magnetosphere
Instruments and Experiments	1. Plasma Analyzer measured the complete three-dimensional energy-per-charge distributions of ions and electrons over the range of 10 V to 30 keV, as well as a retarding potential analyzer for the measurement of very low energy (~0 eV to 25 eV) electrons. 2. Mass Separating Ion Sensor measured simultaneously the distribution functions of ions of up to 10 different masses over an energy range of 0.01 to 12 keV/q. 3. Suprathermal Energy Ionic Charge Analyzer determined the ionic charge stage and mass composition of all major ions from hydrogen through iron over the energy range of 10–300 keV/q. 4. Magnetometer measured magnetic fields with a sensitivity of 0.1 nT. 5. Plasma Wave Spectrometer measured the intensities of the electric fields associated with plasma waves over the range of DC to 5 MHz with two long antennas and of magnetic wave fields from 30 Hz to 1 MHz with two boom-mounted search coils. 6. Lithium/Barium Release Experiments ejected 16 release canisters in pairs, eight with a Li-CuO mixture and eight with a Ba-CuO mixture, which ignited about a kilometer away from the spacecraft to expel hot lithium or barium gas.
Orbit Characteristics:	
Apogee (km)	113,818
Perigee (km)	402
Inclination (deg.)	27.0
Period (min.)	2,653.4
Weight (kg)	705 (including apogee kick motor)
Dimensions	Diameter of 1.8 m; height of 1.3 m
Shape	16 chemical release containers mounted on cylinder
Power Source	Solar array
Prime Contractor	Max Planck Institute for Extraterrestrial Physics under the sponsorship of the Research and Technology Ministry of the Federal Republic of Germany

Table 4–38. United Kingdom Subsatellite Characteristics

Launch Date/Range	August 16, 1984/Cape Canaveral
Date of Reentry	November 1988
Launch Vehicle	Delta 3924
NASA Role	See Table 4–36
Responsible (Lead) Center	Goddard Space Flight Center; satellite provided by Great Britain
Mission Objectives	Keep station with the IRM spacecraft at controllable distances of up to a few hundred miles to measure local disturbances created in the natural space plasma by the injection of tracer ions by the IRM
Instruments and Experiments	1. Ion Analyzer measured ion distribution over the energy range of 10 eV/q to 20 keV/q. 2. Electron Analyzer measured the electron distribution with high time and angular resolution over the energy range of 6 eV to 25 keV. 3. Particle Modulation Analyzer computed auto correlation functions and fast Fourier transform of electron and ion time variations resulting from wave-particle interactions and processed raw pulses from the electron and ion analyzers to reveal any significant resonances in the frequency range of 1 Hz to 1 MHz. 4. Magnetometer measured fields in the range of 0 to 256 nT or 0 to 9192 nT, with a resolution up to 30 pT, from DC to 10 Hz. 5. Plasma Wave Spectrometer measured the electric component of the plasma-wave field in the range of 10 Hz to 2 MHz and the magnetic component in the range of 30 Hz to 20 kHz.
Orbit Characteristics:	
Apogee (km)	113,417
Perigee (km)	1,002
Inclination (deg.)	26.9
Period (min.)	2,659.6
Weight (kg)	77
Dimensions	Diameter of 1 m, height of 0.45 m
Shape	Cylindrical
Power Source	Solar cells
Prime Contractor	Rutherford Appleton and the Mullard Space Science Laboratories under contract to the British Science and Engineering Research Council
Remarks	The satellite became inoperative after 5 months of operation. During that time, it had supported three chemical releases and had met 70 percent of the United Kingdom project objectives.

Table 4–39. Spartan 1 Characteristics

Launch Date/Range	June 17, 1985/Kennedy Space Center, deployed from Shuttle June 20
Date of Reentry	Retrieved June 24, 1985
Launch Vehicle	STS 51-G (*Discovery*)
NASA Role	Project management
Responsible (Lead) Center	Goddard Space Flight Center
Mission Objectives	Launch and retrieve Spartan 1, map the x-ray emissions from the Perseus Center, the nuclear region of the Milky Way galaxy, and the SCO X-2, and obtain engineering test data to prove the Spartan concept
Instruments and Experiments	The scanner observed various cosmic x-ray sources at rates of about 20 arc-sec/sec to provide x-ray data over an energy range of 0.5 keV to 15 keV. These observations were used for studies of emission processes in clusters of galaxies and the exploration of the galactic center.
Orbit Characteristics (same as Shuttle):	
Apogee (km)	391
Perigee (km)	355
Inclination (deg.)	28.5
Period (min.)	92
Weight (kg)	2,051
Dimensions	320 cm by 107 cm by 122 cm
Shape	Rectangular box
Power Source	Silver zinc batteries
Prime Contractor	Built by the Attached Shuttle Payloads Project at Goddard Space Flight Center

Table 4-40. Plasma Diagnostics Package Characteristics

Launch Date/Range	July 29, 1985
Date of Reentry	Retrieved July 29 after 6 hours of operation away from the orbiter; continued observations on-board orbiter throughout mission
Launch Vehicle	STS 51-F (*Challenger*)
NASA Role	Project management
Responsible (Lead) Center	Marshall Space Flight Center (Spacelab 2)
Mission Objectives	• Study orbiter-magneto plasma interactions in terms of density wakes, direct current electric fields, energized plasma, and a variety of possible wave-particle instabilities
	• Provide engine burns in support of the ground radar observations of the plasma depletion experiments for ionospheric and radio astronomical studies
	• Measure fields, waves, and plasma modifications induced by the orbiter and Spacelab subsystems in the payload bay and out to distances of 600 meters
	• Observe natural waves, fields, and plasmas in the unperturbed magnetosphere
	• Assess the Spacelab system performance of active and passive magnetospheric experiments
	• Develop the methods and hardware to operate instruments at the end of the remote manipulator arm and to eject and retrieve small scientific subsatellites
Instruments and Experiments	1. Quadrispherical low-energy proton and electron differential analyzer
	2. Plasma wave analyzer
	3. Electric dipole and magnetic search coil sensors
	4. Direct current electric field meter
	5. Triaxial flux-gate magnetometer
	6. Langmuir probe
	7. Retarding potential analyzer
	8. Differential flux analyzer
	9. Ion mass spectrometer
	10. Cold cathode vacuum gauge
Orbit Characteristics:	
Apogee (km)	321
Perigee (km)	312
Inclination (deg.)	49.5
Period (min.)	90.9
Weight (kg)	407
Dimensions	Diameter of 106.9 cm; height of 140 cm to top of grapple fixture
Shape	Cylindrical with extendible antennas
Power Source	Battery
Principal Investigator	Dr. Louis A. Frank, University of Iowa

Table 4–41. Spartan 203 Characteristics

Launch Date/Range	January 28, 1986/Kennedy Space Center
Date of Reentry	None
Launch Vehicle	STS 51-L (*Challenger*)
NASA Role	Project management
Responsible (Lead) Center	Goddard Space Flight Center
Mission Objectives	Determine the composition of Comet Halley when it was under greatest heating and was, therefore, most active, and look for changes in the composition and structure of the comet as it drew closer to the Sun
Instruments and Experiments	Two ultraviolet spectrometers were to survey Comet Halley in ultraviolet light from 128 nm to 340 nm wavelength. The spectrometers were also to observe the comet close to the perihelion and to look for cometary composition constituents and their rates of change during this highly active period in the cometary life cycle.
Orbit Characteristics	Did not achieve orbit
Weight (kg)	2,041
Dimensions	Carrier: 132 cm by 109 cm by 130 cm
Shape	Rectangular box
Power Source	Silver zinc batteries
Prime Contractor	General Electric-Matsco, Physical Sciences Laboratory at the University of New Mexico
Remarks	Although the Spartan program would continue during the next decade, this opportunity to observe Comet Halley was lost.

Table 4–42. Polar BEAR Characteristics

Launch Date/Range	November 13, 1986/Western Test Range
Date of Reentry	n/a
Launch Vehicle	Scout
NASA Role	Launch services for U.S. Air Force
Responsible (Lead) Center	n/a
Mission Objectives	Place the Air Force P87-1 (Polar BEAR) satellite into an orbit that will enable the successful achievement of Air Force mission objectives
Payload Objectives	Conduct several experiments to study atmospheric effects on electromagnetic propagation
Instruments and Experiments	1. Geophysics experiment photographed the aurora borealis. 2. Defense Nuclear Agency beacon experiment measured distortion of the ionosphere.
Orbit Characteristics	
Apogee (km)	1,014
Perigee (km)	954
Inclination (deg)	89.6
Period (min.)	104.8
Weight (kg)	122.5
Dimensions	n/a
Shape	Cylindrical
Power Source	Solar arrays
Prime Contractor	Applied Physics Laboratory, Johns Hopkins University

Table 4–43. San Marco D/L Characteristics

Launch Date/Range	March 25, 1988/San Marco Equatorial Range in Kenya, Africa
Date of Reentry	December 6, 1988
Launch Vehicle	Scout (launch was conducted by an Italian crew)
NASA Role	Provided an ion velocity instrument, wind/temperature spectrometer, electric field instrument, and Scout launch vehicle for cooperative mission with Italy
Responsible (Lead) Center	NASA Headquarters Office of Space Science and Applications (OSSA) and Goddard Space Flight Center
Mission Objectives	Launch satellite into low-Earth orbit to explore the possible relationship between solar activity and meteorological phenomena and determine the solar influence on low atmosphere phenomena through the thermosphere by obtaining measurements of parameters necessary for the study of dynamic processes occurring in the troposphere, stratosphere, and thermosphere
Instruments and Experiments	1. Neutral Atmosphere Density Experiment (Italy) measured drag forces on the satellite in orbit. 2. Airglow Solar Spectrometer (West Germany) measured equatorial airglow, solar extreme ultraviolet radiation, solar radiation from Earth's surface and from clouds, and the radiation from interplanetary and intergalactic origins reaching the satellite. 3. Wind and Temperature Spectrometer (Goddard) measured neutral winds, neutral particle temperatures, and concentrations of selected gases in the atmosphere. 4. Three-Axis Electric Field Experiment (Goddard) measured the electric field surrounding the spacecraft in orbit. 5. Ion Velocity Instrument (University of Texas) measured the plasma concentration and ion winds surrounding the spacecraft in orbit.
Orbit Characteristics:	
Apogee (km)	614
Perigee (km)	260
Inclination (deg.)	2.9
Period (min.)	99
Weight (kg)	237
Dimensions	96.5 cm diameter
Shape	Spherical
Power Source	Solar cell array
Prime Contractor	Satellite was provided by Centro Ricerche Aerospaziali (Italy)
Remarks	The wind and temperature spectrometer instrumentation system failed after providing approximately 1 week of data. The remaining four experiments operated satisfactorily.

Table 4–44. Chronology of Spacelab Development

Date	Event
Sept. 10, 1971	First documented use of the term "Sortie Can," predecessor to Spacelab, is used. NASA Headquarters Space Station Task Force Director Douglas R. Lord asks Marshall Space Flight Center to begin an in-house design study of a Sortie Can, a manned system to be carried in the Shuttle cargo bay for the conduct of short-duration missions.
Nov. 30–Dec. 3, 1971	The Joint Technical Experts Group meets in Washington.
Feb. 16, 1972	NASA Associate Administrator for Manned Space Flight Dale Myers investigates the Sortie Can and related activities at Marshall and issues new guidelines.
June 14–16, 1972	A delegation from the European Space Conference travels to Washington for a discussion with senior U.S. officials. The European Research and Technology Center (ESTEC) is assigned the task of determining needed resources for Europe to develop the Sortie Module (Lab).
July 31–Aug. 4, 1972	NASA Associate Administrator for Space Science Dr. John E. Naugle heads a Space Shuttle Sortie Workshop at Goddard Space Flight Center.
Aug. 17–18, 1972	NASA Headquarters hosts a meeting to review provisions that might appear in an agency-to-agency agreement that was developed based on earlier agreements between Europe and NASA.
Nov. 8–9, 1972	European space ministers agree to formulate plan for a single European space agency by December that would merge the existing European Space Research Organization (ESRO) and European Launcher Development Organization (ELDO) into the European Space Agency (ESA).
Dec. 20, 1972	At the space ministers' official meeting, the formal development commitment to the Sortie Lab is made.
By Jan. 1973	NASA and Europeans prepare first drafts of an agency-level agreement.
Jan. 9, 1973	ESRO's format of a Memorandum of Understanding (MOU) is discussed by Roy Gibson, ESRO's deputy of administration, and Arnold Frutkin, NASA's Associate Administrator for International Affairs.
Jan. 15–17, 1973	A symposium is held at ESRO's European Space Research Institute (ESRIN) facility in Frascati, Italy, to acquaint European users with the Sortie Lab (Spacelab) concept.
Jan. 18, 1973	The ESRO Council meets and votes to authorize a "Special Project" to develop the Sortie Lab, which the Europeans call Spacelab.
Jan. 23, 1973	Frutkin receives revised MOU, prepared by ESRO.
Feb. 22–23, 1973	NASA and State Department representatives travel to Paris. Although the stated purpose of the meeting is to work on the agency-to-agency agreement, the U.S. team gets its first look at the intra-European agreement, then in draft form, which would firmly commit the European signers to Spacelab development.

Table 4–44 continued

Date	Event
March 23, 1973	The program directors approve the first Spacelab concept document, "Level I Guidelines and Constraints for Program Definition," formulated by NASA. It addresses programmatics, systems, operations, interfaces, user requirements, safety, and resources.
May 1973	The expanded working groups review the findings from the July 1971 Goddard workshop, identify new requirements for the Shuttle and sortie systems, and identify systems and subsystems to be developed in each discipline. They also identify supporting research and technology needs, note changes in policies or procedures to fully exploit the Shuttle, and prepare cost, schedule, and priority rankings for early missions.
May 3–4, 1973	Representatives from Belgium, France, West Germany, Italy, the Netherlands, Spain, and the United Kingdom meet at the U.S. State Department to negotiate the draft intergovernmental agreement and the related draft NASA/ESRO MOU.
July 25, 1973	The Concept Verification Test (CVT) is assembled to simulate high-data-rate experiments emphasizing data compression techniques, including data interaction and on-board processing.
July 30, 1973	The Interim Programme Board for the European Spacelab Programme meets and approves the text of the intergovernmental agreement, the text of the MOU, and a draft budget.
July 31, 1973	The ministers of 11 European countries agree to a "package deal" by the European Space Conference.
Aug. 10, 1973	Belgium, France, West Germany, Switzerland, and the United Kingdom endorse the "Arrangement Between Certain Member States of the European Space Research Organization and the European Space Research Organization Concerning the Execution of the Spacelab Program." Subsequently, Spain, the Netherlands, Denmark, Italy, and later Austria also sign the arrangement.
Aug. 14, 1973	Belgium, France, West Germany, Switzerland, the United Kingdom and the United States sign the intergovernmental agreement titled "Agreement Between the Government of the United States of America and Certain Governments, Members of the European Space Research Organization, for a Cooperative Program Concerning the Development, Procurement, and Use of a Space Laboratory, In Conjunction with the Space Shuttle System." The Netherlands signs on August 18, Spain on September 18, Italy on September 20, and Denmark on September 21.

Table 4–44 continued

Date	Event
Aug. 15, 1973	This is the "magic" date when NASA would have to initiate the program, in the absence of a European undertaking, to have a Sortie Laboratory available for use by 1979. It states a readiness, therefore, to accept a firm European commitment in October and signed agreement by late October–early November, along with immediate initiation of a full-scale project definition effort, as well as an added proviso that the Europeans could withdraw from that commitment by August 15, 1973, if their definition work indicated that the projected target costs would be unacceptably exceeded.
Sept. 7, 1973	The NASA-developed Spacelab Design Requirements are reviewed and approved by NASA Administrator James Fletcher.
Sept. 21, 1973	The second issue of the Guidelines and Constraints Document is signed.
Sept. 24, 1973	In a U.S. Department of State ceremony in Washington, Acting Secretary of State Kenneth Rush and the Honorable Charles Hanin, Belgian science minister and chairman of the European Space Conference, sign a communiqué noting the completion of arrangements for European participation in the Space Shuttle program and marking the start of a new era in U.S.-European space cooperation. NASA Administrator James C. Fletcher and Dr. Alexander Hocker, director general of the ESRO, also sign the MOU to implement this international cooperative project.
Oct. 1973	The NASA Headquarters Sortie Lab Task Force is renamed the Spacelab Program Office, with responsibilities for overall program planning, direction, and evaluation as well as establishing program and technical liaison with ESRO. The name change from Sortie Lab to Spacelab recognizes the right of ESRO, as the sponsoring agency, to choose its preferred title for the program.
Oct. 9–10, 1973	Marshall reviews the preliminary design effort.
Nov. 16, 1973	NASA Administrator Fletcher directs NASA to evaluate the impact of a Shuttle docking module (then required on Shuttle missions carrying more than three crew members) on the mission model and on specific payloads.
Jan. 1974	The NASA administrator agrees with the recommendations not to use a docking module on all Spacelab missions. A general purpose laboratory, much like a Spacelab module, is added to the CVT complex at Marshall.
Early 1974	The Joint User Requirements Group begins informal discussions of the real Spacelab mission. The Joint Spacelab Working Group (JSWG) expresses its concern over the need to use the first missions to verify Spacelab performance.
March 5, 1974	The third version of the Guidelines and Constraints Document is signed and renamed the "Level I Programme Requirements Document."

Table 4–44 continued

Date	Event
March 19, 1974	The JSWG meets and establishes the Spacelab Operations Working Group with the thought that it would have a limited life, possibly through the Critical Design Review. In actuality, the Operations Working Group continues not only beyond that time, but eventually is divided into two groups, one focused on ground operations, the other on flight operations. The Software Coordination Group is also established; its initial focus is on the HAL-S and GOAL languages, which NASA is to furnish to ESRO, but it quickly broadens its scope to include microprogramming. Dr. Ortner of ESRO proposes a joint ESRO/NASA program called the Airborne Science/Spacelab Experiments System Simulation (ASSESS). By May, it is agreed that a joint mission could be authorized under the umbrella of the Spacelab MOU by a simple exchange of letters between the two program directors. The JSWG states that the Spacelab program should dictate the flight configuration and specify the resources available for experiments. It specifies that the first mission would have a long module and a pallet of two segments; 3,000–4,000 kg of weight, 1.5–2.5 kW of electrical power, and approximately 100–150 hours of crew time would be available for experiment activities; and the first mission would be no longer that 7 days.
April 23, 1974	The NASA/ESRO Joint Planning Group, co-chaired by Dr. Gerald Sharp of NASA and Jacques Collet of ESRO, meet to develop guidelines and procedures for selecting the first Spacelab payload.
May 17, 1974	NASA presents an expanded set of constraints for consideration at a JSWG meeting, including constraints imposed by the Shuttle, one of which is a limit of four to five crew members for the first Spacelab mission if it is conducted, as then planned, on the seventh Shuttle flight.
May 20, 1974	First annual review of the Spacelab program is held.
May 29–30, 1974	After it is suggested that the CVT general purpose laboratory be upgraded to make it more like the Spacelab design, a Preliminary Requirements Review for the improved simulator is held. Its completion is planned for mid-1976.
Summer of 1974	Some 60 Europeans, both ESRO and industry representatives of the Spacelab team, embark on a 2-week visit to the United States.
July 1–14, 1974	Fourteen points are approved by the NASA Manned Space Flight Management Council. The configuration now states a one- or two-segment pallet with the long module. Weight and power are unchanged, but the crew size is to be "minimized" and "up to" 100 crew-hours would be available for experiment operations.
July 12, 1974	John Thomas, NASA's chief engineer for the Spacelab Program Office at Marshall, gives the first detailed requirements of the Verification Flight Instrumentation to the JSWG. He presents parameters to be measured, the type of test equipment, power and weight requirements, and summary mission timelines.

Table 4-44 continued

Date	Event
July 15–19, 1974	An integrated life science mission is conducted in the CVT facility. Planned and conducted by Ames Research Center scientists, this test demonstrates candidate experiment protocols, modular organism housing units, and rack-mounted equipment plus radioisotope tracer techniques.
July 22–23, 1974	The Spacelab team visits Johnson Space Center for technical discussions of the primary Shuttle/Spacelab interfaces.
Aug. 8, 1974	A letter from Lord to Stoewer, the ESRO acting program director, projects a joint mission in 1975 to draw up Spacelab design conclusions, study operational concepts, and perform scientific experiments. Marshall issues an Instrument Pointing System (IPS) Requirements Document.
Aug. 26, 1974	Stoewer's confirmation letter states full agreement with Lord's proposal but cautions that ESRO's funding limit for the first mission is 350,000 accounting units (approximately $440,000 at the time). By the end of 1974, planning for the first ASSESS mission is to take shape. A series of five flights on consecutive days would approximate the useful time of a 7-day Spacelab mission.
Sept. 23, 1974	The Joint Planning Group meets. ESRO reports that a call for Spacelab utilization ideas elicited 241 replies, over half of which were new "customers" for space experimentation. The JSWG members discuss the constraints for the second Spacelab mission, the most important one being that it would not be a joint payload. ESRO does not agree to this point. NASA also suggests that a DOD mission might replace the first Spacelab on the first Shuttle operational flight. ESRO objects strongly to this proposal.
Sept. 26, 1974	The new version of the Programme Requirements Document (Revision 1) is signed.
Oct. 21–31, 1974	After receipt of the data package from ERNO on October 21, independent technical teams are set up by ESRO at ESTEC and by NASA at Marshall. The teams conduct their reviews and write Review Item Discrepancies (RIDs). The three baseline documents for this review are: the Program Requirements Document (Level I), the System Requirements Document (Level II), and the Shuttle Payload Accommodations, Volume XIV.
Nov. 7, 1974	The Shuttle/ Spacelab Interface Working Group on Avionics, or, as it is soon called, the Avionics Ad Hoc Group, is established by agreement of the program directors.
Dec. 1974	NASA accepts ESRO's choice of the Mitra 25 computer system.
Dec. 11, 1974	The Joint Planning Group holds its final meeting; its functions would be assumed by line payload organizations.
Jan. 1975	It is agreed that the transfer tunnel would be offset below the orbiter centerline so that lightweight payloads could be mounted on bridging structures above the tunnel if desired.

Table 4-44 continued

Date	Event
March 1975	A second decision establishes the approach to the orbiter end of the tunnel. The Shuttle program would build a removable tunnel adapter, which would be placed between the Spacelab tunnel and the orbiter cabin wall. The adapter would have doors at both ends and a third door at the top where the airlock could be mounted.
May 29–30, 1975	The NASA Preboard "N" chaired by Jack Lee conducts its review of the System Requirements Documents at Marshall. In the meantime, ESA conducts a parallel review.
June 4, 1975	An annual review of the Spacelab program is held. Roy Gibson, director of ESA, and NASA Administrator Fletcher propose to accept the objectives for the first Spacelab payload as presented by the Joint Planning Group, and the group formally dissolves. A review is also presented on the status of the IPS proposal.
June 7, 1975	The ASSESS simulation flights are conducted, successfully completing the program at Ames Research Center. The international crew of five completes a 6-day mission on board the CV 990 Galileo II.
June 9, 1975	The combined ESA/NASA teams meets in Noordwijk to consider the 1,772 RIDs prepared by both agencies.
Aug. 28–29, 1975	ESA Spacelab Programme Director Deloffre and Lord draft a "package deal" that would commit the agencies to develop or fund activities and equipment that have been in question.
Sept. 1975	By this meeting between Lord and Deloffre, plans for go-ahead have fallen apart. ESA has rejected the Dornier proposal (submitted through ERNO as the prime contractor) because of unacceptable schedule and cost risks. ESA has issued RFPs to ERNO, MBB, and Dornier, with a response due December 5.
Sept. 24, 1975	Revision 2 of the Programme Requirements Document is issued.
Sept. 30, 1975	The main contract between ESA and prime contractor VFW Fokker/ERNO is signed in the amount of approximately 600 million Deutschmarks. Over the next 9 months, negotiations between ERNO and its co-contractors are concluded.
Nov. 17–21, 1975	Another CVT simulation is conducted to determine how effectively a team of scientists in orbit, with only moderate experiment operations training, could conduct experiments while being monitored on the ground by principal investigators using two-way voice and downlink-TV contact.
Nov. 18–19, 1975	The Joint Program Integration Committee meets and reviews preliminary management plans for the first mission, Level I constraints, Level II guidelines imposed by the system and verification test requirements, and payload accommodation study results and plans.

Table 4–44 continued

Date	Event
Winter 1975–1976	The ESA team holds subsystem reviews. Also, ESA Spacelab Programme Director Bernard Deloffre works to sign contracts with each member of the consortium, reduce the backlog of engineering change proposals, recover schedule slips, and meet with European and NASA groups to review the program. To improve NASA's visibility into the European contractor effort, Deloffre invites NASA program management to participate in his quarterly reviews at ERNO beginning in September 1975.
By early 1976	ESA receives two proposals for the IPS: a joint bid on the IPS by Dornier and MBB and a bid from ERNO covering integration of the IPS into the Spacelab.
March 1976	Final approval is obtained to conduct ASSESS II as a joint mission sponsored by NASA's Office of Applications and Office of Space Flight and by ESA. The ESA Industrial Policy Committee authorizes Deloffre to proceed with the IPS contracts.
March 4–5, 1976	At the Joint Spacelab Working Group meeting, ESA reports that 110 engineering change proposals have been resolved with ERNO and only 90 are left open. The cost of the changes recently approved is 15 million accounting units (approximately $15 million at that time).
March 17, 1976	NASA's Fletcher, Low, Naugle, Mathews, Yardley, McConnell, Calio, Culbertson, Frutkin, and Lord deliberate the latest ESA proposal on the IPS. They agree to advise ESA that NASA would use an ESA IPS that meets the specification requirements and that NASA's first potential use would be on Spacelab 2.
March 19, 1976	Deloffre reports that his reserves on the program are down to only 5 million accounting units.
March–June 1976	ESA and NASA jointly conduct the Spacelab Requirements Assessment and Reduction Review. This review evaluates program needs and eliminates those items that have crept into the program but could be deleted with a considerable cost saving.
April 1976	ESA establishes a Software Audit Team to assess the software situation and make recommendations.
May 12, 1976	The Software Audit Team presents its preliminary findings to the ESA Spacelab Programme and project managers.
May 26, 1976	NASA (Marshall) issues an RFP for a Spacelab integration contract to secure a contractor to provide support in developing Spacelab hardware that is NASA's responsibility and analytical and hands-on support in the integration and checkout of Spacelab hardware during the system's operational lifetime.
June 2, 1976	The Software Audit Team makes its final presentation to ESA, ERNO, and co-contractors. The group concludes that Spacelab software is not in good shape and that there does not seem to be a structure for improving the situation.

Table 4–44 continued

Date	Event
June 16, 1976	The third annual meeting of the agency heads (Gibson and Fletcher) occurs in Washington, D.C. Discussed is the claim that the logistics requirements have been almost totally neglected in the agreements and contracts to date. Fletcher signs a letter to Gibson concurring with ESA's plans to proceed with IPS development. Fletcher urges that the delivery schedule provide adequate time for integration of payloads and checkout of the combined system for the planned launch date in 1980.
June 18, 1976	A NASA Program Director's Review is held, and Luther Powell of the Marshall project team summarizes activities in support of Preliminary Design Review-A (PDR-A).
June 24–25, 1976	The technical experts team analyzes its planned reviews with ESA at ESTEC and goes to Bremen for the final reviews between ESA and ERNO. By the time the senior NASA representatives arrive on July 1–2, chaos is reigning. PDR-A is a complete disaster. Documentation is inadequate, schedules are slipping, the budget cannot be held, the contractor team is out of control, and the team morale is at an all-time low.
June 28, 1976	NASA distributes the data packages for the Preliminary Operations Requirements Review for ground operations. The purpose of this review is to obtain agreement on ground operations requirements, including integration at Level I, II, and III, logistics, training of ground processing personnel, ground support equipment, facilities, contamination control, and safety.
July 7, 1976	Gibson signs a PDR implementation plan with Hans Hoffman at ERNO for a simple and straightforward approach to PDR-B.
July 15, 1976	A final CVT simulation to employ a high-energy cosmic ray balloon flight experiment is conducted.
July 30, 1976	Further changes are approved to the Programme Requirements Document. The most important ones note the addition of NASA-furnished utility connectors (from the orbiter to Spacelab) and a trace gas analyzer.
Aug. 1976	At the Program Director's Review, John Waters of Johnson Space Center presents a plan to procure a simulator to operate alone or with the Shuttle Mission Simulator and the Mission Control Center at Houston to produce a high-fidelity mission simulation.
Sept. 18, 1976	Gibson and Fletcher meet at Ames Research Center to tackle Spacelab logistics.
Nov. 1, 1976	ESA selects Michel Bignier as director of the Spacelab Programme.
Early Nov. 1976	Bignier and Gibson recognize that Spacelab funding is out of hand and propose descoping the program.
Nov. 22–23, 1976	NASA astronauts Paul Weitz, Ed Gibson, Bill Lenoir, and Joe Kerwin conduct a walkthrough of the Spacelab module at ERNO. They simulate various airlock operations and note further improvements needed.
Dec. 4 and 8, 1976	ESA, ERNO, and NASA hold board meetings, resulting in agreement that PDR-B represents a major turnaround in the program.

Table 4–44 continued

Date	Event
Jan. 14, 1977	NASA Spacelab Deputy Director Jim Harrington states that ESA proposals could save as much as $84 million in the ESA budget but could impose on NASA an additional funding requirement of $26 million to $33 million. Fletcher and Gibson agree on the descoping items for ESA to go to its Spacelab Programme Board for approval.
Jan. 20–24, 1977	Gibson receives approval from the Spacelab Programme Board for all the proposed changes, with one notable exception. The board refuses to accept deletion of the IPS and decides instead to postpone decisions on this part of the program.
Feb. 23, 1977	The Spacelab module, which is produced by the Italian firm Aeritalia, successfully completes a series of limit, proof, and ultimate pressure testing.
March 1977	After many discussions and studies of various options, the NASA administrator decides to proceed with the development of a "hybrid" pallet to be used on several Shuttle orbital flight test (OFT) missions and that would also be available if the Spacelab system is delayed.
March 9, 1977	NASA announces the selection of McDonnell Douglas for the integration effort.
March 16, 1977	The ESA Spacelab Programme Board decides not to cancel the IPS as part of the overall program descoping.
April 1977	ESA Headquarters submits a proposal for a Spacelab Utilization Programme to its managing council. The report addresses three alternative programs for European use of the Spacelab.
April 25–29, 1977	The first formal Crew Station Review is held at ERNO and includes NASA astronauts Bob Parker, Paul Weitz, and Ed Gibson. Working with NASA, ESA, and ERNO specialists in crew habitability, they review the Spacelab design.
May 2, 1977	Bignier writes to Lord that only three engineering model pallets would be flightworthy, the others having been used in the test program in such a manner that they cannot be flown. NASA initially requested four pallets that could be flightworthy for OFT missions.
May 3–4, 1977	The JSWG meets, and Jim Harrington presents a NASA proposal for six preliminary options to meet the NASA requirement of having four pallets for the OFT missions.
May 16, 1977	"Launch" of the ASSESS II occurs. This mission emphasizes the development and exercise of management techniques planned for Spacelab using management participants from NASA and ESA who would be responsible for the Spacelab 1 mission, then scheduled for 1980.
May 30–June 5, 1977	John Yardley, the NASA associate administrator for space flight, visits Hawker-Siddeley Dynamics, ERNO, and Aeritalia to review the status of the program and progress on hardware fabrication.
June 1977	Co-contractor Critical Design Reviews (CCDRs) are held for electrical and mechanical ground support equipment.

Table 4–44 continued

Date	Event
June 16, 1977	ESA signs a fixed-price contract with Dornier for developing the IPS, with a delivery date of June 18, 1980. Dornier would be solely responsible for managing the IPS/Spacelab interface with no subcontract for this function.
June 20– July 12, 1977	The Preliminary Requirements Review for the transfer tunnel, which provides crew access to the module from the orbiter, is conducted.
July 1977	CCDRs are held for the data management subsystem and module structure. The first Electrical System Integration activity, the T800 self-test, is successfully completed. A Preliminary Requirements Review of the transfer tunnel is held, and the design and development of critical elements are initiated.
Aug. 1–19, 1977	A Preliminary Requirements Review for the Verification Flight Instrumentation is conducted.
Sept. 1977	CCDRs are held for crew habitability, system activation and monitoring, thermal control, and electrical power distribution systems. Testing is completed on the command and data management subsystem portion of the Electrical System Integration.
Oct. 1977	NASA drops its idea of using a hybrid pallet as a Spacelab backup.
Oct. 7, 1977	After touring several European government and industry facilities, new NASA Administrator Dr. Robert Frosch meets with Gibson in Paris. The target dates for Spacelabs 1 and 2 are now December 1980 and April 1981, respectively.
Nov. 1977	Reviews are conducted on the life support system, the igloo structure, and the airlock. A subsystem interface compatibility test is also completed.
Nov. 15–16, 1977	ESA expresses concern about the Spacelab reimbursement policy, particularly the high costs, and that ESA is not given preferential treatment by NASA in view of its development role.
Late 1977	The Spacelab payload planners, reacting to experiment proposals for the second mission, recommend a change in Spacelab 2 to fly a large cosmic ray experiment that could use its own independent structural mount to the orbiter.
Dec. 1977	A compatibility test between the command and data management subsystem and the first set of electrical ground support equipment, newly arrived from BTM, is completed. The IPS Preliminary Design Review is held. Concurrent reviews are held at Marshall and ESTEC; the final phase is held at Dornier. Results are encouraging, except for two discrepancies: certain structural elements are found to be made of materials susceptible to stress corrosion, and IPS software requirements needs better definition.
Jan. 23– March 10, 1978	The Software Requirements Review is conducted to define the operational software for the Spacelab flight subsystems and the ground checkout computers. ESA, NASA, and ERNO reach a technical agreement for the first time.

Table 4–44 continued

Date	Event
Jan. 30, 1978	After evaluation of the Spacelab Simulator by Johnson Space Center, a formal contract agreement is signed, and development begins with ERNO to provide the scientific airlock mockup for the simulator and data support to Link.
Feb. 1978	Another Crew Station Review allows the astronauts to review the scientific airlock hardware at Fokker and the improvements to the module at ERNO. Senior NASA and ESA officials meet to discuss the trade of one Spacelab for NASA launch services for European Spacelab missions. The results of this meeting are so encouraging that NASA terminates work related solely to contractual procurement in favor of concentrating on a barter agreement.
Feb. 7–8, 1978	The NASA preboard meets, and the focus is shifted to ESTEC for the joint team meetings starting on February 17.
Feb. 27, 1978	The final phase of the Critical Design Review begins in Bremen.
March 9, 1978	A draft MOU of the barter arrangement is reviewed by NASA and ESA representatives.
May 1978	Information on the planned mounting structure of the new Spacelab 2 configuration is submitted to ESA.
May 8, 1978	NASA administrator Frosch and ESA director general Gibson exchange letters that agree on a set of guidelines and a timetable leading to signature of the MOU to formalize the barter by the end of 1978.
May 16, 1978	ESA sends an RFP to ERNO for a firm evaluation of the cost of the second Spacelab flight unit. A separate request is sent to Dornier for a similar proposal on a second IPS.
June 1978	ESA Project Manager Pfeiffer reports that Electrical System Integration testing has been completed. T004, an assembly test involving the racks and floors of the engineering model of the Spacelab, is completed. McDonnell Douglas reports that it is having problems in both the design and fabrication for the flexible transfer tunnel sections. The Preliminary Design Review for the Verification Flight Instrumentation is completed, but it is not until July and November 1979 that a two-part Critical Design Review is completed for the Verification Flight Instrumentation for Spacelab 1.
June 12–13, 1978	The JSWG reports on user needs for more power, heat rejection, energy, data handling, and a smaller and lighter IPS. Bignier accepts the proposed changes to the Spacelab 2 configuration during the JSWG meeting.
July–Aug. 1978	A Critical Design Review for the OFT pallet system is conducted.
Aug. 1978	NASA and ESA announce the first selection of potential crew members for the early Spacelab missions. Drs. Owen K. Garriott and Robert A.R. Parker are named as mission specialists for the first Spacelab mission.

Table 4–44 continued

Date	Event
Aug. 8, 1978	ESA and NASA introduce their final candidates for the single payload specialist to be provided by each side. ESA has selected Dr. Wubbo Ockels, a Dutch physicist; Dr. Ulf Merbold, a German materials specialist; and Dr. Claude Nicollier, a Swiss astronomer. NASA has selected Byron K. Lichtenberg, a doctoral candidate in bioengineering at MIT, and Dr. Michael Lampton, a physicist at the University of California at Berkeley.
Sept. 14, 1978	A NASA delegation headed by John Yardley and Arnold Frutkin meets with the ESA Spacelab Programme Board to propose the mechanism for NASA to obtain the second Spacelab flight unit in exchange for Shuttle launch services.
Oct. 1978	The newly developed flexible multiplexer/demultiplexer (from the orbiter program) is accepted from Sperry, and the first OFT pallet structure is accepted at British Aerospace.
Oct. 7, 1978	Frosch and Gibson meet for a formal review of the overall Spacelab program. The meeting results in assignments to the Spacelab program directors to prepare a post-delivery change control plan, review an ESA proposal for operational support, and continue the analysis of European source spares. The Spacelab 1 mission is now targeted for June 1981 and Spacelab 2 for December 1981.
Oct. 10–11, 1978	European news media representatives attend a 2-day symposium at ERNO sponsored by the West German minister of research and technology, Volker Hauff. His opening remarks strongly endorse space efforts, Spacelab in particular, and issue an equally strong challenge to demonstrate the payoff for space activities.
Oct. 16 and 27, 1978	ERNO and Dornier submit their proposals for a procurement contract for the second Spacelab. ESA and NASA begin their evaluations.
Oct. 30, 1978	ERNO proposes a new schedule to ESA, which forecasts delivery of the engineering model to NASA in April 1980 and delivery of the flight unit in two installments: July and November 1980.
Nov. 13, 1978	A NASA team joins its ESA counterpart in Europe with the goal to define a procurement contract as early as possible in 1979.
Dec. 4, 1978	The OFT pallet arrives at Kennedy Space Center.
Jan. 1979	The oft-postponed module subsystems test is finally completed. NASA Administrator Frosch formally announces that NASA would proceed with both a free-flying 25-kW power module and an orbiter-attached power extension package to provide up to 15 kW power for a maximum of 20 days. Colin Jones presents a detailed progress review of the IPS to the JSWG. The delivery to Kennedy is now projected for July 1981.
Jan. 16, 1979	NASA applies to the Bureau of Customs of the Treasury Department for duty-free entry of the Spacelab from Europe under the Educational, Scientific, and Cultural Materials Importation Act of 1966.

Table 4–44 continued

Date	Event
March 12, 1979	Bignier and Lord attend the program review at Dornier and observe progress in the assembly and testing of all major hardware elements.
March 29, 1979	A meeting between Frosch and Gibson is held, and NASA proposes the formation of a joint ESA/NASA working group to define the follow-on development program.
By April 1979	Good progress is finally reported in the development of the flexible toroidal sections to be placed at each end of the transfer tunnel, which would minimize the transfer of loads between the tunnel and its adjoining structural elements. The development test program of the tunnel "flex unit" is successfully completed. Two sets of tests have been completed in at Johnson Space Center using European-supplied development components from the Spacelab data system.
May 1979	Preliminary Design Review activities previously terminated because of flex unit development problems are resumed and satisfactorily completed.
June 1979	A System Compatibility Review is held to verify the IPS design qualifications on the basis of testing already performed.
July 4, 1979	NASA and ESA agree to a letter contract for the procurement of essential long-lead items necessary for producing a second Spacelab.
Sept. 1979	The total hardware system of the simulator is shipped to Johnson Space Center and accepted. This includes the crew station, an instructor operator station from which training operations would be controlled, and supporting computer equipment.
Sept. 12, 1979	Bignier writes to Lord expressing serious concern over the escalation of cost of the vertical access kit, then under design review at SENER.
By Oct. 1979	The ESA Spacelab Programme Board indicates its reluctance to approve additional funding for Spacelab improvements in light of cost overruns in the current development program.
Nov. 1979	A two-part Critical Design Review is completed for the Verification Flight Instrumentation for Spacelab 1. MDTSCO has the complete Software Development Facility operational at the IBM Huntsville, Alabama, complex. The facility provides a duplication of the Spacelab system and simulates all the orbiter interfaces and also can model the experiments that would fly on Spacelab. Both pallets are ready for Level IV integration of the payload.
Late 1979	During the NASA administrator's review of the 1981 Office of Space Science budget, the consolidated Spacelab utilization costs raise serious concern about their magnitude. In particular, the administrator states that the costs are not in keeping with the concept of a walk-on laboratory. He calls for formation of a Spacelab Utilization Review Committee to analyze the costs and to make recommendations for making the Spacelab a cost-effective vehicle for science missions. The pallet for the OSS-1 payload is transported from Kennedy to Goddard over the road, using the Payload Environmental Transportation System.

Table 4-44 continued

Date	Event
Jan. 1980	A contract is signed by Marshall (as the procurement agent for NASA) and ESA to purchase the second flight unit at a cost of approximately $184 million. The first assembly test of the racks, floor, and subfloor of the flight unit is completed, a full 2 weeks ahead of the new schedule.
Feb. 1980	Work starts on the long module integration test of the engineering model. Jesse Moore proposes to Lord to modify the Spacelab 2 configuration again to change from a three-pallet train with igloo to a single pallet with igloo plus a two-pallet train. This is accepted as the new configuration for Spacelab 2 unless later loads analyses show the need for further changes.
Feb. 14, 1980	Agency heads meet to review the Spacelab program in Paris. It is noted that, despite considerable progress by both ESA and NASA, the date for the first Spacelab flight has slipped to December 1982.
April 1980	Part I of the Engineering Model Acceptance Review is held. Nine teams evaluated a major portion of the deliverable acceptance data package and some 800 discrepancy notices are written.
Late May 1980	ESA and NASA sign an agreement for procurement of a second IPS for approximately $20 million, scheduled for delivery in the fourth quarter of 1983.
July 1980	The second major test of the flight unit is completed, although special test equipment has to be used to replace a faulty diverter valve.
Oct. 1980	The October monthly program report from ESA and NASA states that the engineering model and flight unit test (including electromagnetic compatibility) was completed on October 1, and with that test, the engineering model system integration program is completed.
Oct. 20, 1980	The Engineering Model Test Review Board gives final approval for full disassembly of the engineering model.
Nov. 4, 1980	The Engineering Model Test Review Board gives final approval for the start of the formal acceptance review, also known as the Engineering Model Acceptance Review Part II.
Nov. 24–25, 1980	The Engineering Model Acceptance Review Part II is successfully completed, with the final board giving permission to ship the hardware to Kennedy.
Nov. 28, 1980	The final segment of the engineering model is rolled out of the ERNO Integration Hall and is transported to Kennedy in three major shipments.
Late 1980	The first pallet is moved to the cargo integration test equipment stand to prepare for a simulated integration with the orbiter.
Dec. 5, 1980	The first shipment of the engineering model is brought to Kennedy on a C5A airplane. It contains the core segment, one pallet, and miscellaneous electrical ground support equipment (EGSE) and mechanical ground support equipment (MGSE), with a total weight of 29.9 metric tons.

Table 4–44 continued

Date	Event
Dec. 8, 1980	The second shipment of the engineering model arrives at Kennedy via a Lufthansa 747 airplane containing two pallets, miscellaneous EGSE and MGSE, and documentation, with a total weight of 36.3 metric tons.
Dec. 13, 1980	The third shipment of the engineering model arrives at Kennedy via a C5A plane containing the experiment segment, two pallets, and miscellaneous EGSE and MGSE, with a total weight of 33.6 metric tons.
Mid-Dec. 1980	The flight unit racks are accepted by NASA and delivered to the SPICE facility in Porz-Wahn.
March 4, 1981	A symbolic turnover of OSTA-1 from Rockwell to Johnson is accomplished.
March 10, 1981	A second turnover of OSTA-1 to Kennedy takes place.
April 8, 1981	ESA project manager Pfeiffer writes to John Thomas, the new NASA Spacelab program manager at Marshall, advising him of the April 3 selection of a new design concept for the IPS. ESA concludes that the existing mechanical design would have failed at several critical sections from the structural loads. The basic electronics concept, however, would be retained.
June 1981	The first part of the Flight Unit 1 Acceptance Review covering EGSE servicers, flight software, and spares is successfully completed. (Flight Unit 1 contains the module.)
June 15, 1981	The modified igloo is returned to ERNO for SABCA, and, after small modifications are made to the igloo support structure, work begins on integrating Flight Unit 2 (which contains the igloo).
June 26, 1981	The quarterly progress meeting at Dornier is held. Dornier presents the details of its new design concept and the results of recent hardware testing. Jim Harrington, NASA program director, summarizes the successful first flight of the Space Shuttle.
July 27, 1981	The first set of Flight Unit 1 hardware is shipped to Kennedy.
July 1981	Dornier's redesign concept of the IPS is given a go-ahead. The first set of EGSE is received by Kennedy. Following the successful completion of the tests in the cargo integration test equipment stand, a payload Certification Review certifies that OSTA-1 is prepared to support the STS-2 Flight Readiness Review and that the integrated payload and carrier are ready for testing with the orbiter. This affirms the operational readiness of the supporting elements of the mission.
Aug. 31, 1981	The report from Pfeiffer states that there are no outstanding technical problems in the first part of Flight Unit 1.
Sept. 1981	A new Preliminary Design Review is held of the IPS.
Nov. 4, 1981	Orbiter processing proceeds normally; the second Shuttle launch occurs. OSTA-1 provides abundant data. From the Spacelab viewpoint, OSTA-1 demonstrates the outstanding performance of the pallet for carrying experiments.

Table 4-44 continued

Date	Event
Nov. 30, 1981	The second part of the Flight Unit 1 Acceptance Review is completed, with the board's decision to approve Flight Unit 1 for shipment to Kennedy. A formal Certificate of Acceptance is signed by the program directors, project managers, and acceptance managers for the two agencies and for the prime contractor.
Dec. 7, 1981	Testing resumes 3 weeks late on the Flight Unit 2 systems.
Dec. 8, 1981	The OFT Pallet Program Manager's Review is conducted at Marshall.
Dec. 15, 1981	The OSS-1 Pallet Pre-Integration Review is conducted at Marshall.
Jan. 1982	A Spacelab 2 Interface Review is held of the IPS. By early 1982, the entire transfer tunnel assemblage is delivered to Kennedy, ready for processing for the first Spacelab mission.
Jan. 5, 1982	The Cargo Readiness Review of the OSS-1 Pallet is held at Kennedy.
Jan. 26–28, 1982	An OSS-1 simulation is conducted at Johnson.
Feb. 1982	The engineering model is powered up to begin tests simulating those to be conducted later with the first flight unit.
March 9, 1982	The Flight Readiness Review for OSS-1 is completed.
March 22, 1982	STS-3 is launched on its successful 7-day mission with the OSS-1 payload in the cargo bay.
March–Oct. 1982	It is agreed that NASA would conduct a Design Certification Review with support from ESA and its prime contractor ERNO to: review the performance and design requirements; determine that design configurations satisfied the requirements; review substantiating data verifying that the requirements had been met; review the major problems encountered during design, manufacturing, and verification and the corrective action taken; and establish the remaining effort necessary to certify flightworthiness.
June 10, 1982	Spacelab 1 faces its first operational review, the Cargo Integration Review for the STS-9 mission, conducted at Johnson. The board concludes that the hardware, software, flight documentation, and flight activities would support the planned launch schedule of September 30, 1983.
June 17, 1982	Agency heads meet in Paris. James E. Beggs has replaced Dr. Frosch as NASA administrator.
July 3, 1982	The final Flight Unit Acceptance Review for Flight Unit 2 is completed with the board meeting.
By July 7, 1982	A new cost review is presented to the administrator by Mike Sander and Jim Harrington. Their presentation focuses on three areas of Spacelab costs: operations, mission management, and instrument development.
July 8, 1982	The second Certificate of Acceptance is signed for Flight Unit 2.
July 29, 1982	The final shipment of large components of Flight Unit 2 is delivered to Kennedy from Hanover. It contains the igloo and the final three pallets, carried by C5A.
Aug. 1982	A Critical Design Review of the redesign of the IPS is held.

Table 4–44 continued

Date	Event
Dec. 6–9, 1982	The Johnson Mission Integration Office under Leonard Nicholson conducts an STS-9 Integration Hardware/Software Review to verify the compatibility of the integrating hardware and software design and orbiter capability against the cargo requirements for STS-9. The overall findings verify that the orbiter payload accommodations would meet the cargo requirements and can support the STS-9 launch schedule
Jan. 13, 1983	The final presentations and NASA Headquarters board review of the Design Certification Review are held.
Jan.–March 1983	The Spacelab 1 system test is conducted, verifying the internal interfaces between the subsystem and the experiment train, including the pallet.
March and April 1983	The experiments are powered up and total system verified in a mission sequence test simulating about 79 hours of the planned 215-hour flight, with the orbiter simulated by ground support equipment and the high-data-rate recording and playback demonstrated.
April 1983	A Design Certification Review on the verification flight tests and Verification Flight Instrumentation is completed.
May 1983	Subsystem integration of the new IPS system begins. The transfer tunnel is integrated to the module and its interfaces verified.
May 17, 1983	The NASA administrator signs a blanket certificate for the duty-free entry of Spacelab and Remote Manipulator System materials.
May 18, 1983	Spacelab is moved to the cargo integration test equipment stand for a higher fidelity simulation of the orbiter interface and use of the Kennedy launch processing system. During this test, the data link to the Payload Operations Control Center is simulated using a domestic satellite in place of the Tracking and Data Relay Satellite System. The cargo integration test equipment test is problem free.
June 17, 1983	Glynn Lunney, manager of the National Space Transportation System program at Johnson, issues the plan for the STS-9 Flight Operations Review to baseline the operations documentation through this management evaluation of the transportation of payload requirements into implementation plans and activities.
June 30, 1983	Lunney chairs the Flight Operations Board meeting at Johnson. The meeting includes a "walkthrough" of the STS-9 flight operations.

Table 4–44 continued

Date	Event
July 25, 1983	John Neilon, manager of NASA's cargo projects office, chairs a meeting of the Cargo Readiness Review Board. The review verifies the readiness of Spacelab 1 and supporting elements for on-line integration with the orbiter, verifies the readiness of the orbiter to receive Spacelab 1, and reviews the Kennedy cargo integration assessment from cargo transfer to the orbiter through mission completion, including identification of any major problems, constraints, or workarounds. The milestone events in the Spacelab program are reviewed, and all objectives are accomplished in three key tests at Kennedy: the integrated systems test, the cargo/orbiter interface test, and the closed loop test from Spacelab to the Mission Control Center and Payload Operations Control Center
Aug. 15, 1983	Spacelab is placed in the payload canister, transferred to the Orbiter Processing Facility, and installed in the orbiter *Columbia*. Three tests are conducted during the next month: the Spacelab/orbiter interface test verifies power, signal, computer-to-computer, hardware/software, and fluid/gas interfaces; the Spacelab/tunnel/orbiter interface test verifies tunnel lighting, air flow, and Verification Flight Instrumentation sensors; and the end-to-end command/data link test verifies the Spacelab/orbiter/Tracking and Data Relay Satellite System/White Sands/Domat/Johnson/Goddard link.
Sept. 23, 1983	The orbiter is moved to the Vehicle Assembly Building.
Sept. 28, 1983	The Shuttle assembly is rolled out to the launch pad, with launch scheduled for September 30.
Sept. 29, 1983	The Shuttle assembly returns to the Vehicle Assembly Building because of a suspect exhaust nozzle on the right solid rocket booster.
Nov. 4, 1983	The orbiter is moved to the Vehicle Assembly Building for a second time.
Nov. 8, 1983	The Shuttle is rolled out again to the pad.
Nov. 28, 1983	Spacelab 1 flies on Shuttle mission STS-9.

Source: Douglas R. Lord, *Spacelab—An International Success Story*, NASA Scientific and Technical Division, NASA, Washington, DC, 1987.

Table 4–45. Spacelab 1 Experiments

Experiment/Number	Principal Investigator	Class	Purpose/Objective	Success	Result
Solidification of Immiscible Alloys, 1ES301	H. Ahlborn, Federal Republic of Germany	Materials Science and Technology	Study alloys immiscible on Earth in a near-zero gravity environment to provide knowledge that may apply to industrial processes on Earth	Yes (Y)	100% of the planned objectives were accomplished using the isothermal heating furnace in the Materials Science Double Rack (MSDR).
Solidification of Technical Alloys, 1ES302	D. Poetschke, Federal Republic of Germany	Materials Science and Technology	Study technical alloys and the solidification process in near-zero gravity to provide knowledge that may apply to industrial processes on Earth	No (N)	This experiment was not performed in orbit because of the isothermal heating facility failure.
Skin Technology, 1ES303	H. Sprenger, Federal Republic of Germany	Materials Science and Technology	Study the casting of metals and composites in a near-zero gravity environment to provide knowledge which may apply to industrial processes on Earth	N	This experiment was not performed in orbit because of the isothermal heating facility failure.
Vacuum Brazing, 1ES304	E. Siegfried, Federal Republic of Germany	Materials Science and Technology	Study vacuum brazing in near-zero gravity to provide knowledge that may apply to industrial processes on Earth	Y	100% of the planned objectives were accomplished using the isothermal heating furnace in the MSDR.
Vacuum Brazing, 1ES305	R. Stickler, Austria	Materials Science and Technology	Study vacuum brazing in near-zero gravity to provide knowledge that may apply to industrial processes on Earth	Y	100% of the planned objectives were accomplished using the isothermal heating furnace in the MSDR.

Table 4-45 continued

Experiment/ Number	Principal Investigator	Class	Purpose/Objective	Success	Result
Emulsions and Dispersion Alloys, 1ES306	H. Ahlborn, Federal Republic of Germany	Materials Science and Technology	Study the influence of surface tension on the separation process in immiscible alloys in a near-zero gravity environment to provide knowledge that may apply to industrial processes on Earth	Y	100% of the planned objectives were accomplished using the isothermal furnace in the MSDR. Fourteen samples of zinc-lead-bismuth alloys were processed.
Reaction Kinetics in Glass, 1ES307	H.G. Frischat, Federal Republic of Germany	Materials Science and Technology	Study reaction kinetics of glass in near-zero gravity to provide knowledge that may apply to industrial processes on Earth	N	This experiment was not performed in orbit because of the isothermal heating facility failure.
Metallic Emulsions of Aluminum-Lead, 1ES309	P.D. Caton, United Kingdom	Materials Science and Technology	Investigate stability and properties, effects of cooling rates, and alloy composition on particle size and distribution in an aluminum-lead system in near-zero gravity to provide new knowledge that may be applied to industrial processes on Earth	Partial (P)	Approximately 25% of the planned objectives were accomplished. The objectives not accomplished were attributed to the isothermal heating furnace failure.
Bubble Reinforced Materials, 1ES311	P. Gondi, Italy	Materials Science and Technology	Study bubble-reinforced materials in a near-zero gravity environment to provide knowledge that may apply to industrial processes on Earth	Y	100% of the planned objectives were accomplished using the isothermal heating furnace in the MSDR.

Table 4–45 continued

Experiment/Number	Principal Investigator	Class	Purpose/Objective	Success	Result
Nucleation Behavior of Silver-Germanium, 1ES312	Y. Malmejac, France	Materials Science and Technology	Study the nucleation behavior of silver-germanium in a near-zero gravity environment to provide knowledge that may apply to industrial processes on Earth	P	Approximately 33% of the planned objectives were accomplished. The objectives not accomplished were attributed to the isothermal heating furnace failure.
Solidification of Near Monotetic Zinc-Lead Alloys, 1ES313	H. Fischmeister, Austria	Materials Science and Technology	Study the lead content, size and distribution of lead particles, temperature and time of the solidification process, and structure of immiscible zinc-lead alloys in a near-zero gravity environment to provide knowledge that may apply to industrial processes on Earth	Y	100% of the planned objectives were accomplished using the isothermal heating furnace in the MSDR.
Dendrite Growth and Microsegregation, 1ES314	H. Fredriksson, Sweden	Materials Science and Technology	Study dendrite growth in near-zero gravity to provide knowledge that may apply to industrial processes on Earth.	Y	100% of the planned objectives were accomplished using the isothermal heating furnace in the MSDR.

Table 4-45 continued

Experiment/Number	Principal Investigator	Class	Purpose/Objective	Success	Result
Composites with Short Fibers and Particles, 1ES315	A. Deruyttere, Belgium	Materials Science and Technology	Melt and allow solidification of various metallic composites to investigate the casting process in a near-zero gravity environment and to study the behavior of solid particles dispersed in a liquid metal	P	Approximately 50% of the planned objectives were accomplished. The objectives not accomplished were attributed to the isothermal heating furnace failure. Several different aluminum and copper composites were used. Resultant enhanced bonding characteristics were observed because of the near-zero gravity environment.
Unidirectional Solidification of Aluminum-Zinc Emulsions, 1ES316	C. Potard, France	Materials Science and Technology	Study the homogenous distribution of aluminum-zinc emulsions and the solidification process in space to determine the structural properties of the solidified alloy	Y	100% of the planned objectives were accomplished in the low-temperature gradient furnace in the MSDR.
Unidirectional Solidification of Aluminum-Aluminum II Copper, and Silver-Germanium Eutectics, 1ES317	Y. Malmejac, France	Materials Science and Technology	Study the homogenous distribution of aluminum-aluminum II copper and silver germanium eutectics and the solidification process in space to determine the structural properties of the solidified alloy	Y	100% of the planned objectives were accomplished in the low-temperature gradient furnace in the MSDR.

Table 4–45 continued

Experiment/Number	Principal Investigator	Class	Purpose/Objective	Success	Result
Growth of Lead Telluride, 1ES318	H. Rodot, France	Materials Science and Technology	Study the homogenous distribution of lead telluride and the solidification process in space to determine the structural properties of the solidified alloy	Y	100% of the planned objectives were accomplished in the low-temperature gradient furnace in the MSDR.
Unidirectional Solidification of Eutectics, 1ES319	K.L. Muller, Federal Republic of Germany	Materials Science and Technology	Study the growth and distribution of tellurium, doped indium antimonide, and nickel antimonide alloy in a near-zero gravity environment	Y	100% of the planned objectives were accomplished in the low-temperature gradient furnace in the MSDR.
Thermodiffusion in Tin Alloys, 1ES320	Y. Malmejac, France	Materials Science and Technology	Study the homogenous distribution of tin alloys and the solidification process in space to determine the structural properties	Y	100% of the planned objectives were accomplished in the low-temperature gradient furnace in the MSDR.
Zone Crystallization of Silicon, 1ES321	R. Nitsche, Federal Republic of Germany	Materials Science and Technology	Study the thermodiffusion effects on silicon crystal growth and the composition changes resulting from the near-zero gravity environment	Y	100% of the planned objectives were accomplished in the mirror heating facility in the MSDR. This experiment produced the first floating zone silicon crystal grown in space.
Traveling Solvent Growth of Cadmium Telluride, 1ES322	H. Jager, Federal Republic of Germany	Materials Science and Technology	Study the thermodiffusion effects on cadmium telluride crystal growth and the composition changes resulting from the near-zero gravity environment	Y	100% of the planned objectives were accomplished in the mirror heating facility in the MSDR.
Traveling Heater Method of III-V Compounds, Indium-Antimony, 1ES323	K.W. Benz, Federal Republic of Germany	Materials Science and Technology	Study the thermodiffusion effects of indium-antimony crystal growth and the composition changes resulting from the near-zero gravity environment	Y	100% of the planned objectives were accomplished in the mirror heating facility in the MSDR.

Table 4-45 continued

Experiment/ Number	Principal Investigator	Class	Purpose/Objective	Success	Result
Crystallization of Silicon Spheres, IES324	Dr. Kolker, Federal Republic of Germany	Materials Science and Technology	Study the thermodiffusion effects on silicon spheres and the composition changes resulting from the near-zero gravity environment	Y	100% of the planned objectives were accomplished in the mirror heating facility in the MSDR.
Unidirectional Solidification of Cast Iron, IES325	T. Luyendijk, The Netherlands	Materials Science and Technology	Study cast iron solidification in a near-zero gravity environment to provide knowledge that may apply to industrial processes on Earth	Y	100% of the planned objectives were accomplished using the isothermal heating furnace in the MSDR.
Oscillation Damping of a Liquid in Natural Levitation, IES326	H. Rodot, France	Materials Science and Technology	Study the phenomena of natural levitation of a liquid, oscillation damping of the liquid, and hydrodynamics of floating liquid zones	Y	100% of the planned objectives were accomplished in the fluid physics module in the MSDR.
Kinetics of Spreading of Liquids on Solids, IES327	J.M. Haynes, United Kingdom	Materials Science and Technology	Study the phenomena of spreading liquids on solids and the kinetics of spreading in a near-zero gravity environment	Y	100% of the planned objectives were accomplished in the fluid physics module in the MSDR.
Free Convection in Low Gravity, IES328	L.G. Napolitano, Italy	Materials Science and Technology	Study the phenomena of free convection in a near-zero gravity environment	Y	100% of the planned objectives were accomplished in the fluid physics module in the MSDR. Other data were also successfully collected.

Table 4–45 continued

Experiment/ Number	Principal Investigator	Class	Purpose/Objective	Success	Result
Capillary Surfaces in Low Gravity, IES329	J.F. Padday, United Kingdom	Materials Science and Technology	Study the phenomena of capillary surfaces in near-zero gravity and the hydrodynamics of floating liquid zones	Y	100% of the planned objectives were accomplished in the fluid physics module in the MSDR.
Coupled Motion of Liquid-Solid Systems in Near-Zero Gravity, IES330	J.P.B. Vreeburg, The Netherlands	Materials Science and Technology	Study the phenomena of coupled motion of a liquid-solid system in near-zero gravity and the hydrodynamics of floating liquid zones	Y	100% of the planned objectives were accomplished in the fluid physics module in the MSDR.
Floating Zone Stability in Zero-Gravity, IES331	I. Da Riva, Spain	Materials Science and Technology	Study the phenomena of floating zone stability of a liquid in near-zero gravity	Y	100% of the planned objectives were accomplished in the fluid physics module in the MSDR. Other data were also successfully collected.
Organic Crystal Growth, IES332	K.F. Nielsen, Denmark	Materials Science and Technology	Study organic crystal growth in near-zero gravity and the effect of weightlessness on the crystals	Y	Approximately 100% of the planned objectives were accomplished.
Growth of Manganese Carbonate, IES333	A. Authier, France	Materials Science and Technology	Study the growth of manganese carbonate in near-zero gravity and the effect of weightlessness on the manganese carbonate	Y	Approximately 100% of the planned objectives were accomplished.
Crystal Growth of Proteins, IES334	W. Littke, Federal Republic of Germany	Materials Science and Technology	Study crystal growth of proteins in near-zero gravity and the effects of weightlessness on the crystals	Y	Approximately 100% of the planned objectives were accomplished.

Table 4–45 continued

Experiment/ Number	Principal Investigator	Class	Purpose/Objective	Success	Result
Self-Diffusion and Inter-Diffusion in Liquid Metals, 1ES335	Dr. Kraatz, Federal Republic of Germany	Materials Science and Technology	Study self-diffusion and inter-diffusion in liquid metals exposed to near-zero gravity and the effect of weightlessness on the liquid metals	Y	Approximately 100% of the planned objectives were accomplished.
Crystal Growth of Mercury Iodide by Physical Vapor Transport, 1ES338	C. Belouet, France	Materials Science and Technology	Study crystal growth of mercury iodide by physical vapor transport in near-zero gravity and the effect of weightlessness on the mercury iodide crystals	Y	Approximately 100% of the planned objectives were accomplished.
Interfacial Instability and Capillary Hysteresis, 1ES339	J.M. Haynes, United Kingdom	Materials Science and Technology	Study interfacial instability and capillary hysteresis in a near-zero gravity environment	Y	100% of the planned objectives were accomplished in the fluid physics module in the MSDR.
Adhesion of Metals, Ultra High Vacuum Chamber, 1ES340	G. Ghersini, Italy	Materials Science and Technology	Study adhesion of metals in an ultrahigh vacuum chamber in a near-zero gravity environment	Y	Approximately 100% of the planned objectives were accomplished.
Tribiological Experiments in Zero-Gravity, 1NT011	C.H.T. Pan, F. Whitaker and R.L. Gause, United States	Materials Science and Technology	Study the wetting and spreading phenomena and fluid distribution patterns in a near-zero gravity environment	Y	100% of the planned objectives were accomplished. Other data were also successfully collected.

Table 4–45 continued

Experiment/ Number	Principal Investigator	Class	Purpose/Objective	Success	Result
An Imaging Spectrometric Observatory, 1NS001	M.R. Torr, United Sates	Atmospheric Physics and Earth Observations	Measure the airglow spectrum in wavelengths ranging from extreme ultraviolet to infrared	P	Approximately 75% to 80% of the planned objectives were accomplished. The Imaging Spectrometric Observatory (ISO) obtained the first broad-band spectrum of dayglow from 300 to 12,800 angstroms. It also obtained a database for a detailed assessment of the Shuttle environment for optical remote sensing in the visible, ultraviolet, and near-infrared ranges. The ISO obtained additional data concurrent with the electron beam firings and neutral releases. All science functional tests were run and the data were recorded. Because of RAU21, HDRR, and TDRSS coverage problems, some data were lost.

Table 4–45 continued

Experiment/ Number	Principal Investigator	Class	Purpose/Objective	Success	Result
Grille Spectrometer. 1ES013	M. Ackerman, Belgium; A. Girard, France	Atmospheric Physics and Earth Observations	Study, on a global scale, the atmosphere between 15 km and 150 km altitude	P	The low percentage of planned objectives accomplished (16%) was from the large beta angle constraint caused by the launch delay from October to November. The first observations of CO_2 in the thermosphere and water and methane in the mesosphere were made. Other gases were also observed. Solar absorption spectra of the atmospheric Earth limb in infrared light at sunset and sunrise were taken with a spectral resolution better than 10.0. Atmospheric absorptions were observed from 12 km to 130 km.
Waves in the Oxygen-Hydrogen Emissive Layer, 1ES014	M. Herse, France	Atmospheric Physics and Earth Observations	Photograph a layer of the high atmosphere to examine cloudlike structures that were observed within that layer	Y	100% of the planned objectives were accomplished. In addition to the planned photography, other measurements were taken.

Table 4-45 continued

Experiment/ Number	Principal Investigator	Class	Purpose/Objective	Success	Result
Investigation on Atmospheric Hydrogen and Deuterium through the Measurement of Their Lyman-Alpha Emission, 1ES017	J.L. Bertaux, France	Atmospheric Physics and Earth Observations	Study various sources of Lyman-Alpha emission in the atmosphere, in interplanetary space, and possibly in the galactic medium	P	Approximately 80% of the planned objectives were accomplished. Deuterium was discovered in the upper atmosphere between 100 km and 150 km. This discovery would allow for determination of the atmospheric eddy diffusion coefficient. The atomic hydrogen vertical profile between 80 km and 250 km was determined, and observations of interplanetary Lyman-Alpha emission were successful.
Metric Camera Experiment, 1EA033	M. Reynolds, Federal Republic of Germany	Atmospheric Physics and Earth Observations	Test the mapping capabilities of high-resolution photography from space	P	Approximately 80% of the planned objectives were accomplished. Although the metric camera experienced a jammed film advance mechanism, an in-flight maintenance procedure was developed to correct the problem. In addition to the planned observations, several targets of opportunity were photographed.

Table 4–45 continued

Experiment/ Number	Principal Investigator	Class	Purpose/Objective	Success	Result
Microwave Remote Sensing Experiment, 1EA034	G. Dieterle, Federal Republic of Germany	Atmospheric Physics and Observations	Develop an all-weather microwave remote-sensing system	P	Approximately 20% of the planned objectives were accomplished because of primary experiment equipment malfunctions. Measurements were taken over the planned target areas using the backup radiometer mode. In addition to the planned objectives, other data were obtained.
Atmospheric Emission Photometric Imaging (AEPI), 1NS003	S.B. Mende, United States	Space Plasma Physics	Observe faint optical emissions associated with natural and artificially induced phenomena (such as auroras) in the upper atmosphere	P	The AEPI operated with the camera in the stowed position because of a hardware failure, but as a result of orbiter maneuvers, the payload specialists were able to successfully complete approximately 65% to 70% of the planned objectives. Principal investigators gathered significant diagnostic data during joint operations with SEPAC. Information was also gathered about the double-layer airglow phenomena.

Table 4–45 continued

Experiment/ Number	Principal Investigator	Class	Purpose/Objective	Success	Result
Space Experiments With Particle Accelerators (SEPAC), 1NS002	T. Obayashi, Japan	Space Plasma Physics	Perform active and interactive perturbation experiments in Earth's ionosphere and magnetosphere	P	Approximately 80% of the planned objectives were accomplished. Vehicle charge neutralization was accomplished by the Magnetoplasma Dynamic Arcjet (MPD). A suspected-but-never-proven beam plasma discharge phenomenon was observed. Scientific experiments were successful except those requiring high-power electron gun firings. At the start of the SEPAC electron beam high-power firing test, the electron beam accelerator shut down and did not come back on-line for the remainder of the flight. Testing determined that vehicle neutralization was only partially achievable using the neutral gas plume. The planned coordination with the Phenomena Induced by Charged Particle Beams experiment (see below) was successful.

Table 4–45 continued

Experiment/ Number	Principal Investigator	Class	Purpose/Objective	Success	Result
Low Energy Electron Flux and Its Reaction to Active Experimentation on Spacelab, 1ES019A	K. Wilhelm, Federal Republic of Germany	Space Plasma Physics	Use artificially accelerated electrons as tracer particles for electric fields parallel to Earth's magnetic field	P	Approximately 90% of the planned objectives were accomplished. Principal investigators detected detailed high-resolution auroras when operating with SEPAC. However, failure of the SEPAC high-power electron beam limited the results.
Direct Current Magnetic Field Vector Measurement, 1ES019B	R. Schmidt, Austria	Space Plasma Physics	Determine the magnetic field surrounding the orbiter during the Spacelab 1 mission	P	Approximately 90% of the planned objectives were accomplished. Only the failure of the SEPAC high-power electron beam limited the results.
Phenomena Induced by Charged Particle Beams, 1ES020	C. Beghin, France	Space Plasma Physics	Study the effects of charged particle beam injections into Earth's upper atmosphere	P	All primary independent objectives were met. The planned coordination experiment with SEPAC was completed successfully. Loss of some data was experienced because of a gas bottle failure and the AEPI camera lock problem. This resulted in 60% of all planned objectives being accomplished.

494 NASA HISTORICAL DATA BOOK

Table 4-45 continued

Experiment/Number	Principal Investigator	Class	Purpose/Objective	Success	Result
Isotopic Stack-Measurement of Heavy Cosmic Ray Isotopes, 1ES024	R. Beaujean, Federal Republic of Germany	Space Plasma Physics	Measure heavy cosmic ray nuclei with a nuclear charge of 3 or more	Y	100% of the planned objectives were accomplished.
Far Ultraviolet Astronomy Using the FAUST Telescope, 1NS005	S. Bowyer, France	Astronomy and Solar Physics	Observe faint ultraviolet emissions from various astronomical sources with higher sensitivity than previously possible	P	Approximately 96% of the planned objectives were accomplished. Other targets were photographed in addition to the planned objective.
Very Wide Field Camera, 1ES022	G. Courtes, France	Astronomy and Solar Physics	Make a general ultraviolet survey of the celestial sphere in a study of large-scale phenomena	Y	100% of the planned objectives were accomplished. All primary targets were photographed, plus additional targets and spectra.
Spectroscopy in X-Ray Astronomy, 1ES023	R. Andresen, The Netherlands	Astronomy and Solar Physics	Study detailed features of cosmic x-ray sources and their variations in time	Y	100% of the planned objectives were accomplished. Measurements of the iron emission from the supernova remnant Cassiopeia A and the iron emission from Cygnus X-3 were taken. Additional measurements were obtained from other targets.

Table 4-45 continued

Experiment/Number	Principal Investigator	Class	Purpose/Objective	Success	Result
Active Cavity Radiometer, 1NA008	R.C. Wilson, United States	Astronomy and Solar Physics	Measure the total solar irradiance and its variation through time with state-of-the-art accuracy and precision	P	Approximately 90% of the planned objectives were accomplished.
Measurement of the Solar Constant, 1ES021	D. Crommelynck, Belgium	Astronomy and Solar Physics	Measure the absolute value of the solar constant with improved accuracy and to detect and measure long-term variations	P	Solar constant measurements were made with undetermined results. 90% of the planned objectives were completed.
Solar Spectrum from 170-3200 Nanometers, 1ES016	G. Thuillier, France	Astronomy and Solar Physics	Measure the energy output in the ultraviolet-to-infrared range of the solar spectrum	Y	100% of the planned objectives were accomplished. All of the scheduled observations were completed, and additional data were obtained.
Effects of Rectilinear Accelerations, Optokinetic, and Caloric Stimulations in Space, 1ES201	R. von Baumgarten, Federal Republic of Germany	Life Sciences	Investigate the vestibular functions of the inner ear, particularly the otolith organs that help maintain balance	P	Approximately 75% of the planned objectives were accomplished. Results were highly successful in the linear threshold, oscillopia, and caloric operations.

Table 4-45 continued

Experiment/Number	Principal Investigator	Class	Purpose/Objective	Success	Result
Vestibular Experiments, 1NS102	L.R. Young, United States	Life Sciences	Study the causes of space motion sickness and to study sensory-motor adaptation to weightlessness	P	Approximately 90% of the planned objectives were accomplished. Several experiments were performed, including the rotating dome, "hop and drop," and provocative testing. Data quality was excellent. Additional exploratory investigations were also done.
Vestibulo-Spinal Reflex Mechanisms, 1NS104	M.F. Reschke, United States	Life Sciences	Observe changes in spinal reflexes and posture during sustained weightlessness	P	Approximately 85% of the planned objectives were accomplished.
The Influence of Space Flight on Erythrokinetics in Man, 1NS103	C.S. Leach, United States	Life Sciences	Measure changes in the circulating red blood cell mass of people exposed to weightlessness	Y	100% of the planned objectives were accomplished.
Measurement of Central Venous Pressure and Determination of Hormones in Blood Serum During Weightlessness, 1ES026 and 1ES032	K. Kirsch, Federal Republic of Germany	Life Sciences	Collect data on changes in the distribution of body fluids and in the balance of water and minerals in the blood	Y	100% of the planned objectives were accomplished. Excellent television coverage of blood work and venous pressure activity was downlinked.

Table 4-45 continued

Experiment/Number	Principal Investigator	Class	Purpose/Objective	Success	Result
Effects of Prolonged Weightlessness on the Humoral Immune Response of Humans, INS105	E.W. Voss, Jr., United States	Life Sciences	Determine the effect of weightlessness on the body's immune response or ability to resist disease	Y	100% of the planned objectives were accomplished.
Effect of Weightlessness on Lymphocyte Proliferation, 1ES031	A. Cogoli, Switzerland	Life Sciences	Study the effect of weightlessness on lymphocyte activation	Y	100% of the planned objectives were accomplished.
Three-Dimensional Ballistro-cardiography in Weightlessness, 1ES028	A. Scano, Italy	Life Sciences	Record a three-dimensional ballistrocardiogram under a unique condition and to compare the results with tracings recorded on the same subject on the ground	Y	100% of the planned objectives were accomplished. Several additional runs were completed.
Personal Miniature Electro-physiological Tape Recorder, 1ES030	H. Green, United Kingdom	Life Sciences	Collect physiological data on a normal man in an abnormal environment as a basis for future studies	Y	100% of the planned objectives were accomplished.
Mass Discrimination During Weightlessness, 1ES025	H. Ross, United Kingdom	Life Sciences	Compare the perception of mass in space with the perception of weight on Earth	P	Approximately 90% of the planned objectives were accomplished. The crew's performance of mass discrimination was significantly poorer in near-zero gravity than in a one-gravity environment.

Table 4-45 continued

Experiment/Number	Principal Investigator	Class	Purpose/Objective	Success	Result
Nutation of Helianthus Annuus in a Microgravity Environment, 1NS101	A.H. Brown, United States	Life Sciences	Observe the growth movements of plants in a near-zero gravity environment	P	Approximately 60% of the planned objectives were accomplished. Experiment camera synchronization problems might have caused some loss of data.
Preliminary Characterization of Persisting Circadian Rhythms During Spaceflight: Neurospora as a Model System, 1NS007	F.M. Sulzman, United States	Life Sciences	Compare the growth of plants cultured in Spacelab and on the ground to test whether circadian rhythms persist in space	Y	100% of the planned objectives were accomplished. The implanted fungus demonstrated circadian growth within a 24-hour period.
Microorganisms and Biomolecules in Hard Space Environment, 1ES029	G. Horneck, Federal Republic of Germany	Life Sciences	Measure the influence of the space environment on various biological specimens	Y	100% of the planned objectives were accomplished.
Radiation Environment Mapping, 1NS006	E.V. Benton, United States	Life Sciences	Measure the cosmic radiation inside Spacelab	Y	100% of the planned objectives were accomplished.
Advanced Biostack Experiment, 1ES027	H. Bucker, Federal Republic of Germany	Life Sciences	Determine the radiobiological importance of cosmic radiation particles of high charge and high energy	Y	100% of the planned objectives were accomplished.

Table 4-46. Spacelab 3 Experiments

Experiment/Number	Principal Investigator	Class	Purpose/Objective	Success	Result
Solution Growth of Crystals in Zero-Gravity, Fluid Experiment System (FES)	R. Lal, United States	Materials Science	Develop a technique for solution crystal growth in a near-zero gravity environment, to characterize the growth environment under orbital conditions and its influence on crystal growth behavior, and to evaluate the properties of the resultant crystal	Yes (Y)	During the first flight of the new fluid experiment system, two triglycine sulfate crystals were successfully grown from a liquid. For the first time, scientists could see in detail the crystal growth process in a microgravity environment and determine the differences between crystal growth on the ground and growth in microgravity where convection effects are negligible. Visual observations by the crew provided real-time descriptions of the crystal and aided investigators on the ground as they controlled the progress of the investigation.
Mercuric Iodide Growth, Vapor Crystal Growth System (VCGS)	W.F. Schnepple, United States	Materials Science	Grow higher quality mercuric iodide crystals in a near-zero gravity environment and to gain an improved understanding of crystal growth by a vapor process	Y	A mercuric iodide crystal measuring 14 mm x 8 mm x 7 mm was successfully grown from a seed crystal 20 times smaller in this new facility by a vapor transport process. The crystal grew at a carefully controlled rate, varying from 1 mm to 3 mm per day.

Table 4–46 continued

Experiment/ Number	Principal Investigator	Class	Purpose/Objective	Success	Result
Mercury Iodide Crystal Growth (MICG)	R. Cadoret, France	Materials Science	Grow near-perfect single crystals of mercury iodide in a near-zero gravity environment at different pressures to analyze the effects of the environment on vapor transport	Y	Six cartridges of mercury iodide material without seed crystals were processed for up to 70 hours at a time, each under different growth conditions.
Dynamics of Rotating and Oscillating Free Drops, Drop Dynamics Module (DDM)	T. Wang, United States	Fluid Mechanics	Perform fundamental experiments to verify that the new facility can acoustically manipulate drops, to test theoretical predictions of drop behavior, and to observe any new phenomena encountered	Y	After initial startup difficulties, this facility underwent significant in-flight maintenance and operated successfully for research in the behavior of free floating drops. For the first time, a principal investigator operated and repaired his own experiment in space as a Spacelab crew member. Interaction between the ground team and flight crew resulted in investigation recovery and the accomplishment of virtually all the intended research.

Table 4–46 continued

Experiment/ Number	Principal Investigator	Class	Purpose/Objective	Success	Result
Geophysical Fluid Flow Cell (GFFC)	J. Hart, United States	Fluid Mechanics	Study fluid motions in a near-zero gravity environment to understand fluid flows in oceans, atmospheres, and stars and test an elaborate new facility for laboratory experiments on geophysical flows	Y	The GFFC facility performed nominally and obtained excellent data. All planned scenarios were performed during an 84-hour period, and the mission added 13 unscheduled scenarios in 18 hours of extra operations. Approximately 46,000 shadow-graph images, which permitted the fluid density gradients to be observed, were recorded on film for postflight analysis.

Table 4–46 continued

Experiment/Number	Principal Investigator	Class	Purpose/Objective	Success	Result
Ames Research Center Life Sciences Payload (ARCLSP)	P. Callahan, J. Tremor, United States	Life Sciences	Perform engineering tests to ensure that the Research Animal Holding Facility was a safe and adequate facility for housing and studying animals in the space environment, observe the animals' reactions to the space environment, and evaluate the operations and procedures for in-flight animal care	Y	The new Research Animal Holding Facility provided a suitable animal habitat. However, there were some difficulties with food and waste containment. Food, water, and activity monitors provided good engineering data about the facility and the status of the animals; they adjusted well to spaceflight and demonstrated their suitability for research in orbit. One of the two primates developed symptoms of space adaptation syndrome but recovered in a manner analogous to human experience. This suggests that nonhuman primates may be good models for vestibular research pertinent to human adaptation of microgravity. The Biotelemetry System provided data on physiological functions of four rodents.

SPACE SCIENCE 503

Table 4–46 continued

Experiment/ Number	Principal Investigator	Class	Purpose/Objective	Success	Result
Autogenic Feedback Training (AFT)	P. Cowlings, United States	Life Sciences	Test a treatment for space adaptation syndrome and to test a technique for training people to control bodily processes voluntarily	Y	In general, the hardware performed nominally and did not interfere with other crew activities.
Urine Monitoring System (UMS)	H. Schneider, United States	Life Sciences	Verify the operation of the UMS in collecting and sampling urine, perform in-flight measurement system calibration, develop and utilize a procedure for monitoring crew water intake using existing orbiter facilities, and verify the system for preparing urine samples for postflight analysis	Y	All planned calibrations and dead volume measurements were performed, but urine samples were collected for only one rather than two crew members as planned.
Very Wide Field Camera (VWFC)	G. Courtes, France	Atmospheric Science and Astronomy	Make an ultraviolet survey of the celestial sphere in a study of large-scale phenomena, such as clouds, within our galaxy	Partial (P)	The VWFC operated nominally on its first deployment but could not be subsequently deployed when the bent latch handle on the scientific airlock precluded further airlock operations. Ground teams assessed the airlock malfunction but determined that in-flight maintenance was inappropriate. During the initial extension into space, the camera acquired its first target and made a 1-minute exposure. However, the five subsequent operations were suspended.
Auroral Imaging Experiment	T. Hallinan, United States	Atmospheric Science and Astronomy	Observe and record the visual characteristics of pulsating and flickering auroras	P	Of the 21 scheduled opportunities for auroral observations, 18 were accomplished, with auroras clearly visible on each.

Table 4–46 continued

Experiment/Number	Principal Investigator	Class	Purpose/Objective	Success	Result
Atmospheric Trace Molecules Spectroscopy (ATMOS)	C.B. Farmer, United States	Atmospheric Science and Astronomy	Obtain fundamental information related to the chemistry and physics of Earth's upper atmosphere using infrared absorption spectroscopy, determine, on a global scale, the compositional structure of the upper atmosphere and its spatial variability, and provide the high-resolution, calibrated spectral information essential for the detailed design of advanced instrumentation for future global monitoring of species critical to atmospheric stability	Y	Although the ATMOS instrument was deactivated earlier than planned, the investigation was one of the most successful of the mission. In 19 3-minute operations, ATMOS obtained 150 independent atmospheric spectra, each of which contained at least 100,000 individual spectral measurements. During five solar calibrations, detailed infrared spectra of the Sun were obtained. Initial examination of the data indicated unexpected evidence about molecular constituents there.
Studies of the Ionization States of Solar and Galactic Cosmic Ray Heavy Nuclei (IONS)	S. Biswas, India	Atmospheric Science and Astronomy	Use a newly designed detector system to determine the composition and intensity of energetic ions emitted from the Sun and other galactic sources toward Earth's atmosphere	Y	IONS provided data on the arrival time and directions in space of cosmic ray particles and hence the magnetic rigidity. During the mission, the instrument initially did not respond to commands to rotate the detector stack. After in-flight maintenance was performed, the instrument operated nominally. The investigation accomplished two-thirds of its operational timeline.

Table 4–47. Spacelab 2 Experiments

Experiment/Number	Principal Investigator	Class	Purpose/Objective	Success	Result
Vitamin D Metabolites and Bone Demineralization, 2SL-01	H.K. Schnoes, United States	Life Sciences	Measure quantitatively the blood levels of biologically active vitamin D metabolites of the Spacelab 2 flight crew members	Yes (Y)	100% of the planned objectives were accomplished.
Interaction of Oxygen and Gravity Influenced Lignification, 2SL-02	J.R. Cowles, United States	Life Sciences	Determine the effect of weightlessness upon lignification and to establish the overall effect of oxygen on lignin formation independent of any gravity effects	Y	100% of the planned objectives were accomplished. The real-time downlinked video from this experiment exceeded expectations in both quantity and quality.
Ejectable Plasma Diagnostics Package (PDP), 2SL-03	L.A. Frank, United States	Plasma Physics	Study natural plasma processes, orbiter-induced plasma processes, and beam plasma physics	Partial (P)	Approximately 82% of the objectives were accomplished. Some of the attached operations on the Remote Manipulator System (RMS) arm and one orbiter fly-around were lost because of low propellant levels. However, the PDP performed flawlessly on the pallet, with the RMS, and as a free-flyer.
Plasma Depletion Experiments for Ionospheric and Radio Astronomical Studies, 2SL-04	M. Mendillo and A.V. DaRosa, United States	Plasma Physics	Study the ionospheric depletions and related effects caused by the exhaust gases from the orbiter Orbital Maneuvering System (OMS) burns	P	50% of the planned objectives were accomplished. Only four of eight planned OMS burns were performed because of low OMS propellant. The Millstone Hill and Arecibo burns created ionospheric holes deeper and wider than expected.

Table 4–47 continued

Experiment/ Number	Principal Investigator	Class	Purpose/Objective	Success	Result
Small Helium-Cooled Infrared Telescope, 2SL-05	G.G. Fazio, United States	Infrared Astronomy	Study the diffuse emission and extended sources in the infrared sky, the measurement of the natural and spacecraft-induced infrared background, and the determination of suitable procedures and techniques for the in-space use of superfluid helium and cryogenic telescopes	P	The infrared telescope operated well throughout the mission but did not achieve its primary objective of an all-sky survey. During the first viewing period, many of the detectors were quickly saturated by a strong source of mysterious origin. A survey of the instrument with the RMS camera before payload deactivation revealed apparent debris within the sun shade. A section of the galaxy was mapped in shorter wavelengths. These few minutes of data represent a new and valuable complement to the infrared astronomical satellite data. Evaluations were also made of the dewar system, and several attempts were made to alter the state of the superfluid helium to observe the fluid dynamic behavior.

Table 4–47 continued

Experiment/ Number	Principal Investigator	Class	Purpose/Objective	Success	Result
Elemental Composition and Energy Spectra of Cosmic Ray Nuclei, 2SL-06	P. Meyer and D. Muller, United States	High-Energy Physics	Determine the abundance distributions of elements and isotopes in the cosmic radiation, study the composition of cosmic rays at high energies, investigate the role of a galactic halo in particle confinement, and determine whether the relative abundancies of different source nuclei change with energy	Y	100% of the planned objectives were accomplished.
Hard X-Ray Imaging of Clusters of Galaxies and Other Extended X-Ray Sources, 2SL-07	A.P. Willmore, United Kingdom	High-Energy Physics	Use x-ray measurements to observe a component of galaxies and study their temperatures and mass distribution, understand the properties of intergalactic gas emitted from clusters of galaxies, use the x-ray observations to determine the spectrum and distribution of gigaelectron volt electrons in the clusters of galaxies, and use x-ray observations to demonstrate the differences between clusters of galaxies	P	Approximately 90% of the planned objectives were accomplished. The dual x-ray telescope operated well throughout the mission with very good image quality, detector sensitivity, and stability. Images of point sources and extended sources were successfully reconstructed from downlinked data.

Table 4–47 continued

Experiment/ Number	Principal Investigator	Class	Purpose/Objective	Success	Result
Solar Magnetic and Velocity Field Measurement System, 2SL-08	A.M. Title, United States	Solar Physics	Measure magnetic and velocity fields in the solar atmosphere, follow the evolution of solar magnetic structure over several days, study magnetic field changes associated with transient events, and provide a test of the pointing accuracy and stability of the IPS	Y	The Solar Optical Universal Polarimeter started its observations late in the mission after an unexplained shutdown on the first day and an equally unexplained startup on the next-to-the-last day of the mission. (See the "Mission Anomalies" section.) Thereafter, the instrument performed almost perfectly to observe the strength, structure, and evolution of magnetic fields in the solar atmosphere.
Solar Coronal Helium Abundance Spacelab Experiment, 2SL-09	A.H. Gabriel and J.L. Culhane, United Kingdom	Solar Physics	Determine accurately the abundance of helium in the solar atmosphere	P	Approximately 66% of the objectives were accomplished. Early mission observing time was lost because of difficulties with the IPS. However, spectral scans of the limb of the solar disc were achieved. In addition, the instrument was used in a mapping mode to study and make images of the structure of the Sun's corona.

Table 4-47 continued

Experiment/ Number	Principal Investigator	Class	Purpose/Objective	Success	Result
Solar Ultraviolet High Resolution Telescope and Spectrograph, 2SL-10	G.E. Brueckner, United States	Solar Physics	Make spectral scans and images of the solar disc and, particularly, record rapidly changing solar features	P	Approximately 60% of the objectives were accomplished. The resolution of the telescope was very good, but IPS pointing difficulties compromised the early data. Downlink television from the instrument revealed the birth of a spicule, which was never witnessed before.
Solar Ultraviolet Spectral Irradiance Monitor, 2SL-11	G.E. Brueckner, United States	Atmospheric Physics	Improve the accuracy of knowledge of the absolute solar fluxes, provide a highly accurate traceability of solar fluxes, and measure the variability of solar fluxes	P	Approximately 50% of the planned objectives were accomplished. The experiment made spectral scans of the Sun with excellent accuracy, verified by calibration and alignment checks.
Vehicle Charging and Potential, 2SL-14	P.M. Banks, United States	Plasma Physics	Investigate electron beam interactions in space plasma, vehicle charging processes, and electromagnetic wave generation processes	P	Approximately 72% of the objectives were accomplished. Television images of beam and aurora activities were not permitted because of bright moonlight. Joint operations and observations were accomplished, primarily with the PDP and with the nearby Dynamics Explorer satellites. The electron generator was fired more than 200 times. Instrument performance was nearly perfect.

Table 4–47 continued

Experiment/Number	Principal Investigator	Class	Purpose/Objective	Success	Result
Properties of Superfluid Helium in Zero-Gravity, 2SL-13	P.V. Mason, United States	Technology	Determine the fluid and thermal properties that are required for the design of planned space experiments using superfluid helium as a cryogen, advance scientific understanding of the interactions between superfluid and normal liquid helium, and demonstrate the use of superfluid helium as a cryogen in zero-gravity	P	Approximately 88% of the objectives were accomplished. The dewar or cryosat performed as expected during the mission. The existence of quantized surface waves in thin films of helium was clearly established, and several hundred recordings were made across a range of temperatures. Bulk thermal dynamics measurements of temperature variations within the dewar were quite successful. However, bulk fluid dynamics measurements were prevented by sensors, which remained frozen throughout the mission.

Table 4-47 continued

Experiment/ Number	Principal Investigator	Class	Purpose/Objective	Success	Result
Protein Crystal Growth	C.E. Bugg, United States	Technology	Develop hardware and procedures for growing proteins and other organic crystals by two methods in the orbiter during the low-gravity portion of the mission	Y	Generally, hardware for both methods worked as planned. Postflight analysis showed minor modification in the flight hardware was needed, and a means of holding the hardware during activation, crystal growth, deactivation, and photography was desirable. The dialysis method produced three large tetragonal lysozyme crystals with average dimensions of 1.3 mm x 0.65 mm x 0.65 mm. The solution growth methods produced small crystals of lysozyme, alpha-2 interferon, and bacterial purine nucleoside phosphorylase.

Table 4–48. Spacelab D-1 Experiments

Experiment/Number	Investigator/Sponsor	Class
Floating Zone Hydrodynamics, FPM 04	J. Da Riva, U. Madrid, Spain	Fluid Physics
Capillary Experiments in Low Gravity Fields, FPM 06	J.F. Padday, Kodak, Ltd., Harrow, United Kingdom	Fluid Physics
Forced Liquid Motions, FPM 08	J.P.B. Vreeburg, NAL, Amsterdam, The Netherlands	Fluid Physics
Oberflachenspannung (Surface Tension Studies), HOL 03	D. Neuhaus, DFVLR, Koln, Germany	Fluid Physics
Maragonikonvektion Im Offenen Boot (Marangoni Convection), MKB 00	D. Schwabe, U. Gieben, Germany	Fluid Physics
Marangoni Flows, FPM 07	L. Napolitano, U. Neapel, Italy	Fluid Physics
Marangoni Convection in Gas-Liquid Mass Transfer, FPM 01	A.A.H. Drinkenburg, U. Groningen, The Netherlands	Fluid Physics
Convection in Nonisothermal Binary Mixture Presenting a Surface Tension Minimum as a Function of Temperature, FPM 05	J.C. Legros, U. Brussels, Belgium	Fluid Physics
Blasentransport (Bubble Transport), HOL 01	A. Bewersdorff, DFVLR, Koln, Germany	Fluid Physics
Selbst- und Interdiffusion (Self- and Inter-Diffusion), HTT 00	K.H. Kraatz, H. Wever, G. Frohberg, TU Berlin, Germany	Fluid Physics
Thermal Diffusion, GHF 01	J. Dupuy, U. Lyon, France	Fluid Physics
Interdiffusion, IDS 00	W. Merkens, TH Aachen, Germany	Fluid Physics
Homogenitat von Glasern (Homogeneity of Glasses), IHF 05	Chr. Frischat, TU Clausthal, Germany	Fluid Physics
Diffusion of Liquid Zinc and Lead, GPRF 2	R.B. Pond, Marvalaud Inc., United States	Fluid Physics
Thermomigration of Cobalt in Tin, GHF 07	J.P. Praizey, CEN, Grenoble, France	Fluid Physics
Warmekapazitat am Kritischen Punkt (Heat Capacity Near Critical Point), HPT 00	J. Straub, TU Munchen, Germany	Fluid Physics
Phasenbildung am Kritischen Punkt (Phase Separation Near Critical Point), HOL 02	H. Klein, DFVLR, Koln, Germany	Fluid Physics
GETS, HOL 04	A. Ecker, TH Aachen, Germany	Solidification

Table 4–48 continued

Experiment/Number	Investigator/Sponsor	Class
Al-Cu, Phasengrenzflachendiffusion (Aluminum-Copper Phase Boundary Diffusion), GFQ 01	H.M. Tensi, TU Munchen, Germany	Solidification
Erstarrungskonvektion (Solidification Dynamics), GFQ 02	S. Rex, TH Aachen, Germany	Solidification
Dendritic Solidification of Aluminum-Copper Alloys, GHF 04	J.J. Favier, D. Camel, CEN, Grenoble, France	Solidification
Cellular Morphology in Lead Thallium Alloys, GHF 02	B. Billia/J. Favier, U. Marseilles, France	Solidification
InSb-NiSb-Eutektikum (Indium Antimonide-Nickel Antimonide Eutectics), ELI 04	G. Muller, U. Erlangen-Nurnberg, Germany	Solidification
Containerless Melting of Glass, SAAL	D.E. Day, U. Missouri-Rolla, United States	Solidification
Suspensionserstarrung (Solidification of Suspensions), IHF 02	J. Potschke, Krupp-Forschungsinst Essen, Germany	Solidification
Teilchen vor Schmelz und Erstarrungsfront (Particle Behavior at Solidification Fronts), IHF 06	D. Langbein, Battelle-Inst., Frankfurt, Germany	Solidification
Stutzhauttechnologie (Skin Technology), IHF 03	H. Sprenger, MAN, Munchen, Germany	Solidification
Liquid Skin Casting of Cast Iron, IHF 07	H. Sprenger, MAN, Munchen, Germany	Solidification
Erstarrung eutsktischer Legierungen (Solidification of Eutectic Alloys), IHF 09	Y. Malmejac, CEN, Grenoble, France	Solidification
Erstarrung von Verbundmaterialien (Solidification of Composite Materials), IHF 08	A. Deruyttere, U. Leuven, Germany	Solidification
Shmelzzonenzuchtung Si (Silicon-Crystal Growth by Floating Zone Technique), MHF 01	R. Nitsche, A. Croll, U. Freiburg/Br., Germany	Solidification
Si-Kugel (Melting of Silicon Sphere), MHF 04	H. Kolker, Wacker-Chemie, Munchen, Germany	Solidification
Doped Indium Antimonide and Gallium Indium Antimonide, GHF 03	C. Potard, CEN, Grenoble, France	Solidification
Traveling Heater Method (GaSb), MHF 02	K.W. Benz, U. Stuttgart, Germany	Solidification

Table 4–48 continued

Experiment/Number	Investigator/Sponsor	Class
Traveling Heater Method (CdTe), MHF 03	R. Schonholz, R. Freiburg/Br., Germany	Solidification
Traveling Heater Method (InP), ELI 01	K.W. Benz, U. Stuttgart, Germany	Solidification
Traveling Heater Method (PbSnTe), ELI 02	M. Harr, Battelle-Inst, Frankfurt, Germany	Solidification
Gaszonenzuchtung CdTe (Vapor Growth of Cadmium), ELI 03	M. Bruder, U. Freiburg/Br., Germany	Solidification
Ge/Gel4 Chemical Growth, GHF 05	J.C. Launay, U. Bordeaux, France	Solidification
Ge-12 Vapor Phase, GHF 06	J.C. Launay, U. Bordeaux, France	Solidification
Vapor Growth of Alloy-Type Crystal, GPRF 4	H. Wiedemeier, Rensalear Polytechnic Institute, Troy, New York, United States	Solidification
Semiconductor Materials, GPRF 5	R.K. Crouch, Langley R.C., United States	Solidification
Proteinkristalle (Protein Crystals), CRY 00	W. Littke, R. Freiburg/Br., Germany	Solidification
Separation Nichmischbarer Legierungen (Separation of Immiscible Alloys), IHF 01	H. Ahlborn, U. Hamburg, Germany	Solidification
Separation of Immiscible Liquids, FPM 03	D. Langbein, Battelle-Inst., Frankfurt, Germany	Solidification
Separation of Fluid Phases, FPM 02	R. Naehle, DFVLR, Koln, Germany	Solidification
Liquid Phase Miscibility Gap Materials, GPRF 3	H.S. Gelles, Columbus, Ohio, United States	Solidification
Ostwaldreifung (Ostwald Ripening), IHF 04	H. Fischmeister, MPI, Stuttgart, Germany	Solidification
Human Lymphocyte Activation, BR 32CH	A. Cogoli, ETH Zurich, Switzerland	Biology
Cell Proliferation, BR 21 F	H. Planel, U. Toulouse, France	Biology
Mammalian Cell Polarization, BR 48 F	M. Bouteille, U. Paris, France	Biology
Circadian Rhythm, BR 27 D	D. Mergenhagen, U. Hamburg, Germany	Biology
Antibacterial Activity, BR 58 F	R. Tixador, U. Toulouse	Biology
Growth and Differentiation of Bacillus Subtilis, BR 28 D	H.D. Mennigmann, U. Frankfurt, Germany	Biology
Effect of Microgravity in Interaction Between Cells, BR 07 I	O. Ciferri, U. Pavia, Italy	Biology
Cell Cycle and Protoplasmic Streaming, BR 16 D	V. Sovick, DFVLR, Koln, Germany	Biology

Table 4-48 continued

Experiment/Number	Investigator/Sponsor	Class
Dosimetric Mapping Inside Biorack, BR 19 D	H. Bucker, DFLVR, Koln, Germany	Biology
Froschstatolith (Frog Statoliths), STA 00	J. Neubert, DFVLR, Koln, Germany	Biology
Dorsoventral Axis in Developing Embryos, BR 52NL	G. Ubbels, U. Utrecht, The Netherlands	Biology
Distribution of Cytoplasmic Determinants in the Drosophilia Melanogaster Egg, BR 15 E	R. Marco, U. Madrid, Spain	Biology
Embryogenesis and Organogenesis of Caracausius Morosus, BR 18 D	H. Bucker, DFVLR, Koln, Germany	Biology
Graviperzeption (Gravi-Perception), BOT 01	D. Volkmann, U. Bonn, Germany	Biology
Geotropismus (Geotropism), BOT 02	J. Gross, U. Tubingen, Germany	Biology
Differenzierung (Differentiation of Plant Cells), BOT 03	R.R. Theimer, U. Munchen, Germany	Biology
Statocyte Polarity and Geotropic Response, BR 39 F	G. Perbal, U. Paris, France	Biology
Vestibular Research,VS-ES 201	R.V. Baumgarten, U. Mainz, Germany	Medicine
Vestibular Research,VS-NS 102	L. Young, MIT, Cambridge, Massachusetts, United States	Medicine
Zentraler Venendruck (Central Venous Pressure), ZVD 00	K. Kirsch, FU Berlin, Germany	Medicine
Tonometer, TOM 00	J. Draeger, U. Hamburg, Germany	Medicine
Body Impedance Measurement, BIM 300	F. Baisch, DFVLR, Koln, Germany	Medicine
Mass Discrimination, ROS 230	H.E. Ross, U. Stirling, United Kingdom	Medicine
Spatial Description in Space, LAN 200 SDS 00	A.D. Friederici, MPI, Nijmegen, The Netherlands	Medicine
Gesture and Speech in Microgravity, LAN 200GPS 00	A.D. Friederici, MPI, Nijmegen, The Netherlands	Medicine
Reaktionszeitmessung (Determination of Reaction Time), JUF 250	M. Hoschek/J. Hund, Muhltal, Germany	Medicine
Uhrensynchronisation (Clock Synchronization), NX-USY 00	S. Starker, DFVLR, Oberpfaffenhofen, Germany	Navigation
Einwegenthfenungsmessung, NX-EWE 00	D. Rother, SEL, Stuttgart, Germany	Navigation

Table 4–49. OSS-1 Investigations

Investigation	Principal Investigator	Institution
Contamination Monitor Package measured the buildup of molecular and gas contaminants in the orbiter environment to determine how molecular contamination affects instrument performance.	J. Triolo	Goddard Space Flight Center/U.S. Air Force
Microabrasion Foil Experiment measured the numbers, chemistry, and density of micrometeorites encountered by spacecraft in near-Earth orbit.	J.A.M. McDonnell	University of Kent, England
Vehicle Charging and Potential Experiment measured the electrical characteristics of the orbiter, including its interactions with the natural plasma environment of the ionosphere and the disturbances that result from the active emission of electrons.	P. Banks	Utah State University
Shuttle-Spacelab Induced Atmosphere provided data on the extent that dust particles and volatile materials evaporating from the orbiter produced a local "cloud" or "plume" in the "sky" through which astronomical observations could be made.	J. Weinberg	University of Florida
Solar Flare X-Ray Polarimeter measured x-rays emitted during solar flare activities on the Sun.	R. Novick	Columbia University
Solar Ultraviolet Spectral Irradiance Monitor was designed to establish a new and more accurate base of solar ultraviolet irradiance measurements over a wide wavelength region.	G. Brueckner	Naval Research Laboratory
Plant Growth Unit demonstrated the effect of near weightlessness on the quantity and rate of lignin formation in different plant species during early stages of development and tested the hypothesis that, under microgravity, lignin might be reduced, causing the plants to lose strength and droop rather than stand erect.	J.R. Cowles	University of Houston

Table 4–49 continued

Investigation	Principal Investigator	Institution
Thermal Canister Experiment determined the ability of a device using controllable heat pipes to maintain simulated instruments at several temperature levels in thermal loads.	S. Ollendorf	Goddard Space Flight Center
Plasma Diagnostics Package studied the interaction of the orbiter with its surrounding environment, tested the capabilities of the Shuttle's Remote Manipulator System, and carried out experiments in conjunction with the Fast Pulse Electron Generator of the Vehicle Charging and Potential Experiment, also on the OSS-1 payload pallet. The package was deployed for more than 20 hours and was maneuvered at the end of the 15.2-meter RMS. (See also Table 4–40.)	S. Shawhan	University of Iowa

Table 4–50. Hubble Space Telescope Development

Date	Event
1940	Astronomer R.S. Richardson speculates on the possibility of a 300-inch telescope placed on the Moon's surface.
1960/1961	The requests for proposal (RFP) for the Orbiting Astronomical Observatory spacecraft and the astronomical instruments to be flown aboard them are issued.
1962	The National Academy of Sciences recommends the construction of a large space telescope.
1965	The National Academy of Sciences establishes a committee to define the scientific objectives for a proposed large space telescope.
1968	The first astronomical observatory, the Orbiting Astronomical Observatory-1, is launched.
1972	The National Academy of Sciences again recommends a large orbiting optical telescope as a realistic and desirable goal.
1973	NASA establishes a small scientific and engineering steering committee headed by Dr. C. Robert O'Dell of the University of Chicago to determine which scientific objectives would be feasible for a proposed space telescope.
1975	The European Space Agency becomes involved in the project.
1977	NASA selects a group of 60 scientists from 38 institutions to participate in the design and development of the proposed space telescope.
June 17, 1977	NASA issues the Project Approval Document for the space telescope. The primary project objective is to "develop and operate a large, high-quality optical telescope system in space which is unique in its usefulness to the international science community. The overall scientific objectives…are to gain a significant increase in our understanding of the university—past, present, and future—through observations of celestial objects and events...."
Oct. 19, 1977	NASA awards the contract for the primary mirror to Perkin-Elmer of Danbury, Connecticut.
1978	Congress appropriates funds for the development of the space telescope.
April 25, 1978	Marshall Space Flight Center is designated as the lead center for the design, development, and construction of the telescope. Goddard Space Flight Center is chosen to lead the development of the scientific instruments and ground control center.
Dec. 1978	Rough grinding operation begins at Perkin-Elmer in Wilton, Connecticut.
1979	
Jan. 20, 1979	Money requests for space science program increase 20 percent ($100 million), which includes money for the space telescope.
Feb. 1979	Debate over which institute NASA should choose to develop the space telescope takes place. (John Hopkins University is chosen.)

Table 4–50 continued

Date	Event
May 29, 1979	The decision is made to have Fairchild Space & Electronics Company modify the communications and data handling module it developed for NASA's Multimission Modular Spacecraft for use on the space telescope.
June 1979	Marshall Space Flight Center decides that the alternative sensor was receiving little management attention at the Jet Propulsion Laboratory and the space telescope was unlikely to be ready for a 1983 launch.
July 1979	Marshall Space Flight Center compiles its Program Operating Plan for fiscal year 1980; Lockheed and Perkin-Elmer overshot the cost for the space telescope by millions of dollars of the original budgeted adjusted program's reserves.
Nov. 18, 1979	Five states compete for the space telescope: Maryland, New Jersey, Illinois, Colorado, and California. Competing groups include University Research Association, Associated Universities, Inc. (AUI), and Association of Universities for Research and Astronomy (AURA). AUI wants the project at Princeton; AURA wants it at Johns Hopkins University.
Dec. 14, 1979	Goddard Space Flight Center releases the Space Telescope Science Institute RFP. Proposals are due March 3, 1980.
1980	
Feb. 13, 1980	Dr. F.A. Speer, manager of the High Energy Astronomy Observatory program at Marshall Space Flight Center, is named manager of the space telescope project for Marshall.
Feb. 21, 1980	NASA Associate Administrator Dr. Thomas A. Mutch informs Congress that the space telescope can be completed within its "originally estimated costs." NASA estimates space telescope development costs at $530 million, with another $600 million allotted for operation of the system over a 17-year period. Mutch says progress toward launch in December 1983 "continues to be excellent."
May 29, 1980	NASA announces the selection of Ford Aerospace to negotiate a contract for overall system design engineering on preliminary operations requirements and the test support system for the space telescope.
Sept. 18, 1980	NASA officials admit to space telescope cost and schedule problems in hearing before the House Science and Technology subcommittee.
1981	
Jan. 6, 1981	A.M. Lovelace, NASA associate administrator/general manager, submits a revised space telescope cost and schedule estimate. The launch period is revised to the first half of 1985, and the estimated development cost at launch is $700 million to $750 million (in 1982 dollars).

Table 4–50 continued

Date	Event
Jan. 16, 1981	NASA selects AURA for final negotiation of a contract to establish, operate, and maintain the Space Telescope Science Institute. It will be located at Johns Hopkins University. The contractor's estimate of the cost of the 5-year contract is $24 million, plus additional funds to support a guest observer and archival research program.
April 29, 1981	Perkin-Elmer completes polishing of the 2.4-meter primary mirror (see events dated November 1990).
April 30, 1981	Goddard Space Flight Center awards the contract for the management of the Space Telescope Science Institute to AURA. The period of performance for the $40.4 million contract extends through 1986. The institute will be located at Johns Hopkins University.
Oct. 23, 1981	Space telescope's "main ring" is delivered to Perkin-Elmer Corp. from Exelco Corp., which fabricated the ring over a period of 18 months.
Dec. 10, 1981	Perkin-Elmer finishes putting an aluminum coating 3 millionths of an inch thick on the primary mirror.
1982	
Jan. 26, 1982	Congress increases space telescope funding by $2 million to $121.5 million.
March 1982	The Critical Design Review of the space telescope's support systems module is completed, and the design is declared ready for manufacturing.
March 28, 1982	A report from the House Appropriations Committee states that the space telescope would cost $200 million more and reach orbit a year later than expected because of difficulties in development. The report blames delays and cost overruns on NASA for understaffing the program by 50 percent in its early development and on Perkin-Elmer for failing to properly plan for a project of the technical and manufacturing difficulty of the space telescope. Also, unremovable dust on the primary mirror after 15 months in a Perkin-Elmer "clean room" had lowered its reflecting power by 20 to 30 percent.
1983	
Feb. 4, 1983	NASA Administrator Beggs tells the House Science and Technology Committee that technical problems in developing the electronics and guidance and pointing system of the optical telescope assembly of the space telescope will delay the launch of the telescope and increase costs.
March 24, 1983	NASA Administrator Beggs tells House subcommittee that the space telescope has problems in a number of areas—the latching mechanism, the fine guidance sensor system, and the primary mirror—that are likely to result in cost overruns of $200 million or more and at least a 12- to 18-month delay. Beggs says that the primary mirror is coated with dust after sitting in a clean room for a year and may not be able to be cleaned without harming its surface. Its capability could be limited to 70 or 80 percent.

Table 4–50 continued

Date	Event
March 25, 1983	The preliminary report by the Investigations and Survey Staff of the House Appropriations subcommittee states that the space telescope will overrun its costs by $200 million, boosting its overall cost to $1 billion.
April 13, 1983	NASA names James B. Odom as manager of Marshall Space Flight Center's space telescope project.
April 26, 1983	James Welch, NASA's director of space telescope development, states that NASA may accept the dirty primary mirror because a current study indicates that the mirror would be within the acceptable range and would meet the original specifications in the contract. Also, NASA has decided to coat the sticking latching mechanism with tungsten carbide rather than redesign it.
June 15, 1983	Dr. William Lucas, Marshall Space Flight Center director, tells the House Space subcommittee that NASA estimates that telescope project costs will increase $300 million to $400 million to approximately $1.1 billion to $1.2 billion, and it expects to be able to launch in June 1986. He states that technical problems "are now understood and resolution is in hand."
June 15, 1983	Administrator Beggs acknowledges that, in retrospect, NASA made some errors in planning and running the space telescope program, but that the instrument has not been compromised.
Oct. 5, 1983	The space telescope is officially renamed the Edwin P. Hubble Space Telescope.
Nov. 17, 1983	NASA submits a report to Congress on proposed action that would augment efforts planned for the space telescope development by $30.0 million above the authorized and appropriated amount, for a revised FY 1984 level of $195.6 million.
Dec. 22, 1983	Space telescope officials are cautiously optimistic that the serious problems that surfaced on the space telescope over the last year have been solved and that the instrument can be launched on schedule in 1986.
1984	
April 2, 1984	The estimated cost of the space telescope has risen to $1.175 million. NASA Administrator Beggs states that Lockheed will lose some of its award fees because of poor workmanship problems.
April 30, 1984	NASA reports that tests of the fine guidance sensors have demonstrated that the telescope will meet stringent pointing and tracking requirements.
May 14, 1984	The idea surfaces of refurbishing the space telescope in space.
May 31, 1984	The five science instruments to fly on the space telescope complete acceptance testing at Goddard Space Flight Center: high-resolution spectrograph, faint-object spectrograph, wide-field/planetary camera, faint-object camera, high-speed photometer.
July 12, 1984	Technicians at Perkin-Elmer clean the primary mirror. NASA states that cleaning of the primary mirror has confirmed that the observatory will have the very best optical system possible.

Table 4–50 continued

Date	Event
Dec. 6, 1984	Goddard Space Flight Center's Telescope Operations Control Center satisfactorily conducts command and telemetry tests with the Hubble Space Telescope at Lockheed Missile and Space Corporation. This is the first of seven assembly and verification tests.
1985	
Jan. 17–18, 1985	A workshop by the Space Telescope Science Institute is held to give scientists an opportunity to present their recommendations for key projects for the space telescope.
Feb. 1, 1985	The National Society of Professional Engineers presents an award to Perkin-Elmer Corp. for its development of the Hubble Space Telescope's optical telescope assembly.
July 8, 1985	Lockheed Missiles and Space Co. reports that it has completed assembly of the primary structure for the Hubble Space Telescope.
July 19, 1985	Goddard Space Flight Center releases the RFP for design and fabrication of an Imaging Spectrograph for the space telescope. Proposals are due September 17.
Dec. 5, 1985	NASA selects three scientific investigations for the space telescope to lead to the development of one or two advanced scientific instruments for Hubble.
1986	
Jan. 26, 1986	The destruction of *Challenger* delays the launch of Hubble and other missions.
Feb. 27, 1986	Hubble completes acoustic and dynamic and vibrational response tests. The tests indicate that it can endure the launch environment.
May 2– June 30, 1986	Thermal-vacuum testing is conducted.
May 21, 1986	The last elements of Hubble—the solar arrays—are delivered to Lockheed Missiles and Space Co. (Sunnyvale, California) for integration into the main telescope structure.
May 27, 1986	Hubble successfully completes the thermal-vacuum testing in the Lockheed thermal-vacuum chamber.
Aug. 7, 1986	NASA and the Space Telescope Science Institute in Baltimore announce that 19 U.S. amateur astronomers will be allowed to make observations with Hubble. This decision is to show gratitude to the amateur astronomers for their help with telescopes for the last 400 years.
Aug. 8, 1986	Hubble successfully completes 2 months of rigorous testing.
1987	
March 17, 1987	Hubble starts a 3-day ground system test involving the five instruments that will be carried on board: wide field and planetary camera, high-resolution spectrograph, faint object spectrograph, high-speed photometer, and faint object camera.
Aug. 31– Sept. 4, 1987	Goddard Space Flight Center's Space Telescope Operations Control Center, Marshall Space Flight Center, and the Space Flight Telescope Science Institute conduct a joint orbital verification test.

Table 4–50 continued

Date	Event
Sept. 9, 1987	Hubble completes the reevaluation of Failure Mode and Effects Analysis (FMEA). This reevaluation of the FMEA/Critical Items List/hazard analysis is directed by the Space Telescope Development Division as part of NASA's strategy to return the Space Shuttle to flight status.
1988	
Feb. 10, 1988	Fred S. Wojtalik is appointed manager of the Hubble project at Marshall Space Flight Center.
March 31, 1988	The draft Program Approval Document for Hubble is completed. The draft contains the objectives of Hubble, the technical plan, including the experiments and descriptions, and the systems performance requirements.
June 20, 1988	NASA begins the fourth ground system test (GST-4) of Hubble. This will be the longest ground test to date, lasting 5 1/2 days, and also the most sophisticated because all of the six instruments will be used in their various operational modes; the new instrument is the fine-guidance astrometer.
July 24, 1988	Hubble completes the GST-4 tests successfully, except for a timing incompatibility between the science instruments and the computer. The problem is to be corrected by adjusting the software.
August 31, 1988	NASA delays launch of Hubble from June 1989 to February 1990.
1989	
July 19, 1989	The Space Telescope Science Institute completes its selection of the first science observation proposals to be carried out using Hubble. Among the 162 accepted proposals (out of 556 submitted) are plans to search for black holes in neighboring galaxies, to survey the dense cores of globular star clusters, to better see the most distant galaxies in the universe, to probe the core of the Milky Way, and to search for neutron stars that may trigger bizarre gamma-ray bursts.
Oct. 1989	A modified Air Force C-5A Galaxy transports the Hubble Space Telescope from Lockheed in California to its launch site at the Kennedy Space Center in Florida.
1990	
Jan. 19, 1990	NASA delays the Hubble launch to replace O-rings.
Feb. 5–7, 1990	Confidence testing is held.
Feb. 10, 1990	End-to-end communications test run using Tracking and Data Relay Satellite-East is concluded to interconnect the payload interfaces of *Discovery* in its hangar, Hubble in the Vertical Processing Facility, and the Space Telescope Operations Control Center at Goddard Space Flight Center.
Feb. 13, 1990	The final confidence test is held.
Feb. 15, 1990	Closeout operations begin.
Feb. 17, 1990	Functional testing of Hubble's science instruments is completed.
March 29, 1990	Hubble is installed in the Space Shuttle orbiter *Discovery*'s payload bay.
April 24, 1990	Hubble is launched on STS-31.

Table 4–50 continued

Date	Event
June 21, 1990	Hubble's project manager announces the telescope's inability to focus properly.
July 2, 1990	The Hubble Space Telescope Optical Systems Board of Investigation is formed under the chairmanship of Dr. Lew Allen of the Jet Propulsion Laboratory.
Oct. 16, 1990	Responsibility for the Hubble project (except for the optical system failure questions) is transferred from Marshall to Goddard.
Nov. 1990	The Board of Investigation releases findings, which conclude that a spherical aberration was caused by a flawed measuring device that was used to test the primary mirror at the manufacturer's facility.
Dec. 2, 1993	The Hubble Repair Mission on STS-61 installs corrective lenses and replaces solar panels.

Table 4–51. Ulysses Historical Summary

	Spacecraft	Launch Vehicle/ Upper Stage	Launch Date
October 1978 Project Start	1 NASA spacecraft 1 ESA spacecraft	Single STS/IUS (3-stage launch)	1983 launch
April 1980		Split launches: 1 NASA, 1 ESA	Launch deferred to 1985
February 1981	NASA spacecraft "slowdown"	Launch vehicle changed to STS/Centaur	Launch deferred to 1986
September 1981	U.S. spacecraft canceled		
January 1982		Launch vehicle changed to STS/IUS (2-stage)	
July 1982		Launch vehicle changed to STS/Centaur	
January 1986		*Challenger* accident	Launch deferred indefinitely
June 1986		STS/Centaur program canceled	
November 1986		IUS/PAM-S upper stage procurement decision	
			Launch date selected: October 1990

INDEX

A

Aaron, John, 373
Abrahamson, James A., 17, 18
Active Magnetospheric Particle Tracer Explorer, AMPTE, 373, 386, 388
Advanced Communications Technology Satellite, ACTS, 50
Advanced Launch System, 56
Advanced Solid Rocket Motor, ASRM, 57, 58
Aerojet Strategic Propulsion Company, 222
Aerojet TechSystems, 53
Air Force, U.S., 17, 24, 28, 29, 169, 179, 221, 225, 375, 385, 386
Aldrich, Arnold J., 19, 111, 196, 197, 200
Aldrin, Edwin "Buzz," 125
Allen, Joseph P., 178
Ames Research Center, 118, 179
Anderson, Paul G., 118
Apogee and Maneuvering System, AMS, 51
Apollo, 107, 108, 115, 125, 134, 135, 165
Arizona, University of, 366
Armstrong, Neil, 125
Atlantic Ocean, 33, 138, 156
Atlantic Research Corporation, 222
Atlantis, 125, 153, 182, 232, 233, 235, 405
Atlas, 13, 16, 22, 24, 25, 26, 27, 28, 29, 47, 51, 375
Aurora, satellite, 28
Australia, 179
Austria, 150

B

Ball Aerospace, 380
Beggs, James M., 54, 370
Belgian, 150
Big Bang, 404, 410
Black, David C., 118
Bluford, Guion S., 173
Boeing Aerospace Corporation, 52, 53, 241, 242, 402
Brand, Vance D., 174
Briggs, Geoffrey, 371
British Science and Engineering Research Council, 399, 400, 408
Bundeministerium für Forschung und Technologie, BMFT, 408, 409

C

California Institute of Technology, 402
California, University of, 402, 406
Canada, 170, 173, 241, 243, 244, 250
Canadian Space Agency, 250
Capital Development Plan, 241
Captain Cook, 125
Carnegie Institute, 367
Castor, II, IV, 31
Centaur, 16, 17, 22, 24, 26, 29, 51, 57, 217, 375, 417, 418, 420
Centre Spatial Guyanais, 252
Centro Ricerche Aerospaziali, 392
Challenger, 3, 14, 19, 28, 41, 43, 45, 52, 107, 108, 111, 115, 122, 123, 125, 128, 139, 153, 171, 172, 174, 175, 176, 177, 185, 186, 187, 188, 190, 191, 195, 196, 199, 203, 212, 217, 218, 235, 364, 368, 390, 403, 409, 410, 417, 418, 420
Chang-Diaz, Franklin R., 183
Charge Composition Explorer, CCE, 386, 387, 388
Chicago, University of, 402, 419
Clean Air Act of 1977, 379
Coast Guard, U.S., 190
Collins, Michael, 125
Colorado, University of, 380, 390
Columbia, 107, 108, 122, 125, 153, 155, 156, 157, 161, 162, 163, 164, 165, 167, 168, 169, 183, 232, 233, 235
Columbus, 251, 252
Combined Release and Radiation Effects Satellite, CRRES, 29
Comet Giacobini-Zinner, 368, 402, 421
Comet IRAS-Araki-Alcock, 383
Commerce Business Daily, 27
Compton, Arthur Holly, 405
Compton Gamma Ray Observatory, CGRO, 401, 404, 405, 406
Compton Telescope, COMPTEL, 405
Congress, U.S., 3, 4, 14, 20, 21, 22, 56, 57, 119, 120, 121, 188, 189, 212, 239, 241, 242, 246, 248, 363, 373, 374, 417
Copernicus, Orbiting Astronomical Observatory, 402
Cosmic Background Explorer, COBE, 409, 410
Crew Activity Plan, 141

Crippen, Robert L., 112, 167, 172, 214, 215, 222
Critical Design Review, 44–45, 56
Critical Item List, CIL, 210, 214, 215
Culbertson, Philip E., 18, 114

D

Data and Design Analysis Task Force, 51-L, 211
Delta, 13, 16, 21, 22, 24, 25, 26, 27, 28, 30, 31, 49, 51, 368, 379, 382, 383, 384, 399, 406, 407, 409, 410
Denmark, 150
Department of Defense, DOD, 19, 27, 28, 33, 48, 53, 56, 57, 122, 131, 179, 182, 186, 218, 227, 385, 391
Design Certification Review, 45
Deutsche Forschungs Versuchsanstat für Luft und Raumfahrt, DFVLR, 393 (see German Aerospace Research Establishment)
Diaz, Alphonso V., 118, 119
Discovery, 125, 147, 153, 177, 178, 185, 187, 232, 233, 235, 236, 237, 238, 388
Douglas Aircraft Company, 31
Drop Dynamics Module, 395
Dutch additional experiment, DAX, 382, 383
Dynamics Explorer 1, 378, 379

E

Earth, 33, 34, 48, 52, 53, 55, 56, 107, 108, 125, 126, 129, 141, 143, 148, 152, 157, 161, 163, 165, 167, 168, 171, 174, 177, 178, 180, 238, 239, 243, 246, 252, 254, 363, 364, 366, 369, 370, 372, 373, 374, 375, 379, 383, 385, 388, 395, 396, 399, 401, 402, 404, 406, 410, 411, 413, 414, 416, 418, 419, 420, 421
Eaton, Peter, 19
Edelson, Burton I., 371
Edwards Air Force Base, California, 34, 136, 142, 169, 171, 204, 208, 210, 216, 225, 238
Ellington Field, 145
Endeavour, 126, 153
Enterprise, 108
European Space Agency, ESA, 26, 109, 148, 150, 173, 174, 241, 243, 244, 252, 368, 384, 394, 399, 400, 401, 402, 405, 408, 419, 420
European X-Ray Observatory Satellite, EXOSAT, 26, 384
Evolution Management Council, 246
Expendable Launch Vehicles, ELV, 13, 14, 15, 18, 19, 20, 21, 22, 24, 25, 27, 28, 33, 47, 49, 217, 368

Extravehicular activity, EVA, 146, 147, 171, 174, 181, 239, 250
Extreme Ultraviolet Explorer, EUVE, 368, 401, 406, 407

F

Feynman, Richard, 199
Finarelli, Margaret, 118
Fisher, William F., 181
Fisk, Lennard A., 371
Fitts, Jerry, 17
Fletcher, James C., 115, 116, 120, 212, 218
Flight Readiness Firing, 133, 236, 237
Flight Readiness Review, 112, 195, 196, 199, 207, 215, 223
France, 150
Freedom, Space Station, 242, 243, 250
Fullerton, Gordon, 162

G

Galaxy satellite, 27
Galileo, 52, 218, 368, 369, 383, 417, 418, 419
Gardner, Dale A., 178
Garn, Senator Jake, 180
German Aerospace Research Establishment, DFVLR, 182 (see Deutsche Forschungs Versuchsanstat fur Luft und Raumfahrt)
German Federal Ministry of Research and Technology, 182
Germany, Federal Republic of, 150, 172, 173, 405, 407, 418
Gemini, 142
General Accounting Office, 6
General Dynamics, 29, 52
General Electric, GE, 52, 53, 242
General purpose computers, GPC, 127, 128, 233
Geostationary Operational Environmental Satellite, GOES, 25, 26, 28
Get-Away Special, 151, 152, 169, 180, 182
Gibson, Robert L. "Hoot," 174
Glaser, Harold, 371
Goddard Space Flight Center, 31, 116, 117, 143, 144, 151, 180, 240, 248, 368, 369, 371, 390, 392, 399, 402, 403, 406, 409
Graham, William, 189
Greenbelt, Maryland, 248
Gregory, Frederick, 224
Griggs, David, 147, 179
Grumman Corporation, 242
Grumman Gulfstream II, 146
Guastaferro, Angelo, 371
Gunn, Charles R., 15

INDEX

H

Halley's Comet, 183, 184, 190, 390, 401, 416, 421
Halpern, Richard E., 118
Hercules Inc., 222
Herman, Daniel H., 118, 371
High Energy Astronomy Observatories, HEAO, 365, 366, 375, 376, 409
High-pressure fuel turbopump, HPFT, 36
Hilat, 385, 391
Hinners, Noel W., 369, 372
Hitchhiker, 183
Hodge, John D., 18, 114, 116
Hoffman, Jeffrey, 147, 179
Honeywell, Inc., 51, 52
Hubble, Edwin P., 367
Hubble Space Telescope, 218, 367, 368, 401, 402, 403, 404
Hudson, Henry, 125
Hughes Aircraft, 26, 178, 416
Huntsville, Alabama, 247
Hutchinson, Neil, 240

I

Indian Ocean, 33, 137, 238
Inertial Upper Stage, IUS, 17, 48, 49
Infrared Astronomy Satellite, IRAS, 26, 381, 382, 383
Insat, 26, 172, 173
Intelsat, 26
Intercontinental ballistic missile, ICBM, 29, 30
Intermediate range ballistic missile, 31
International Cometary Explorer, ICE, 402, 421
International Halley Watch, 421
International Solar Polar Mission, 52, 419, 420
International Sun-Earth Explorer, ISEE, 401, 402, 415
International Ultraviolet Explorer, IUE, 399, 400, 401
Intravehicular activity, IVA, 147
Ion Release Module, IRM, 386, 387, 388
Italian Commissione per le Ricerche Spaziali, 391
Italy, 150, 391
Itek, 402

J

Japan, 241, 243, 244, 252, 401
Japanese Experiment Module, JEM, 253
Jarvis, Gregory, 190

Jet Propulsion Laboratory, JPL, 143, 371, 380, 383, 406, 416
Johns Hopkins University, 403
Johnson Space Center, JSC, 34, 50, 109, 110, 111, 112, 113, 114, 115, 116, 117, 123, 143, 144, 145, 155, 189, 196, 217, 223, 239, 240, 248
Jupiter, 365, 366, 396, 411, 413, 417

K

Keel, Alton G., Jr., 188
Kennedy Space Center, KSC, 29, 34, 37, 50, 109, 110, 112, 113, 116, 117, 122, 125, 133, 135, 136, 137, 141, 142, 143, 155, 169, 176, 189, 190, 196, 200, 203, 204, 208, 210, 216, 217, 224, 225, 232, 235, 241, 251, 369
Kenya, 392
Kilminster, Joseph, 196
Kodak, 402
Kohrs, Richard H., 111, 112
Kourou, French Guiana, 252

L

Landsat, 26
Langley Research Center, 33, 246, 247
Laser initial navigation system, LINS, 50, 51
Launch Abort Panel, 225
Launch Control Center, 134, 135, 236
Launch Processing System, 134, 135
Leasat, 147, 177, 179, 181
Leicester University, United Kingdom, 408
Lewis Research Center, 30, 52, 53, 115, 116, 117, 118, 240, 249, 250, 383
Lightweight external tank, LWT, 37
Lockheed, 241, 402
Long Duration Exposure Facility, 177
LTV Corporation, 29, 33
Lunney, Glynn S., 111

M

Magellan, 369, 416, 417
Magnetic satellite, Magsat, 24
Mahon, Joseph B., 16, 18, 19
Main engine cutoff, MECO, 34, 40, 137, 139
Main propulsion system, MPS, 34, 35
Manipulator Foot Restraint, 174
Manned Maneuvering Unit, 174, 175, 176
Manufacturing Review, 56
Mariner, 365, 390, 421
Mars, 29, 30, 247, 365, 366, 368, 372, 413, 414, 420

Mars Observer, 50, 420, 421
Marshall Space Flight Center, MSFC, 34, 37, 44, 50, 52, 55, 109, 110, 112, 113, 116, 117, 142, 143, 144, 189, 196, 197, 200, 204, 207, 210, 212, 217, 240, 247, 248, 402
Martin, Franklin D., 117, 371, 373
Martin Marietta Corporation, 29, 33, 34, 51, 53, 113, 242, 402, 416
Max Planck Institut fuer Extraterrestrische Physik, MPE, 408
McAuliffe, S. Christa, 190
McCandless, Bruce, II, 174
McDonald, Franklin B., 372
McDonnell Douglas, 29, 49, 169, 178, 242
McNair, Ronald E., 174, 190
Memorandum of Understanding, MOU, 241, 243, 252, 391, 392, 393, 408
Merbold, Ulf, 173, 394
Mercury, 29, 144, 365
Mexico, 180
Milky Way, 401
Mission Control Center, 139, 141, 142, 143, 146, 227
Mission Operations Reports, MOR, 154
Mobile Launcher Platform, 134, 135
Moon, 29, 30, 372, 402
Moore, Jesse W., 18, 196, 371
Moore, R. Gilbert, 151
Morocco, Ben Guerir, 225
Moser, Thomas L., 115, 117, 118
Mulloy, Lawrence B., 196
Mulroney, Brian, 250
Musgrave, F. Story, 172
Mutch, Thomas A., 369

N

NASA Advisory Council, 20
NASA Authorization Act, 379
National Academy of Sciences, 402, 418
National Advisory Committee for Aeronautics, NACA, 3
National Aeronautics and Space Act, 3
National Aeronautics and Space Administration, NASA, 3, 4, 5, 6, 19–33, 36, 37, 41, 43, 44, 47, 48, 50–55, 58, 107–123, 134, 135, 136, 143, 144, 145, 150–156, 164, 166, 171, 174, 177–191, 195, 197–201, 205, 208–228, 233, 235, 238, 239, 240–247, 252, 254, 255, 363–379, 388, 391, 393, 406, 408
National Aerospace Plane, 3
National Air and Space Museum, 367
National Oceanic and Atmospheric Administration, NOAA, 24, 25, 26, 27, 28, 244

National Research Council, NRC, 120, 206, 207, 214, 215, 216, 219, 222, 223
National Research Council of Canada, 175
National Science Teachers Association, NSTA, 152
National Space Development Agency, 252
National Space Policy, 27, 246
National Space Technology Laboratories, 109, 144
National Space Transportation System, NSTS, 109, 110, 111, 112, 113, 121, 222, 223, 225, 228, 230
National Transportation Safety Board, 190
National Weather Service, 29
Naugle, John E., 372
Naval Research Laboratory, 180, 388, 399
Navy, U.S., 33, 125
Neptune, 367, 368, 411, 414, 415
Netherlands, 150, 367, 381, 405
Neutral Buoyancy Laboratory, 248
New York Times, 240
Nicogossian, Arnauld, 371
Nixon, Richard, 121
Nobel Prize, 405
Norris, Theodrick B., 371
NOVA-II satellite, 29

O

O'Connor, Bryan, 223
O'Dell, C. Robert, 402
Odom, James B., 119
Office of Management and Budget, OMB, 4, 6, 21, 363
Office of Technology Assessment, 120
Onizuka, Ellison, 190
Orbital Flight Test, OFT, 154, 155, 156, 157, 158, 159, 160, 161, 162, 164, 166
Orbital maneuvering system, OMS, 34, 35, 123, 124, 134, 137, 138, 139, 141, 156, 157, 163, 230, 231, 235, 236, 237
Orbital maneuvering vehicle, OMV, 54, 55,
Orbital Sciences Corporation, OSC, 50
Orbital Transfer Vehicle, OTV, 52, 53, 54
Orbiter Processing Facility, 141, 142, 227, 235, 236, 237
Orbiting Solar Observatory, 365
Oscar, satellite, 391

P

Pacific Ocean, 33
Palapa, 172, 174, 178
Paules, Granville, 118
Payload Assist Module, PAM, 49, 50, 51, 170, 174, 177, 179, 183

INDEX 531

Payload Flight Test Article, PFTA, 172, 173
Payload Operations Control Center, POCC, 142, 143
Pellerin, Charles J., 371
Perkin-Elmer, 402
Peterson, Donald H., 172
Phillips, Samuel C., 115, 117, 214
Pioneer, 365, 366, 367, 414, 415, 416
Plasma Diagnostics Package, PDP, 161, 162, 389
Pluto, 365, 367, 404
Polar Beacon Experiments and Auroral Research satellite, Polar BEAR, 391
Pratt & Whitney, 52, 53
Preliminary Design Review, 44, 45, 55, 214
Preliminary Requirements Review, 55
Program Requirements Review, 242

R

Raney, William P., 119
RCA, 24, 25, 26
Reaction control system, RCS, 124, 128, 134, 137, 140, 141, 160, 161, 167, 230, 231, 235, 236
Reagan, Ronald, 13, 26, 43, 56, 109, 114, 120, 187, 205, 212, 238, 240, 250, 254, 255
Redesigned Solid Rocket Motor, RSRM, 43, 44, 45, 47, 57, 199
Redmond, Thomas W., 111
Remote Manipulator System, RMS, 144, 145, 147, 161, 168, 171, 172, 175, 177, 179, 239, 250
Request for Proposal, RFP, 239, 241, 243
Research Animal Holding Facility, 396
Resnik, Judith A., 190
Reston, Virginia, 248
Return-to-launch-site, RTLS, 137, 138
Ride, Sally, 172, 178
Ritchey Chretien Telescope, 400, 403
Rocketdyne, 30, 51, 53, 204, 242
Rockwell International, 31, 33, 36, 51, 125, 126, 189, 197, 198, 242
Rodney, George A., 113, 215, 224
Roentgen Satellite, ROSAT, 401, 407, 408, 409
Rogers Commission, 108, 111, 113, 115, 187
Rogers, William P., 187, 188
Rome, University of, 393
Ross, Jerry L., 183

S

Sander, Michael, 371
Satcom, 182, 183
Satellite Business Systems, SBS, 25
Saturn, 107, 135, 365, 366, 367, 411, 415

Saudi Arabia, 180
SCATHA satellite, 24
Scobee, Francis R., 190
Scout, 13, 16, 22, 26, 27, 28, 33, 391, 393
Seddon, M. Rhea, 179
Senegal, 156
Shuttle Mission Simulator, 146,
Shuttle Pallet Satellite, SPAS, 384
Shuttle Student Involvement Program, SSIP, 152, 184
Simpson, J.A., 419
Skylab, 107, 134, 135
Slay, Alton, 223
Smith, Bradford, 366
Smith, Michael J., 190
Smithsonian Astrophysical Observatory, 409
Smithsonian Institute, 367
Soffen, Gerald, 371
Solar Maximum Satellite, Solar Max, 139, 174, 175, 176, 177, 366, 377, 378, 406
Solar Mesospheric Explorer, 379, 380
Solid Rocket Booster, SRB, 17, 22, 23, 33, 34, 39, 40, 41, 42, 43, 110, 112, 123, 126, 134, 156, 166, 172, 192, 193, 194, 203, 227, 236, 237
Solid Rocket Motor, SRM, 42, 44, 45, 48, 49, 57, 170, 192, 205, 206, 212, 214, 219, 221, 222
Solid Spinning Upper Stage, SSUS, 17
SOOS-3 satellite, 29
Soviet Union, 401
Soyuz, 108, 135
Space Industries, Inc., 241
Space Shuttle, 13, 14, 15, 16, 18, 19, 20, 21, 22, 23, 28, 33, 34, 37, 38, 43, 48, 49, 52, 54, 55, 56, 57, 107, 108, 109, 110, 111, 112, 121, 122, 123, 124, 125, 126, 127, 131, 132, 134, 135, 139, 140, 141, 143, 144, 145, 146, 147, 150, 152, 153, 154, 155, 157, 158, 159, 160, 162, 163, 164, 165, 166, 169, 170, 172, 173, 174, 175, 176, 177, 178, 181, 182, 183, 184, 185, 186, 187, 188, 189, 190, 191, 192, 195, 198, 199, 200, 202, 203, 204, 205, 206, 207, 211, 214, 215, 216, 217, 218, 219, 222, 223, 224, 225, 227, 233, 234, 235, 236, 237, 238, 239, 240, 250, 251, 253, 364, 367, 368, 371, 373, 374, 375, 378, 384, 389, 390, 394, 395, 399, 401, 404, 406, 408, 409, 410, 420
Space Shuttle Main Engine, SSME, 22, 23, 34, 35, 36, 39, 40, 123, 124, 127, 136, 137, 144, 191, 204, 230, 234
Space Station, 107, 114, 115, 116, 117, 118, 119, 120, 121, 238, 239, 240, 241, 242, 243, 244, 246, 247, 248, 252, 253, 254, 255

Space Station Control Center, 248
Space Station Development Plan, 241
Space Station Training Facility, 249
Space Station User's Handbook, 246
Space Systems Automated Integration and Assembly Facility, 248
Space Telescope Advisory Committee, 403
Space Telescope Science Institute, 402
Space transfer vehicle, STV, 57
Space Transportation System, STS, 13, 14, 26, 50, 51, 52, 56, 108, 112, 121, 122, 123, 148, 152, 153, 154, 155, 166, 168, 170, 177, 186, 201, 206, 218, 222, 23, 24, 226, 227, 228, 229, 230, 248, 367
Spacelab, 109, 147, 148, 150, 154, 162, 173, 174, 180, 181, 182, 251, 364, 366, 368, 371, 389, 394, 395, 397, 399
Spain, 150, 156
Spartan, 180, 190, 388
Spring, Sherwood C., 183
Stennis Space Center, 36, 37, 109
Stever, H. Guyford, 214, 222
Stewart, Robert L., 174
Stofan, Andrew J., 115, 116, 119, 369, 371
Strategic Defense Initiative, 181
Stratospheric Aerosol and Gas Experiment, SAGE, 24
STS-1, 157, 158, 163, 164, 165, 166, 167, 168, 172, 393
STS-2, 156, 157, 158, 160, 164, 165, 166, 167, 168, 169, 367
STS-3, 151, 153, 156, 157, 158, 162, 164, 165, 168, 169, 170, 389
STS-4, 155, 156, 157, 158, 160, 162, 163, 165, 166, 169
STS-5, 37, 151, 156, 157, 165, 166, 170, 171, 174
STS-6, 37, 171, 172, 173
STS-7, 172, 398
STS-8, 172, 173
STS-9, 173, 181, 194, 394
STS-26, 36, 37, 41, 47, 49, 185, 186, 187, 221, 235
STS-27, 186, 235
STS-30, 417
STS-32, 177
STS-34, 418
STS 41-B, 174, 178
STS 41-C, 139, 174, 176, 378
STS 41-D, 177, 178
STS 41-F, 177, 388
STS 41-G, 55, 154, 178
STS 51-A, 178, 182
STS 51-B, 180, 196, 394, 395
STS 51-C, 179, 196
STS 51-D, 147, 179, 181

STS 51-F, 181, 389, 396
STS 51-G, 180, 181, 394
STS 51-I, 181, 182, 235
STS 51-J, 182
STS 51-L, 46, 47, 108, 128, 135, 184, 185, 187, 189, 190, 191, 192, 194, 196, 197, 198, 199, 200, 202, 204, 211
STS 61-A, 182, 394, 398, 403
STS 61-B, 49, 182, 183
STS 61-C, 49, 183, 232, 235
Sun, 126, 169, 177, 365, 366, 367, 374, 377, 378, 379, 381, 386, 390, 402, 410, 419
Switzerland, 150
Syncom, 147, 177, 179, 181
System Design Review, 230

T

Teacher in Space Project, 184, 190
Technical and Management Information System, TMIS, 116, 241
Teledyne, 52
Telesat, 26, 170, 172, 178, 179
Telstar, 177
Tethered Satellite System, TSS, 55, 56
Thiokol Corporation, 33, 43, 44, 47, 122, 188, 189, 196, 197, 198, 199, 222
Thomas, James, 196
Thompson, Robert F., 111
Titan, 30, 48, 50
Tracking and Data Relay Satellite, TDRS, 127, 142, 152, 171, 172, 173, 184, 185, 190, 218, 235, 236, 403
Transfer Orbit Stage, TOS, 50, 51
Transpace Carriers, Inc., 27
Truly, Richard H., 19, 111, 112, 209, 210, 211, 227
TRW, Inc., 55

U

Uhuru, 365
UK-6/Ariel, 24
Ulysses, 218, 368, 419, 420
United Kingdom, UK, 150, 367, 375, 382, 387, 388, 405, 407, 408
United States, U.S., 14, 25, 26, 27, 122, 135, 150, 173, 174, 178, 241, 243, 252, 375, 382, 405, 407, 408, 418, 420, 421
United Technologies Corporation, 222
Uranus, 368, 411, 413, 414, 415
Utah State University, 152, 169

V

Vandenberg Air Force Base, California, 28, 33, 34, 109, 136, 176, 235
Vanguard, 31, 33
Van Hoften, James D.A., 181
Van Renssalaer, Frank, 17
Vehicle Assembly Building, 133, 134, 142, 235, 236
Venus, 29, 126, 365, 367, 369, 413, 416, 418
Viking, 30, 365, 366, 367, 414, 421
Vought Corporation, 33
Voyager, 365, 366, 367, 368, 401, 411, 413, 414

W

Walker, Charles, 178
Wallops Flight Facility, 27, 33
Wallops Island, Virginia, 392
Weeks, L. Michael, 16
Weiss, Stanley I., 16
Welch, James C., 371
Westar satellite, 26, 174, 178
Western Union, 24, 26, 174
White House, 187
White Sands, New Mexico, 136
Wisconsin, University of, 402
Woods Hole Oceanographic Institute, 125

X

X-ray Timing Explorer, XTE, 407

Y

Yardley, John F., 15, 16, 151
Young, John W., 167

Z

Zeiss, Carl, Company, 408

ABOUT THE COMPILER

Judy A. Rumerman is a professional technical writer who has written or contributed to numerous documents for the National Aeronautics and Space Administration. She has been the author of documents covering various spaceflight missions, the internal workings of NASA's Goddard Space Flight Center, and other material used for training. She was also the compiler of *U.S. Human Spaceflight: A Record of Achievement, 1961–1998,* a monograph for the NASA History Office detailing NASA's human spaceflight missions.

Ms. Rumerman has degrees from the University of Michigan and George Washington University. She grew up in Detroit and presently resides in Silver Spring, Maryland.

THE NASA HISTORY SERIES

Reference Works, NASA SP-4000

Grimwood, James M. *Project Mercury: A Chronology* (NASA SP-4001, 1963).
Grimwood, James M., and Hacker, Barton C., with Vorzimmer, Peter J. *Project Gemini Technology and Operations: A Chronology* (NASA SP-4002, 1969).
Link, Mae Mills. *Space Medicine in Project Mercury* (NASA SP-4003, 1965).
Astronautics and Aeronautics, 1963: Chronology of Science, Technology, and Policy (NASA SP-4004, 1964).
Astronautics and Aeronautics, 1964: Chronology of Science, Technology, and Policy (NASA SP-4005, 1965).
Astronautics and Aeronautics, 1965: Chronology of Science, Technology, and Policy (NASA SP-4006, 1966).
Astronautics and Aeronautics, 1966: Chronology of Science, Technology, and Policy (NASA SP-4007, 1967).
Astronautics and Aeronautics, 1967: Chronology of Science, Technology, and Policy (NASA SP-4008, 1968).
Ertel, Ivan D., and Morse, Mary Louise. *The Apollo Spacecraft: A Chronology, Volume I, Through November 7, 1962* (NASA SP-4009, 1969).
Morse, Mary Louise, and Bays, Jean Kernahan. *The Apollo Spacecraft: A Chronology, Volume II, November 8, 1962–September 30, 1964* (NASA SP-4009, 1973).
Brooks, Courtney G., and Ertel, Ivan D. *The Apollo Spacecraft: A Chronology, Volume III, October 1, 1964–January 20, 1966* (NASA SP-4009, 1973).
Ertel, Ivan D., and Newkirk, Roland W., with Brooks, Courtney G. *The Apollo Spacecraft: A Chronology, Volume IV, January 21, 1966–July 13, 1974* (NASA SP-4009, 1978).
Astronautics and Aeronautics, 1968: Chronology of Science, Technology, and Policy (NASA SP-4010, 1969).
Newkirk, Roland W., and Ertel, Ivan D., with Brooks, Courtney G. *Skylab: A Chronology* (NASA SP-4011, 1977).
Van Nimmen, Jane, and Bruno, Leonard C., with Rosholt, Robert L. *NASA Historical Data Book, Volume I: NASA Resources, 1958–1968* (NASA SP-4012, 1976; rep. ed. 1988).
Ezell, Linda Neuman. *NASA Historical Data Book, Volume II: Programs and Projects, 1958–1968* (NASA SP-4012, 1988).
Ezell, Linda Neuman. *NASA Historical Data Book, Volume III: Programs and Projects, 1969–1978* (NASA SP-4012, 1988).
Gawdiak, Ihor Y., with Fedor, Helen, compilers. *NASA Historical Data Book, Volume IV: NASA Resources, 1969–1978* (NASA SP-4012, 1994).
Astronautics and Aeronautics, 1969: Chronology of Science, Technology, and Policy (NASA SP-4014, 1970).

Astronautics and Aeronautics, 1970: Chronology of Science, Technology, and Policy (NASA SP-4015, 1972).
Astronautics and Aeronautics, 1971: Chronology of Science, Technology, and Policy (NASA SP-4016, 1972).
Astronautics and Aeronautics, 1972: Chronology of Science, Technology, and Policy (NASA SP-4017, 1974).
Astronautics and Aeronautics, 1973: Chronology of Science, Technology, and Policy (NASA SP-4018, 1975).
Astronautics and Aeronautics, 1974: Chronology of Science, Technology, and Policy (NASA SP-4019, 1977).
Astronautics and Aeronautics, 1975: Chronology of Science, Technology, and Policy (NASA SP-4020, 1979).
Astronautics and Aeronautics, 1976: Chronology of Science, Technology, and Policy (NASA SP-4021, 1984).
Astronautics and Aeronautics, 1977: Chronology of Science, Technology, and Policy (NASA SP-4022, 1986).
Astronautics and Aeronautics, 1978: Chronology of Science, Technology, and Policy (NASA SP-4023, 1986).
Astronautics and Aeronautics, 1979–1984: Chronology of Science, Technology, and Policy (NASA SP-4024, 1988).
Astronautics and Aeronautics, 1985: Chronology of Science, Technology, and Policy (NASA SP-4025, 1990).
Noordung, Hermann. *The Problem of Space Travel: The Rocket Motor.* Stuhlinger, Ernst, and Hunley, J.D., with Garland, Jennifer, editors (NASA SP-4026, 1995).
Astronautics and Aeronautics, 1986–1990: A Chronology (NASA SP-4027, 1997).

Management Histories, NASA SP-4100

Rosholt, Robert L. *An Administrative History of NASA, 1958–1963* (NASA SP-4101, 1966).
Levine, Arnold S. *Managing NASA in the Apollo Era (NASA SP-4102, 1982).*
Roland, Alex. *Model Research: The National Advisory Committee for Aeronautics, 1915–1958* (NASA SP-4103, 1985).
Fries, Sylvia D. *NASA Engineers and the Age of Apollo* (NASA SP-4104, 1992).
Glennan, T. Keith. *The Birth of NASA: The Diary of T. Keith Glennan.* Hunley, J.D., editor (NASA SP-4105, 1993).
Seamans, Robert C., Jr. *Aiming at Targets: The Autobiography of Robert C. Seamans, Jr.* (NASA SP-4106, 1996)

Project Histories, NASA SP-4200

Swenson, Loyd S., Jr., Grimwood, James M., and Alexander, Charles C. *This New Ocean: A History of Project Mercury* (NASA SP-4201, 1966).
Green, Constance McL., and Lomask, Milton. *Vanguard: A History* (NASA SP-4202, 1970; rep. ed. Smithsonian Institution Press, 1971).

Hacker, Barton C., and Grimwood, James M. *On Shoulders of Titans: A History of Project Gemini* (NASA SP-4203, 1977).

Benson, Charles D. and Faherty, William Barnaby. *Moonport: A History of Apollo Launch Facilities and Operations* (NASA SP-4204, 1978).

Brooks, Courtney G., Grimwood, James M., and Swenson, Loyd S., Jr. *Chariots for Apollo: A History of Manned Lunar Spacecraft* (NASA SP-4205, 1979).

Bilstein, Roger E. *Stages to Saturn: A Technological History of the Apollo/Saturn Launch Vehicles* (NASA SP-4206, 1980).

SP-4207 not published.

Compton, W. David, and Benson, Charles D. *Living and Working in Space: A History of Skylab* (NASA SP-4208, 1983).

Ezell, Edward Clinton, and Ezell, Linda Neuman. *The Partnership: A History of the Apollo-Soyuz Test Project* (NASA SP-4209, 1978).

Hall, R. Cargill. *Lunar Impact: A History of Project Ranger* (NASA SP-4210, 1977).

Newell, Homer E. *Beyond the Atmosphere: Early Years of Space Science* (NASA SP-4211, 1980).

Ezell, Edward Clinton, and Ezell, Linda Neuman. *On Mars: Exploration of the Red Planet, 1958–1978* (NASA SP-4212, 1984).

Pitts, John A. *The Human Factor: Biomedicine in the Manned Space Program to 1980* (NASA SP-4213, 1985).

Compton, W. David. *Where No Man Has Gone Before: A History of Apollo Lunar Exploration Missions* (NASA SP-4214, 1989).

Naugle, John E. *First Among Equals: The Selection of NASA Space Science Experiments* (NASA SP-4215, 1991).

Wallace, Lane E. *Airborne Trailblazer: Two Decades with NASA Langley's Boeing 737 Flying Laboratory* (NASA SP-4216, 1994).

Butrica, Andrew J., editor. *Beyond the Ionosphere: Fifty Years of Satellite Communication* (NASA SP-4217, 1997).

Butrica, Andrews J. *To See the Unseen: A History of Planetary Radar Astronomy* (NASA SP-4218, 1996).

Mack, Pamela E. Editor. *From Engineering Science to Big Science: The NACA and NASA Collier Trophy Research Project Winners* (NASA SP-4219, 1998).

Reed, R. Dale, with Lister, Darlene. *Wingless Flight: The Lifting Body Story* (NASA SP-4220, 1997).

Heppenheimer, T.A. *The Space Shuttle Decision: NASA's Quest for a Reusable Space Vehicle* (NASA SP-4221, 1999).

Center Histories, NASA SP-4300

Rosenthal, Alfred. *Venture into Space: Early Years of Goddard Space Flight Center* (NASA SP-4301, 1985).

Hartman, Edwin, P. *Adventures in Research: A History of Ames Research Center, 1940–1965* (NASA SP-4302, 1970).

Hallion, Richard P. *On the Frontier: Flight Research at Dryden, 1946–1981* (NASA SP- 4303, 1984).

Muenger, Elizabeth A. *Searching the Horizon: A History of Ames Research Center, 1940–1976* (NASA SP-4304, 1985).
Hansen, James R. *Engineer in Charge: A History of the Langley Aeronautical Laboratory, 1917–1958* (NASA SP-4305, 1987).
Dawson, Virginia P. *Engines and Innovation: Lewis Laboratory and American Propulsion Technology* (NASA SP-4306, 1991).
Dethloff, Henry C. *"Suddenly Tomorrow Came . . .": History of the Johnson Space Center* (NASA SP-4307, 1993).
Hansen, James R. *Spaceflight Revolution: NASA Langley Research Center from Sputnik to Apollo* (NASA SP-4308, 1995).
Wallace, Lane E. *Flights of Discovery: 50 Years at the NASA Dryden Flight Research Center* (NASA SP-4309, 1996).
Herring, Mack R. *Way Station to Space: A History of the John C. Stennis Space Center* (NASA SP-4310, 1997).
Wallace, Harold D., Jr. *Wallops Station and the Creation of the American Space Program* (NASA SP-4311, 1997).
Wallace, Lane E. *Dreams, Hopes, Realities: NASA's Goddard Space Flight Center's First Forty Years* (NASA SP-4312, 1999).

General Histories, NASA SP-4400

Corliss, William R. *NASA Sounding Rockets, 1958–1968: A Historical Summary* (NASA SP-4401, 1971).
Wells, Helen T., Whiteley, Susan H., and Karegeannes, Carrie. *Origins of NASA Names* (NASA SP-4402, 1976).
Anderson, Frank W., Jr. *Orders of Magnitude: A History of NACA and NASA, 1915–1980* (NASA SP-4403, 1981).
Sloop, John L. *Liquid Hydrogen as a Propulsion Fuel, 1945–1959* (NASA SP-4404, 1978).
Roland, Alex. *A Spacefaring People: Perspectives on Early Spaceflight* (NASA SP-4405, 1985).
Bilstein, Roger E. *Orders of Magnitude: A History of the NACA and NASA, 1915–1990* (NASA SP-4406, 1989).
Logsdon, John M., editor, with Lear, Linda J., Warren-Findley, Jannelle, Williamson, Ray A., and Day, Dwayne A. *Exploring the Unknown: Selected Documents in the History of the U.S. Civil Space Program, Volume I: Organizing for Exploration* (NASA SP-4407, 1995).
Logsdon, John M., editor, with Day, Dwayne A., and Launius, Roger D. *Exploring the Unknown: Selected Documents in the History of the U.S. Civil Space Program, Volume II: External Relationships* (NASA SP-4407, 1996).
Logsdon, John M., editor, with Launius, Roger D., Onkst, David H., and Garber, Stephen J. *Exploring the Unknown: Selected Documents in the History of the U.S. Civil Space Program, Volume III: Using Space* (NASA SP-4407, 1998).